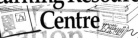
Kingston College
Kingston Hall Road
Kingston upon Thames
KT1 2AQ

MARKETING MICHELIN

Stephen L. Harp

MARKETING **Michelin**

Advertising & Cultural Identity

in Twentieth-Century France

The Johns Hopkins University Press

Baltimore & London

The Johns Hopkins University Press

2715 North Charles Street

Baltimore, Maryland 21218-4363

www.press.jhu.edu

Library of Congress Cataloging-in-Publication Data

Harp, Stephen L.

Marketing Michelin : advertising and cultural identity in

twentieth-century France / Stephen L. Harp.

p. cm.

Includes bibliographical references and index.

ISBN 0-8018-6651-0 (hardcover : perm. paper)

1. Advertising—Tires—France—History. 2. Pneu Michelin

(Firm)—History.   I. Title.

HF6161.T55 H37 2001

338.7'67832'0944—dc21

00-011288

A catalog record for this book is available from the British Library.

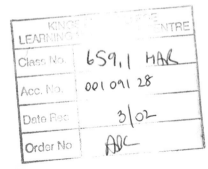

To Sarah and Marie,

and to Lisa: jeg elsker dig.

Contents

AS A GOOD MANY HISTORIANS have now pointed out, it seems inevitable that a historian of France who teaches in Akron, Ohio, once the world's "rubber city," would sooner or later write a book about Michelin. That may well be true, but my interest in French culture and civilization has long been personal as much as professional, at least to the extent that any academic can distinguish between the two.

During my first serious consideration of things French, as an undergraduate studying in Strasbourg in the mid-1980s, I was fascinated by just how different the French seemed to be. I concentrated on those aspects that now seem superficial, almost entertaining. Whereas the male members of my own family measured vacations in how many hours and miles were traveled in a day, French men and women recounted what region they had explored that year and what they had eaten. It seemed that every corner of France had its regional culinary specialities and wines that were just waiting to be savored. I adored the Michelin green guides for their detail and their photographs as I dutifully trooped across France on vacation. In the guidebooks as in my conversations with French people, I was also amazed by French consciousness of the past, particularly as regarded the First and Second World Wars, although as a student of history I was surprised by how much their common understanding of the wars differed from textbook interpretations I had learned in European history classes. I was further amazed that the French state generously paid people to have large families. Moreover, like many Americans, I thought images of the United States were seriously skewed and simplistic, and I was a little defensive. Above all, despite my own interest in history, I took differences between French and other nationalities to be immutable, as if they had always existed in their present state and always would. There was, I presumed, something essentially "French," distinguishing the inhabitants of the hexagon from the rest of us. It took me years to figure out that such constructions of "French" or "American" cultures were in many respects of relatively recent vintage, created and re-created by people, leaders, and companies often having something to gain.

In a sense, I have since my undergraduate days been preoccupied with permutations of what I consider to be a fundamental question in French history: What has made France French? If there is not something uniquely French or British or German, why then is historiography largely divided into the histories of nation-states? Clearly, the drawing of borders, the legal definition of citizenship, and many other actions by centralized governments do not even begin to tell the story. I first tackled the question by writing a comparative history of nation-building in France and Germany that focused on the primary schools of Alsace-Lorraine. For the nineteenth century, there were few institutions more crucial in the spread and maintenance of the idea of the nation-state and all of its trappings. Yet I was nagged by the fact that schools seemed less central to the articulation of national cultures in the twentieth century. Despite the ongoing importance of schooling in the twentieth century, it was obvious that powerful new forces, not controlled by the nation-state, were at work. In particular, the rise of large and small businesses attempting to sell their products seemed to have an increasing role in the definition of what France was assumed to be.

This book is a case study of Michelin's role in reflecting and helping to create French national culture in the twentieth century. It is *a*, and not *the*, cultural history of twentieth-century France, and it concentrates on those cultural spheres in which Michelin was most active. There is, of course, a huge and ever growing literature in cultural studies and cultural history that I have mined for approaches to the evidence. This book, however, is not intended to be a critique of existing theory but instead an attempt by a workaday historian interested in historical agency to examine, in specific ways and in concrete language destined for general readers and students as well as colleagues, how specific aspects of French culture came to be defined in the twentieth century. As in any book, the reader will not find a complete answer in these pages, but I hope to have made a convincing argument that one cannot answer the question of how French people have seen their country and themselves as French without taking into account the history of French and multinational businesses that did so much to convince people to become consumers.

It is absolutely humbling to think of the support that so many people and institutions have given me in order to facilitate research and writing.

The University of Akron funded two of three summers of research in the form of a Faculty Research Grant and a special Carnegie II incentive grant from the Provost's Office. A University Teachers' Fellowship from the National Endowment for the Humanities allowed me to take the academic year 1999–2000 to write the first draft; I cannot begin to describe how crucial the NEH's support has been. I also owe special thanks to my chair and my dean, Walter Hixson and Roger Creel, who made it possible for me to take the NEH grant a year before my sabbatical. The university generously granted my sabbatical year in 2000–2001, so that I could refine the manuscript and begin work on a related project. Although I have missed students in large and small ways, I am very grateful to have had two years of virtually uninterrupted time in which to research and write. I will return to the classroom charged with more new ideas than I had ever hoped to have.

At meetings of the American Historical Association, French Historical Studies, and the Western Society for French History, at special conferences in Brussels and Oslo, and in correspondence, a host of historians have generously offered their comments. I owe more than they probably realize to Seth Armus, Connie Bouchard, William Chew, Sarah Curtis, Venita Datta, Nancy Edwards, Patrick Fridenson, Ellen Furlough, Sam Goodfellow, Bertram Gordon, Isabelle Gournay, Kolleen Guy, William Keylor, Cheryl Koos, Richard Kuisel, James Laux, Shanny Peer, Dan Sherman, and Patrick Young. Bob Brugger, acquisitions editor of the press, similarly offered very sound advice for revisions while dealing patiently with my constant queries. Antoine Champeaux graciously sent me a copy of his master's thesis.

Shelley Baranowski, William Cohen, Sarah Curtis, Carol Harrison, Dan Nelson, and an anonymous reader for the press carefully read the entire manuscript, and Dan Sherman expertly evaluated chapter 3. Their comments were clear and helpful; I hope subsequent revisions have done justice to their fine critiques. Bill, my (supposedly former) teacher of so many years now, and Shelley taught me to ask the big questions in European history by providing such good models. They, like Ellen Furlough, tirelessly wrote letters of recommendation and helped to refine my ideas. Connie Bouchard and Elizabeth Mancke offered their ever sound judgment in how to write better proposals. Finally, many years ago Janina Traxler taught me good French and convinced me to spend a year in France as an undergraduate, giving me an interest in things French that has aged like a good cheese.

In a very practical sense, various archives and libraries made possible the research for this work. In Clermont-Ferrand, the staffs of the Archives Departementales du Puy-de-Dôme and the Bibliothèque Municipale et Universitaire de Clermont-Ferrand were incredibly cheerful in helping me to locate materials. In Paris, the Archives Nationales, the Bibliothèque Nationale de France, the Bibliothèque Forney, the Bibliothèque Trocadéro, the Bibliothèque Historique de la Ville de Paris, the library of the Institut National d'Etudes Démographiques, and the university libraries of Paris VII and Paris X offered assistance to yet another demanding foreign scholar in a hurry to see as much as possible in summers that are far too short. Closer to home, the library of the Stevens Institute of Technology photocopied and sent correspondence between the Michelins and Frederick Winslow Taylor, saving me a trip to Hoboken. Finally, the Interlibrary Loan Service at the University of Akron managed, in the capable hands of Sarah Akers and John Ball, to deal with an exceedingly large volume of requests from a faculty member who was already a heavy user.

Friends and family provided much moral support. After years of suffering through my "obsessive-compulsive" behavior, George Boudreau and Sarah Curtis deserve much thanks. Nicole Junger and Myriam and Frédéric Thareau opened their homes for dinners and evenings that I could never forget. My parents, Sara and Larry Gotshall, Greg and Barb Harp, offered both love and its very practical manifestation in the form of child care, often for a week at a time.

This book is dedicated to my wife, Lisa Bansen-Harp, and our daughters, Sarah and Marie. Lisa's knowledge of modern France has improved this book in many ways as she refined and transformed my ideas with her ability to ask such good questions. She took time from her own research projects to spend portions of summers alone with two small children and had the amazing good humor to remain supportive when *I* whined about being miserably alone in Clermont-Ferrand and Paris. Above all, she deserves thanks for enduring what must have seemed interminable discussions about everything I prepared to write, was writing, and worried about afterward. It is no exaggeration to say that without her this book would not have been researched, much less written. Sarah and Marie have offered diversion, perspective, and even tolerance of my preoccupations. A book dedication is an almost laughable compensation for having put up with my rather single-minded intensity the past several years.

In so many ways, this book belongs more to the above people and institutions than it does to me. I can only hope that it seems worthy of their efforts.

MARKETING MICHELIN

INTRODUCTION

IN 1914, on the eve of World War I, Jean Arren published a book entitled *Sa majesté la publicité*, written to describe the "modern power named advertising." Noting that advertising had become "a science" to which "we are all subjected," Arren had two objectives: to explain to the public what modern advertising was and, in particular, to inform industrialists and merchants who "will find useful suggestions." The book had two parts. In the first half, Arren noted the general techniques companies might use to sell their products. The quality of the product, the price, the quantity produced for the market as a whole, the age and experience of the firm, a novel product, the perceived personality that the buyer could adopt by purchasing the product, the overall success of the company, patriotism, hygiene, and "human passions" were all potential topics for companies hoping to create demand for their products. In the second half of the book, Arren recounted successful advertising campaigns for a whole series of international products, including Johnnie Walker scotch,

Menier chocolate, Thermos bottles, Coleman's mustard, and the Remington typewriter. But by far the longest description considered the advertising of the Michelin tire company, which had, according to Arren, successfully employed virtually all of the methods that Arren laid out in the first part of the book.[1]

Arren allowed André Michelin himself—the Michelin brother in charge of advertising for the firm—to present what Arren called Michelin's "direct" advertising. This was the only article in which Arren had a company spokesman pen his own description. André explained early Michelin circulars sent to people considered likely to buy tires, the use of posters, the centrality of races to prove the strength of Michelin tires, a weekly newspaper article, and the creation of the company icon, Bibendum (known as the Michelin Man in English). Then Arren noted in summation that "advertising in France is now for Bibendum [meaning Michelin] a luxury. It is now known everywhere and by everyone. That is why Michelin advertising has become what one calls indirect. It now includes sacrifices that the Michelin company makes for expenses in the general interest, for the study of public issues, and for the generous providing of touristic and technical information to all."[2] Finally, Arren wrote that Michelin's other efforts, its sponsorship of aviation, its work to force the numbering of roads in France, the Michelin guides, and the company's tourist office were all examples of Michelin's indirect advertising on behalf of the general good of France.

As Arren sensed, Michelin had the foresight to identify several different developments in early-twentieth-century France and to associate the company name with France itself. In a period of growing automobile travel, the birth of aviation, intense nationalism, the dissemination of gastronomy, French preoccupation with the birthrate, and the reconsideration of the roles of women and men, elites and masses, Europeans and non-Europeans, Michelin managed to link its name with each of these phenomena. In the process, Michelin reflected and helped, in large and small ways, to create what Charles DeGaulle later called, "a certain idea of France." Michelin's efforts to develop brand recognition and to increase the demand for tires had an effect that went beyond advertising a product. Michelin's advertisements and public outreach of various kinds, both like and unlike those of other companies in the twentieth century, reflected cultural assumptions in society at large and implicitly confirmed, adapted, or rejected them.

The Michelin tire company is an obvious choice for a case study of the role of business in the formulation of French culture in the early twentieth century. As one of the larger French manufacturing firms before World War I (with more than four thousand employees, roughly the size of Renault or Peugeot), as a firm whose fortune was recent, resulting from the explosion of demand for first bicycle and then automobile tires, the company had both the resources and an interest in promoting its products through widespread advertising. Before World War II, Michelin was seemingly omnipresent in both overt and more subtle considerations of what France was supposed to be. In several respects, Michelin provides a window through which we can examine the construction of French national identity before 1940.

In one sense, Michelin's role in the articulation of French culture was subtle, as the company reflected widespread assumptions in early-twentieth-century France and reinforced them by force of repetition. As aptly noted by Jean Arren, Michelin's advertising in the form of Bibendum, created at roughly the same time as Aunt Jemima and well before the Green Giant, the Pillsbury Dough Boy, Mickey Mouse, or Ronald McDonald, signaled an innovation in international advertising. Yet Bibendum also mirrored widespread explicit and implicit cultural notions in the Belle Epoque. The evolution of Bibendum allows for a sustained analysis of the prescribed roles for men and women, of the place of race in metropolitan and imperial France, and of the perceived differences between social classes before the First World War. As technical advisor to fellow upper-class men during the golden era of middle-class feminism in the form of the suffrage movement, Bibendum explained how cars, and women, were to be handled. After World War I, Bibendum appeared to become a father, reflecting Michelin's increasing preoccupation with the French birthrate. Quite consistently, Bibendum revealed what French men were assumed to be doing or needed to be doing, reinforcing key early-twentieth-century notions of the respective roles of men, handlers of things technical, and women, passive subordinates and producers of babies.

Similarly, Michelin's very public efforts to increase the French birthrate promoted a widespread notion in interwar France that France needed more babies. While the Left and Right differed as to their preferences for the kind of policies that the country should adopt, there was relative agreement that

France could not continue, as contemporaries put it, to "stagnate" demographically without being dominated by Germany. Michelin became the single most vocal advocate of a higher birthrate among French companies. Building on its own efforts to promote more births with *allocations familiales* (family allowances offered to those employees having children), Michelin fervently argued that other companies needed to do the same. Michelin even offered a cash prize for the pamphlet most effectively arguing the pronatalist cause and heavily bankrolled the most important pronatalist association, the Alliance Nationale pour l'Accroissement de la Population Française (the National Alliance for the Growth of the French Population). Here Michelin managed to associate its name with pronatalism, a position without much controversy in interwar France, and one viewed above all as a national issue. In the process, the company reaffirmed that many of France's problems might be solved if only French women bore more children, a cultural construction of incredible tenacity in twentieth-century France.

Before World War I, Michelin's sponsorship of French aviation accepted and confirmed the idea that French aeronautics made France truly "modern." In the first decade of the twentieth century, Michelin offered the "Prix Michelin," which established various feats that aspiring aviators would accomplish in return for a trophy and a huge cash prize. To an even greater extent than automobile races, competition for the prizes received widespread (and free) reporting in the national and regional presses, not only disseminating Michelin's name across France but placing it in a context of progress. By 1911, the same year as the second Moroccan crisis, thus in an increasingly bellicose climate, Michelin established a prize for aviators who could successfully drop mock-bombs on a target area. By offering it only to French pilots, Michelin thus encouraged military preparedness, a theme that reappeared in its publications both before and after World War I. France would obviously have been an early leader in aviation without Michelin, and the importance of aviation to the construction of French identity would have occurred in some form anyway, but the company's efforts nevertheless reinforced the importance of flight for the French nation, reflecting others' ideas and offering them increased credibility.

Like the airplane, the United States represented modernity in interwar France, and Michelin's advocacy of mass production and mass consumption mirrored the myriad interwar discussions about the preservation of French national identity in the face of growing American economic and cultural hegemony. By publishing millions of copies of brochures advocating the application of the ideas of Frederick Winslow Taylor and Henry Ford to French industrial production and retailing, Michelin became one of the foremost French promoters of what was (and is) often called the "Americanization" of France. As a tire maker, Michelin wanted to increase the market for tires, which meant increasing production of cars and their sales. Michelin publicly associated itself with the United States in favoring the "modernization" of French production and distribution. Here too, Michelin recapitulated a widespread sense in interwar France that France's future looked a lot like the contemporary United States while at the same time reminding French people that efforts to compete with the United States economically were the only option for escaping the American onslaught, thus for saving the French nation.

In promoting tourism, Michelin did more to create, as well as to reflect, a vision of modern France. Although still the leisure activity of a small elite of Frenchmen, automobile tourism became foremost on Michelin's agenda for France before 1914. An analysis of the red guides, which began to appear in 1900, reveals the conditions of touring in prewar France and the centrality of differences of class in the guidebooks' appeals to early tourists. Moreover, the creation of Michelin's tourist office in 1908, of the Michelin maps in 1910, and of a system for the numbering of French roads in 1912 reveals the expansion of tourism as well as the company's efforts to shape it. Because Michelin dominated the French market in pneumatic tires before World War I, the company realized that the primary means for increasing sales was to foster automobile traffic. On the eve of the war, Michelin played a fundamental role in defining the dynamics of automobile tourism, a role rivaled only by the nonprofit Touring Club de France.

Even during the war the company continued its promotion of automobile tourism by publishing the best-known guides to the battlefields, which began to appear in 1917. Like other advocates for tourism, Michelin recognized the tragic possibilities of the Great War, expecting that the postwar period would bring a deluge of domestic and foreign tourists to northeastern France. Michelin worked to encourage interwar tourism, and at the same time interpreted the war for the French public. The company's perspective reinforced the contemporary idea that the war was purely defensive, unavoidable, and fought in the only manner possible. The only cause of soldiers' suffering supposedly was the Germans' attack. In this, the company reflected a relative consensus after the war and contributed some of the earliest narrative accounts of it. Although, as in the case of other advertising, no direct connection can be made between Michelin's tire sales and sales of the guidebooks, it is nevertheless true that the guidebooks also had the effect of fusing the Michelin name with French nationalism. By contributing the proceeds of the guidebooks to the pronatalist Alliance Nationale pour l'Accroissement de la Population Française, Michelin did a further patriotic deed and simultaneously tightened the link between Michelin and France.

After the First World War, Michelin continued and expanded its prewar efforts to encourage automobile tourism. Participating fully in a much broader redefinition of French tourism to include regional touring and gastronomy, the red guides grew in size and in scope, reflecting not only the increasing number of restaurants but also an ever-growing interest in the quality of such establishments. Gastronomy was not new to interwar Europe, but the interest in fine dining in rural France most certainly grew. For Michelin, promotion of gastronomy culminated in the late 1920s and early 1930s with the famous star system for rating restaurants across France. Moreover, Michelin, like other advocates of fine dining, closely associated gastronomy with the French nation, helping to establish that a fundamental part of French culture consisted of fine food and fine wine, discovered as one toured the French provinces. In creating regional, eventually green, guides Michelin also quite adroitly tied the name of Michelin with the French nation as the guidebooks focused on the wonderful diversity of French regions, waiting to be "found" and consumed by French people, particularly wealthy Parisians on vacation.

In sum, Michelin's direct marketing of its tires and its indirect marketing of its causes allowed the company to associate itself with the French nation in a variety of ways. Having reflected and helped to formulate what France was supposed to be in the twentieth century, Michelin's marketing offers an interesting case study for the examination of the role of business in maintaining and creating national identities in the twentieth century.

MICHELIN THE COMPANY: A BRIEF HISTORY

By the time that Jean Arren was writing, in 1914, Michelin had established itself on the French industrial landscape. In 1830, Edouard Daubrée had founded a small sugar refining company in the region of Auvergne, deep in the mountains of the Massif Central in south-central France. Daubrée's Scottish wife Elizabeth (née Pugh-Barker) was the niece of Charles McIntosh (who eventually became famous for his rubberized overcoats), and Michelin legend has it that she convinced her husband of the necessity and advantages of producing rubber balls like those her uncle had made for her as a child. In 1832, in the Auvergnat city of Clermont-Ferrand known for its medieval bishopric, for its unique gothic cathedral in black, and as the hometown of Blaise Pascal, Daubrée and Aristide Barbier became partners in a reorganized business that began to produce rubber products. After Charles Goodyear discovered in the 1830s the vulcanization of rubber, the process of adding sulfur to the raw rubber in order to make it malleable, the business grew as the producer of a whole series of rubber products for use in industry and agriculture. The firm, known as Barbier et Daubrée and later as Jean-Gilbert Bideau et Compagnie, was floundering by the 1880s, eventually challenged in part by younger firms established by two other local rubber producers, Jean-Baptiste Torrilhon and Raymond Bergougnan.[3]

In 1886, André Michelin, grandson of Aristide Barbier, left Paris for the family's provincial base in Clermont-Ferrand in order to take control of the troubled family business. An engineer out of the Ecole Centrale des Arts et Manufactures, he had also studied architecture at the Ecole des Beaux-Arts and worked as a cartographer for the Interior Ministry as well as an entrepreneur.[4] André Michelin, interested in so much and drawn to Paris far more than Clermont-Ferrand, did not succeed in reviving the firm single-handedly. In 1889, however, his younger brother Edouard, who had thus far devoted himself entirely to painting as a student at the Ecole des Beaux-Arts, took control of the troubled concern, and the firm became known as Michelin et Compagnie. From 1889 until his death in 1940, Edouard ran the company in Clermont-Ferrand. His older brother André, the Parisian man about town, handled advertising for the firm in Paris until his death in 1931.[5]

Like that of other rubber companies during the Second Industrial Revolution, the future of the Michelin company rested on product innovation, in particular the company's ability to tap into the growing market for bicycle and later automobile tires. In 1891, just three years after John Boyd Dunlop had patented the pneumatic bicycle tire in lieu of older solid rubber varieties, Michelin pioneered the development of pneumatic bicycle tires that were not glued to the rim, as Dunlop's were. Michelin became the first to realize the potential use of pneumatic tires for early automobiles and embarked on their production as early as 1895. The company achieved and maintained dominance of the French tire market by 1900, allowing it to set tire prices in France, the single largest market for tires before 1905 and then second only to the United States. Primary competitors, particularly the British firm Dunlop and the German firm Continental, remained competitive within France primarily by underselling Michelin.[6] Moreover, by 1914, Michelin had successfully established subsidiaries in Britain, Germany, Italy, and the United States, which serve as evidence of the firm's international competitiveness before the war.

When compared with its American competitors, who supplied tires to quickly expanding American automobile manufacturers as well as a growing replacement tire market, Michelin fell behind during the First World War. Although protected from imports during the war, the formerly open market in raw rubber became subject to controls of the French state, limiting Michelin's access to raw materials. Michelin's tire production was not given a high priority because pneumatic tires were inadequate for needs at the front, and Michelin never produced solid rubber tires (*bandages pleins*). Motorized transport, particularly that of supply trucks, ran on solid rubber tires, produced by Michelin's competitor in Clermont-Ferrand, Bergougnan. Michelin began to produce various other rubberized articles for the war effort.[7] The company also built a new plant in Clermont in order to build Breguet airplanes.[8] The construction of airplanes, whatever their utility for the war, dovetailed nicely with Michelin's prewar sponsorship of aviation, making it possible for the company to continue to associate itself with the French national cause as the company perceived it. Although the number of employees increased by about 50 percent in the course of World War I, the war did not allow Michelin to keep up with the technical innovations and economies of scale in tire production occurring in the United States.

By the end of the war, although retaining for a time its market share, Michelin had lost much of its technological and pricing edge to American manufacturers, whose normal business operations had not been as interrupted by the war and whose own competitiveness had been honed against each other. One American firm, Goodrich, had built a plant in France before the war and joined Dunlop in rivaling Michelin within France, particularly in the 1920s and 1930s.[9] The German firm Continental, forbidden from selling tires in France during the war, did not return to the French market after the war.[10]

Although Michelin remained the primary producer for the French market, by the 1920s Michelin had lost its ability to set prices and worked instead merely to try to maintain market share against Goodrich and particularly Dunlop.[11]

Except during the war, Edouard Michelin concentrated the firm's production in tire manufacturing. Whereas Michelin's French competitors, including Bergougnan and Torrilhon, relied on a more diversified production of various rubber goods for both industrial and household uses, Michelin focused on tires. In fact, Michelin was even more narrow in its specialization because it produced only pneumatic tires, compared with Bergougnan's solid rubber tires. In the 1920s, Michelin's international competitors were also more diversified, although US Rubber and Goodrich had broader product lines than Goodyear or Firestone, whose fortunes were primarily made in automobile tire manufacturing.[12] Although Michelin never abandoned bicycle tire production, the real mass market in early-twentieth-century France, the profit margins were limited for bicycle tires.[13] In short, Michelin gambled that pneumatic automobile tires (including truck and bus as well as car tires) were the future and decided to specialize in their production. This forced the company to remain at the technological cutting edge of tire research and development. Michelin thus had an extremely strong vested interest in the propagation of the automobile in France. The competitiveness of the early tire industry only intensified Michelin's interest in increasing productivity, thus cutting production costs, and taking advantages of the economies of scale in producing more tires for less as the market for automobile tires grew.

Michelin's industrial strategy had implications for its advertising. By placing all of its eggs in one basket, Michelin needed to secure the basket. As Michelin put it in words reminiscent of Andrew Carnegie, "Companies that scatter their efforts scatter their brains as well. The proverb that claims that one should not put all of one's eggs in the same basket is radically false for industry. I say put all of your eggs in the same basket and keep an eye on the basket."[14] The company needed, more than anything, for the French market for tires to grow as well as for Michelin's proportionate share to increase. Well aware of Henry Ford's Model T, Michelin held out the United States as the model of what a tire market might be. As we will see, Michelin's advertising assumed that increasing the consumption of tires was the primary means of increasing its own profits, particularly given Michelin's dominance of the French tire market. Its leading role among industrialists as an advocate for automobiles and for automobile tourism is thus hardly surprising, however inventive it may have been. Moreover, as the single major French producer of pneumatic tires competing primarily against foreign firms, Michelin could employ with aplomb the advertising strategy that Jean Arren called "patriotism." Michelin tires were French tires; they supposedly embodied the quality that the French bourgeoisie associated with French luxury goods.[15] Michelin's fusion of its own image with that of France was possible because of its early dominance of the French market as the only major French participant—it was a "first mover," to use the terminology of historians of business—and its marketing did much to preserve that dominance.

By traditional measures, Michelin was neither atypical nor typical of French businesses. Its rapid expansion, large size, and innovative marketing made it exceptional compared to French companies taken collectively, but not necessarily compared to automobile-related firms; Edouard Michelin, who ran the company in Clermont-Ferrand, was one of France's most successful entrepreneurs, as was Louis Renault.[16] André Michelin, who handled marketing in Paris, was unquestionably one of France's most adept advertisers, as Jean Arren obviously realized, but in the 1920s André Citroën's marketing tactics placed him alongside André Michelin as a genius in branding a product.[17] The Michelin company obviously did not fit the profile of a small French firm, unwilling to expand, caricatured after World War II.[18] Yet it practiced the same self-financing (*autofinancement*) for which French firms have faced such derision from historians. Michelin was never publicly owned or staffed by an outside cadre of managers.[19] Michelin was and remains a family-controlled firm, an organization far more widespread among large French firms during the twentieth century than among American or German manufacturers, the explicit models for many business historians.[20] Interestingly, the very legal organization of the Michelin company, a *société en commandite par actions*—also once considered a hopeless organizational framework by leading American historian David Landes—may have actually been one of the causes of the company's success in the twentieth century.[21] As a *société en commandite par actions*, Michelin could operate in relative secrecy. The Michelin family has controlled a substantial amount of the shares, allowing the company greater flexibility than Michelin's competitors enjoyed, thus using an "old-fashioned" structure to become one of the world's largest and most competitive tire makers.

The appeal of Michelin as a case study is not the extent to which the company was or was not "typical" either of French firms or of large businesses in Britain, Germany, or the United States. Its interest lies in the diversity and breadth of Michelin's efforts both to sell its products and to advocate the brothers' causes, actions that were in most respects utterly inseparable for a family firm. Although the dearth of studies of the cultural impact of other French companies precludes at present any deep comparative assessment of Michelin's own role in helping to create a notion of modern France, the company's marketing offers a unique opportunity to explore issues of fundamental importance to students of French culture in the twentieth century. As the rich primary and secondary sources of this book reveal, Michelin participated fully in the collective articulations of what France should become and what it should remain.

CULTURE AND BUSINESS IN TWENTIETH-CENTURY FRANCE

This book takes for granted that there existed in early-twentieth-century France a set of societal assumptions and several commonly held references that we can usefully call French culture. Of course, this is not to say that everyone always made the same assumptions or had the same information, but that at least on some level there was a sort of coherence despite diversity. In considering culture, this work relies on William Sewell's recent description of the "thin coherence" of culture. As Sewell puts it, culture is simultaneously a "system," as early structuralists, notably Claude Lévi-Strauss, maintained, and a set of "practices," actions or words that themselves in essence create culture, a position usually associated with the poststructuralism of Jacques Derrida and Michel Foucault.[22] Even practices that alter, undermine, or reject the perceived cultural status quo are inseparable from the culture that they are attempting to change, and practices that merely reinforce the status quo are a reinforcement of the system (although they too are altering it as every rearticulation must by necessity do). In early-twentieth-century France, there was a widely shared set of notions about what France was and what it was not, a set of expectations that determined in people's own minds what their options were in the face of the material reality of this world.

This approach to culture, defined as both system and practices, highlights historical agency. The early structuralist approach in the field of cultural anthropology took for granted that every practice or ritual of a given society

1

THE MAKING OF THE MICHELIN MAN

The Birth and Life of Bibendum in the Belle Epoque

IN *Sa majesté la publicité* (1914), Jean Arren noted with intense admiration the creation of Bibendum, the Michelin Man, in 1898, a figure that became a well-established company icon for Michelin long before the war. And while many newer companies in early-twentieth-century France later employed symbols to identify themselves—Citroën's double chevron and Peugeot's lion became two of the best known in the related industry of automobile production—relatively few companies had successful anthropomorphic symbols of their businesses. In the early twentieth century, the closest potential rival for Bibendum in his ability to substitute entirely for the company name itself was probably Nectar, created for Nicolas, the wine distributor, in 1922. Arren's admiration was thus well-placed; Bibendum was a coup in prewar advertising, unrivaled by Michelin's competitors and few other French—or even international—firms.[1]

Since 1914, Bibendum has not only continued to represent Michelin but Arren's interpretation itself has also seen no substantive reevaluation. Most people and most historical works consider Bibendum a cute, sometimes humorous, even brilliant example of a company icon without ever noting the extent to which this symbol is as loaded with potential meaning as any other symbol of modern France.[2] This chapter sets out to prove by example the oversight of Marc Martin's recent statement that the study of advertising images and their evolution "is not . . . the work of a historian."[3] In several important respects, Bibendum revealed and humorously reinforced gender, racial, and class hierarchies in early-twentieth-century France. Moreover, although like those of any cultural artifact, representations of him can be extremely ambiguous and even contradictory, a few features remained almost entirely consistent. Bibendum was a white, upper-class French man, often a veritable man about town (*mondain*), who could advise or dominate fellow men, "conquer" women, and control racial inferiors. He embodied, in several important respects, strong assumptions in prewar France about what well-off men should be.

BEFORE BIBENDUM

Although Bibendum first appeared in 1898, a mere nine years after the reestablishment of the firm as the Michelin tire company, the new icon was not the company's first effort to advertise its products. In fact, because of the fundamentally nonessential nature of tires in the late nineteenth century, like the bicycles and the automobiles that used them, advertising was inseparable from production because without the creation of demand there might have been little demand. That is, new industries such as tire or automobile manufacturing more often adopted the supposedly "modern" fusion of production and distribution out of necessity than did older industries that were less immediately reliant on advertising for their sales, benefiting from established markets. In addition, such vertical integration of newer firms resulted from the inadequacy of traditional distribution networks for their products; to oversimplify, who in France would sell tires and how would people know of their existence? For Michelin, advertising became, from the very beginning, nearly as crucial as production itself.

In 1891, as the unverifiable company's version goes, a local cart carrying a bicycle and a cyclist arrived at the Michelin works in Clermont-Ferrand. At a time when the Touring Club de France, the national cycling association, had only recently been founded, a cyclist was a rarity in provincial France and particularly so in Auvergne. This bicycle was, however, even rarer than most because it was equipped with pneumatic, rather than solid rubber, tires. The production of pneumatic tires had only been patented by John Boyd Dunlop three years earlier. Edouard Michelin and his employees set to work to repair the tires, discovering that they were glued to the rim, which made repairs both difficult and slow. Taken with the potential of pneumatic tires, Michelin set out to build a better mousetrap, developing and then patenting the first detachable pneumatic tire (the inner tube and the tire were not glued onto the rim). Although Dunlop challenged Michelin in court for patent infringement, Michelin won the suit because its lawyers, including later prime minister and president Raymond Poincaré, successfully argued that a patent, which had since expired, had been filed for pneumatic tires in the 1840s.[4]

For Michelin, as for Dunlop and other early tire producers, however, it was essential to reach potential consumers and convince them of the superiority of their products. In September 1891, the Michelin brothers convinced Charles Terront, an early cyclist of some repute, to equip his bicycle with Michelin tires for the race from Paris to Brest (on the coast of Brittany) and back, a competition created and financed by the mass Parisian daily *Le Petit Journal* in order to increase its own circulation figures. The contest was one of several bicycle, automobile, airplane, and even dirigible contests sponsored by major dailies before the First World War.[5] Because of the prevalence of horseshoe nails, glass, and other sharp objects that did not interfere with horse-drawn traffic, the relative neglect of French roads since the advent of the railway, and the fragility of early tires, winning the Paris-Brest depended as much on the reparation of tires as on the strength of the cyclist. Terront won the race, in part because Michelin tires could be changed more rapidly than those glued to the rim. Michelin tires thus received acclaim in its most likely market, the "sportsmen" (as they were called in French at the time) and bicycle tourists, not to mention the readers of *Le Petit Journal,* who followed the competition.

In 1895, Michelin produced the first pneumatic tire for automobiles. Unable to persuade any automobile manufacturer to equip its cars with Michelin tires in the automobile manufacturer Comte (later Marquis) de Dion's race from Paris to Bordeaux and back in 1895, the company fashioned its own automobiles and used its own tires.[6] Two of the company's three cars never returned to Paris. The third was driven by the Michelin brothers themselves. The Éclair ("lightning bolt," named for its zigzagging due to the lack of a differential) barely finished the race and was disqualified for not meeting the hundred-hour limit.[7] The Eclair allowed the company, however, to proclaim that pneumatic tires were the tires of the future because they had at least made the entire journey.[8] Although the company began the sales of pneumatic tires for automobiles in 1896, sales figures in the 1890s rested on the success of bicycle, not automobile, tires. In the 1890s, due in part to the interest generated by the Paris-Brest competition as well as other races, the company experienced phenomenal growth. Michelin became immediately successful, with the number of employees climbing from 72 to 268 by 1894. Total receipts climbed to 460,000 francs in 1891, to 2 million in 1895, and to 6 million in 1899. By 1906 the amount of total sales was approximately 37 million francs, this in a period of low inflation.[9]

Bicycle and automobile races were for early tire and automobile makers ideal forums for promoting their goods. Initially, because the use to which racers subjected their tires was not substantially different from that of other drivers, the tires available for sale and those for races did not differ markedly. Producers thus reached their likely market, people interested in touring by bicycle or automobile. In the first decade of the twentieth century, bicycle and automobile racing remained the primary proving ground for companies' products. Races across Europe, as well as the Vanderbilt and Daytona races in the United States, provided Michelin opportunities to point out that the winners had Michelin tires.[10] The Coupe Gordon-Bennett, which organized racers across Europe by country, exploited growing nationalism to sell James Gordon-Bennett's English-language *New York Herald*, which was published in Paris, and allowed Michelin to associate its tires with French interests.[11] After the advent of Bibendum in 1898, Michelin regularly used him in the press after races to proclaim the superiority of Michelin tires, making it impossible to separate marketing strategies into neat categories. Racing emerged before Bibendum, but he became omnipresent in Michelin's efforts to get mileage out of its cycling and automotive victories until the company withdrew from racing in 1912.[12]

Le Petit Journal's sponsorship of the early Paris-Brest race already revealed the link between early races and writing about them. Races sold newspapers as well as tires, and advertisements in newspapers about race results also helped to sell tires. There were strong connections among newspapers, racing, manufacturers, and tourists. Because tire and automobile producers needed above all to create a market for their products, newspapers, like the races they sponsored, offered the possibility of reaching potential buyers. Manufacturers appealed, in print, to early tourists by using both printed handbills and newspaper advertisements.

André Michelin, who was in charge of the firm's advertising, quite brilliantly used the printed word to promote Michelin tires. A successful race provided the evidence for a whole series of advertising formats and venues. For instance, when Terront arrived back in Paris in the Paris-Brest race, Michelin had already printed one-page handbills for distribution to the crowd. They included a description of the advantages of Michelin tires and ended with the sort of self-consciously pretentious, humorous historical anecdote that implicitly tied Michelin to the French past, a strategy that became one of the norms of Michelin advertising before World War I. As the handbill put it, "On the fourteenth of July, Louis XVI, learning from the mouth of Lafayette of the taking of the Bastille, cried out, 'But alas, it is a revolt!—No, sire, responded the marquis, IT IS A REVOLUTION.' We have confidence that the cycling public will say about our tire, 'Is it a refinement?—No, IT IS A REVOLUTION' [capital letters in the original]."[13] In the succeeding weeks, the firm sponsored a short statement from Terront assuring the readers of *Vélo Sport* that he won Paris-Brest as a result of the ease of use of Michelin detachable tires. *Le Petit Journal* itself published a letter of thanks from the Michelin company to the newspaper in which Michelin announced its sponsorship of a race from Clermont-Ferrand to Paris.[14]

In addition to more traditional advertisements in large Parisian dailies, the company and its competitors used the nascent sporting press to reach a target audience. In 1900, Michelin joined forces with automobile manufacturers Dion-Bouton, Adolphe Clément, and the Baron de Zuylen to finance the establishment of the famed cyclist Henri Desgrange's new publication, *L'Auto-Vélo*.[15] The new daily resulted from the frustration of the manufacturers and promoters of the automobile with Pierre Giffard, who had organized the Paris-Brest race for *Le Petit Journal* in 1891. Along with Paul Rousseau, Giffard had founded the daily newspaper *Le Vélo* in 1892, which had a circulation of eighty thousand by 1895. During the Dreyfus Affair, Giffard energetically supported the cause of Dreyfus in both *Le Vélo* and *Le Petit Journal*, angering much of his wealthy constituency, upset by his Dreyfusard stance and by his willingness to allow polemics into what should have remained, in their eyes, a "neutral" sporting newspaper.[16] When the Comte de Dion partic- ipated in the attack on President Loubet in 1899 at the Auteuil races, *Le Vélo* criticized the automobile manufacturer.[17] Even if the manufacturers' sympa- thies had not been anti-Dreyfusard, companies stood to lose part of their mar- ket if their products were associated with one side of the conflict and not the other. Potential auto buyers existed across the political spectrum, and given the elite ownership of automobiles before World War I, a good many buyers were anti-Dreyfus. Michelin, by helping to finance *L'Auto-Vélo*, obviously recognized the value of a daily devoted entirely to automobile and bicycling enthusiasts, the only real market for a firm that limited itself to the production of tires and tire-related products.

Without entirely abandoning other publications, Michelin successfully re- inforced the connections among the written word, racing, touring, and tire manufacturing in the pages of the new *L'Auto-Vélo*. Within five months of the first edition of *L'Auto-Vélo* on 16 October 1900, the company pioneered an advertising strategy quite different from short messages frequent in earlier printed advertisements. On 11 March 1901, the first "Lundi de Michelin" ap- peared. Each Monday until the outbreak of the war, Michelin penned a short article, seldom more than five hundred words in length but quite substantive compared to earlier advertisements, addressed to its clients and potential cli- ents.[18] Initially, *L'Auto-Vélo* had a circulation of about twenty thousand, but by 1903 Michelin began to reach the ever-growing readership of *L'Auto* (which was forced to drop "Vélo" from the title after losing a case brought by *Le Vélo*) when Henri Desgrange sponsored the first Tour de France bicycle race in 1903, and circulation reached fifty thousand, a number that doubled during the Tour each summer.[19] Michelin could thus reach a very specific audience, the cyclists and early automobilists most likely to buy tires.

Michelin used the "Lundis" to further set itself apart from other tire mak- ers. Claiming that its sole purpose was to provide "service to the client," the same reason offered for the launching of the red guides in 1900, Michelin used the articles to inform, even educate tire users. Tires were of course fragile at the time, but they were also quite new; knowledge of how to store, change, and preserve the life of tires was not part of the *culture générale* in the way that basic care of horses was. Potential and actual buyers needed to be trained lest they unnecessarily ruin these expensive novelties (costing in 1901 99 francs each at a time when a male laborer in the provinces would have earned 3 francs or so daily, depending on his skills), blame the damage on the company, or swear off cycling or driving entirely.[20] Because tires were not necessary, just as bicycles and cars themselves were not, the company had obvious reasons to argue against turn-of-the-century technophobia. The company also needed to inform automobilists of garages where they could buy Michelin tires, a necessity when tire repair and replacement were fundamental parts of driving.

Initially, the "Lundis" were very technical, written with enthusiasts in mind, and Michelin made relatively few attempts at humor. Bibendum, who eventually embodied both the company and its humorous approach, did not appear at the outset. The first "Lundi" featured a list of *stockistes*, the small business owners who had contracted to carry a full range of Michelin tires and, eventually, tire-related products. These dealers could be found in Angoulême, Biarritz, Bordeaux, Marseille, Nice, Pau, and Toulouse, locations that mostly serve as a reminder of how thoroughly the early development of the automobile depended on tourism of the rich to the seaside.[21] The second "Lundi" described the purpose of the "Lundis," when noting that at least forty-five of the one hundred tires sent back to the factory for repair (a regular practice before the war, particularly since few designated repair facilities existed) resulted from defective installation of the tube in the tire. "Given these conditions, we believe that most of these errors are because the driver [*chauffeur*, who could be the owner or a chauffeur] was not well informed about the way to use a tire and we decided to publish . . . a series of articles destined to give all of the information that we consider useful. We would be very happy if, in addition to our experience, experienced drivers would join us. And we will do our best to publish interesting letters sent to us."[22] Michelin thus stood to profit from the "Lundis" in an additional fashion; by receiving what would eventually become voluminous mail from cyclists and drivers, Michelin conducted what companies would eventually and more systematically call "market research."

Alongside large, more standard advertisements that Dunlop and Continental ran in *L'Auto-Vélo*, Michelin presented itself as the cyclist's and particularly the driver's technical advisor, a role that Bibendum would eventually acquire. For example, the company advised drivers to check their alignment (by eyeballing it), to keep tubes out of the light and sheltered from extreme temperatures before their use, to keep sufficient air in tires, to use plenty of talcum powder on the tubes when changing tires, to check the rims, to keep grease away from tires, and to brake cars slowly.[23] Intermittently, the company's advice extended to information about Michelin products for changing tires, Michelin's "exercise" contraption of heavy rubber bands and pulleys, the Michelin red guides, the *stockistes,* Michelin's racing victories, information about the yearly automobile shows, and the injustice of Britain excluding Michelin tires from its market. Michelin thus advertised its products while educating and gathering information about potential clients writing in with technical questions. The early Touring Club de France, the large nonprofit promoter of tourism in France, was incredibly participatory, with people writing constantly to share information and correct what they perceived to be misinformation in the pages of the group's *Revue.* Michelin quite adeptly attempted to tap into that world of tourist participation, both to serve it and to profit from it.

Like other tire, bicycle, and automobile manufacturers, Michelin participated in the yearly salons, or shows, of bicycles and automobiles. After the early inclusion of automobiles at the salons devoted primarily to bicycles in 1894, 1895, and 1897, the Automobile Club de France began its annual salons for the automobile (and secondarily bicycles) in the Grand Palais adjacent to the Tuileries Gardens in 1898.[24] Connoting earlier art exhibitions known as salons, themselves connoting those aristocratic meetings of the Old Regime, these shows were crucial for marketing before the First World War. Given the early dominance of Parisian buyers, the salons allowed manufacturers (also predominantly concentrated near Paris) to show their new products to likely clients who could find a chassis, select the motor they wished to have installed, and choose accessories. At the salon in 1898, manufacturers received orders for 5 million francs' worth of merchandise.[25] Clients could place orders for automobiles to be delivered in the late spring, when the touring year began. Michelin had initially had its own booth, exhibiting its tires, rims, and equipment for changing tires.[26] More important, the tire manufacturers used the salons for boasting rights. Tire makers ran ads touting the number of cars at the salons equipped with its tires. Michelin used its "Lundis" to prove the company's reputation for quality among manufacturers. In 1912, an advertisement with a drawing of a huge Bibendum sitting on top of the Grand Palais announced that 63 percent of all automobiles at the salon had Michelin tires.[27]

In the first decade of the twentieth century, Michelin increasingly adopted illustrations and used humor to lighten the tone of advertisements. Historical, literary, and artistic references to both France and the classical world abounded, making the "Lundis" less technical and more accessible, at least for a well-educated French elite. Bibendum was crucial to the transformation. By 1909, an illustration of Bibendum writing appeared at the top of each "Lundi," and the humor was often his.[28] Bibendum allowed Michelin to seem unique in the face of the competition not only by his presence but also by not being one of the Michelin brothers himself. Playful, boastful, even risqué humor one might expect among the *mondain* or perhaps among bourgeois men in private could be employed in public without being inappropriate. The creation of Bibendum enabled the company to further capitalize on its early advertising with antics appropriate to a company icon but perhaps unseemly for prewar French industrialists.

THE BIRTH OF BIBENDUM

In the late 1890s, André Michelin turned to colorful illustrations, particularly posters, in order to market tires. Beginning in the late 1860s, Jules Chéret literally created the boldly colored poster advertising a product or a business. By the 1880s, posters, although advertisements, were recognized as a legitimate art form in France, as witnessed by the admission of Chéret's work at the Universal Exposition of 1889 and the government's award of the Légion d'honneur to Chéret.[29] In the 1890s, he pioneered the beautiful, vibrantly colored posters generally associated with poster art in turn-of-the-century France, for which Henri de Toulouse-Lautrec and, to a lesser extent, Edouard Manet are sometimes known.[30] Although newspapers continued to consume the largest chunk of advertising expenditure in France, posters became increasingly important, particularly within large cities and especially in Paris.[31] Moreover, the artistic poster was, and is, closely associated with France, where it originated, making it an ideal medium for firms, including Michelin, that attempted to associate their names with French traditions.

André Michelin contracted Marius Rossillon, who signed his name O'Galop, to draw a poster that became the first representation of Bibendum in 1898. Both the company and popular historians have told and retold the history of this poster. As the legend goes, in 1894, the two Michelin brothers attended the Universal and Colonial Exposition in Lyon. At the Michelin booth, an employee had stacked two tall piles of tires at the entry. Edouard, the younger brother who ran the company in Clermont-Ferrand, is reputed to have said to his older brother André that one pile looked like a man. In early 1898, André met with O'Galop, who showed André some of his work. André was quite interested in one design that O'Galop had originally done for a brewery in Munich. It featured a large man who was supposed to be Gambrinus, the legendary king who had invented brewing. The large king held a beer stein while announcing in Latin, "Nunc est bibendum [Now it is time to drink]." The expression was one that the Epicurean Horace had placed in the words of Marc Antony after the battle of Actium, in 31 B. C. André, remembering his brother's comment about the man made of tires and himself long accustomed to referring to how the pneumatic tire could swallow (*avaler*) or drink (*boire*) the obstacle (*l'obstacle*), had O'Galop redraw the poster to substitute the man of tires for Gambrinus. The man did not, however, have a name. In July 1898, at the time of the Paris-Amsterdam-Paris race, the driver Léon Théry, who did not know Latin, yelled to André, "voilà Bibendum, vive Bibendum," thus equating the man in the poster not only with André, but naming the man of tires in the poster "Bibendum."[32]

O'Galop's poster itself tells us much about the marketing of tires in turn-of-the-century France and about the social divisions within France. First, Bibendum's size is remarkable; he is bigger, much bigger than his competing tire men Continental and Dunlop. Obviously, he still has air in his tires whereas his competitors have developed small holes, leading to their deflated, flaccid state. It is important to keep in mind that at the time weight was a sign of wealth, and French newspapers also carried legions of advertisements for various potions men could take to become larger, both in the sense of muscular development and in terms of overall size.[33] We should not anachronistically project the preoccupation with flat stomachs into turn-of-the-century France. Thus, Bibendum represents upper social strata and upper-middle-class men in particular, the only people wealthy enough to buy both abundant rich food and automobile tires. It goes without saying that Bibendum appears as a man dominating other men, and not as a woman.

Bibendum originates from the aristocracy or upper middle class. The pince-nez is an obvious sign, as are the rings on his fingers and his overall size. Most telling, however, is the glass of champagne. At a time when middle-class reformers worked to promote temperance among the working classes, champagne was becoming an ever more important symbol of status among middle classes who associated the drink with the aristocracy and bourgeoisie.[34] Bibendum is not drinking red wine, itself a symbol of France but available to

1. "Now it is time to drink; that is, cheers. The Michelin tire drinks the obstacle." The well-off white Bibendum outdrinks his smaller, thus lesser, competitors Dunlop and Continental. Set in a restaurant or a banquet
hall, the scene is representative of turn-of-the-century dining norms, as few women had the opportunity to
experience in public the pleasures of French gastronomy. Color poster by O'Galop, 1898.

all classes in some form, although the spilled glass on the right indicates that perhaps the head of Continental to his left had been drinking red wine. As interesting are the contents of Bibendum's glass, which are shards of glass and nails in this first poster, although horseshoes with the nails still attached are featured in later editions of the poster. These are the pieces of glass and nails quite prevalent on French roads, harmless for horse-drawn traffic, but perilous for cyclists and particularly automobilists, given the speed and greater lack of proximity to the road of the latter. At any rate, the nails and glass are termed *obstacles,* supposedly because André Michelin had earlier noted how well Michelin pneumatic tires could "*boire l'obstacle* [drink the obstacle]."[35] But André Michelin was not alone. The importance of surmounting *obstacles* of the road is a constant theme in the *Revue du Touring Club de France,* the forceful nonprofit advocate for the bicycle and then the automobile. Use of the term *obstacle* reinforced a link both between automobiles and horses, but of elite horsemanship in particular. *Obstacle* was and is the word used in the French steeplechase. The *course d'obstacles* consisted of the hedges, barriers, and pools of water over which showmen of horses needed to jump. The aristocratic connotations are obvious. Of course, hierarchies of class involve differentiation as part of identity. André Michelin's account of the naming of Bibendum is here telling because it points out the social differentiation between many car owners and race drivers; it was because Théry did not know Latin, that learned language of the educated bourgeoisie, that he considered Bibendum the name of the man rather than understanding its accurate Latin meaning.[36]

Later adornments gave Bibendum a certain social standing. Although in the first poster of 1898 Bibendum had not yet acquired his cigar—that ultimate symbol of the well-off man in early-twentieth-century France—O'Galop added one to Bibendum's hand in later reeditions of the poster. In fact, through the 1920s, the vast majority of portrayals of Bibendum included a cigar. As is so often the case, a symbol of class cannot always be separated from one of gender; the cigar simultaneously reinforced Bibendum's masculine image at a time when only the "New Woman" smoked, and when she—like George Sand back in the 1840s—was berated for such masculine behavior. Moreover, Bibendum's occasional appearance in a fur coat placed him squarely among France's wealthy.[37]

Advertisements for Michelin's Exerciser similarly show the importance of class in marketing. In 1901, the Exerciseur Michelin first appeared. Aside from later toys and advertising trinkets, it was one of the few products that Michelin produced which was not closely related to tires. The exerciser was a combination of rubberized cables and pulleys that could be attached to an interior door for indoor exercise. The exerciser was available in four versions with different tension levels, ranging from an 8 franc version for "women and children," a 9 franc version for "men and young people," a 10 franc version for "athletes," and a 12 franc version for "Hercules."[38] Advertisements for the exerciser took for granted that the user had an essentially sedentary lifestyle; it was a piece of equipment destined for the bourgeois and middle classes generally, not for peasants or workers, even those willing to spend a few days' salary for the contraption.

Michelin held a competition among artists who drew posters of the exerciser that the company could then use for advertising.[39] In the posters, Bibendum remained strong and barrel-chested, although his waistline became smaller. More telling were the suggestions of class difference made in some of the posters. One poster in particular read "The Michelin tire repels all attacks because it uses the Michelin exerciser." The illustration shows Bibendum, appropriately big, successfully defending himself against three small robbers, who are dressed in working-class clothing and have the thin forms common among working-class men in the nineteenth century. One of the men has been flattened by Bibendum, and the remaining two find their knives completely ineffective against the sharp punch and kick delivered by Bibendum, dressed for a night on the town in a top hat. Bibendum is also carrying a cane, which he does not even need to repulse the attack.[40] In short, the Michelin exerciser could strengthen an upper-class man facing threats from the working class.

Michelin also worked to associate the company and Bibendum with royalty and the international aristocracy. Given the appeal of many royal and aristocratic trappings for wealthy bourgeoisie in nineteenth-century Europe, the strategy is not surprising, although it does reflect Michelin's assumptions about the desires of its potential clients. Michelin regularly made the claim that the company produced the highest quality tires in Europe, playing on the French bourgeoisie's association of early French industrial goods with artisanal *articles de luxe*. Before World War I, even the automotive and tire industries were organized as large combinations of smaller workshops, so the claim of "artisanal production" was not far-fetched. In one poster done by O'Galop for the British subsidiary in 1905, "Sir Bibendum" holds a jousting stick topped with

2. "The Michelin tire defies all attacks because he uses the Michelin exerciser." Bibendum, the wealthy man about town, fends off robbers dressed as working-class men and armed with knives, attempting to puncture him. The image reinforced notions of difference between the upper classes and the "dangerous, laboring classes," to use the nineteenth-century formulation. Color poster, 1905.

a glass of champagne filled with *obstacles* in one hand. In the other he holds a shield, which features his four symbols: a glass of champagne with *obstacles*, his pince-nez, cigars, and the cross-section of a tire pressured but not punctured by a nail. The caption, a paraphrase from Tennyson, read "My strength is as the strength of ten / Because my *rubber's* pure ["rubber" emphasized with blue rather than red print in the original]."[41] The company also frequently pointed out in "Lundis," in brochures, and on postcards that "the Michelin tire is not only the king of tires, it is also the tire of kings." Like Louis XIV, Michelin proclaimed that there were no longer any Pyrenees when the Spanish king bought Michelin tires[42] and that even the German emperor had bought a set of Michelins.[43] One brochure featured a picture of King Edward VII in his car and reminded the reader that the kings of Britain, Belgium, and Italy, as well as the czar, equipped their cars with Michelin tires.[44] Even the nobility, notably the French automobile manufacturer the marquis de Dion, as well as the Russian prince Wladimir Orloff, used Michelin tires.[45] Buying Michelin tires, whose slightly higher prices reinforced an image of quality, could confer status.[46]

As historians have recently begun to focus on class not as a strictly social category in the Marxist sense but as a cultural category, historians have increasingly considered class through the eyes of the people who made class distinctions to describe the society in which they lived.[47] Companies participated in the process. On the one hand, it was obvious to all that only the wealthy could afford an automobile in prewar France, and Michelin clearly wanted to capture that lucrative market. By appealing to bourgeois who could afford to buy tires with advertising that appealed to their sensibilities, and thus set them apart from their social inferiors, Michelin might sell them tires. Like other products, Michelin tires could help to establish or confirm social identity. On the other hand, as was clear in so many of the company's actions as well as its pronouncements, Michelin wanted above all to increase the eventual market in tires, hence the advertisements and "Lundi" articles devoted to cyclists, who not only bought tires in the short term but were also likely users of the automobile in the longer term. Here, too, however, distinctions made between social groups in prewar France could be useful for sales. As long as the middle and eventually the lower middle classes were not actually insulted, these groups could, by buying Michelin tires for their automobiles or even their bicycles, attain the much-vaunted "quality" of a "French product," a pre-

3. The same year that Michelin created the Michelin Tyre Co. of Great Britain, Bibendum appeared as a British knight. By referring to an author of British literature, even paraphrasing Tennyson, Michelin placed Bibendum solidly within the context of British high culture, much as Bibendum's references to French literature appealed to wealthy Frenchmen. Color poster by O'Galop, 1905.

occupation among nineteenth-century bourgeois consumers in France that Michelin brilliantly exploited.[48]

The presentation of Bibendum was not, however, the only way that one might use assumptions about class in order to sell a product. Nectar (as in the nectar of the fruit of the grape), the icon that wine distributor Nicolas began to use in 1922, provides a startling contrast, showing not a different notion of class but a different way of reinforcing class difference. The Nicolas business, which had expanded rapidly in the nineteenth and early twentieth centuries, consisted of opening shops where fine wines might be bought, thus avoiding the buyers' need to work directly with vintners at a time of growing wine consumption. One of the primary advantages of buying wine at Nicolas was that the distributor then delivered the purchases. Nectar, who was supposedly modeled after an actual Nicolas employee by the artist Dransy, embodied the wine deliverer. He generally carried more than a dozen bottles in each hand, sometimes white wines in one hand, red in the other. Whereas Bibendum was young, strong, and healthy with a chest shaped like a barrel, Nectar was older, seemed overburdened, and had a narrow chest. Whereas Bibendum was pure white, Nectar—who much more closely resembled an actual man—wore the typical blue of a working-class man. Both men drank heavily. But whereas Bibendum, with his constant thirst, suffered no ill effects, Nectar had a red face with a large red nose.[49] In buying Michelin tires, one might fashion oneself as a strong, rich, bourgeois superman of sorts, resembling Bibendum; in buying wine chez Nicolas, one might assert superiority differently, contrasting one's own privileged position with that of the lowly deliverer. In both cases, class was clearly operative within the marketing strategy, but it worked in very different ways. Both figures, nevertheless, reinforced the notion that there were real differences between the classes in the minds of contemporaries, even though an absolutely accurate re-creation of those social categories has proven so elusive for historians.[50]

Bibendum was not only well-off; he was a well-off Frenchman. Perhaps the least obvious quality in O'Galop's first poster is Bibendum's nationality. In the poster, Bibendum is holding a glass of champagne, a product assumed in France to be exclusive to France. The stein of beer in O'Galop's original design would no doubt have seemed inappropriate for a French audience. Bibendum is above all French, whereas his primary competitors in the sales of pneumatic (as opposed to solid rubber) tires were Dunlop and Continental. On the left of the poster is a caricature of Harvey du Cros, head of Dunlop (called "pneu y" in later editions) and on the right a caricature of the head of Continental, "pneu x." Upon close inspection, the "x" looks remarkably like a German iron cross. Although in various national markets Bibendum later found himself transformed into an Englishman, a German, an Italian, and so on, he was primarily French.

Later prewar appearances of Bibendum confirmed his nationality and tightened the link between Bibendum and France. Often referred to as *notre Bibendum national* (our national Bibendum), Bibendum could drink all of his competitors under the table. In one newspaper advertisement, Bibendum drank his glass of *obstacles,* winning the Gordon-Bennett in 1905, while his competitors "pneu x" and "pneu y" lie flat on the ground, exhausted and deflated. "Pneu y" is literally under the small table next to a Bibendum holding a champagne bottle in one hand, drinking from the other, and leaning back grandly in his chair.[51] "Lundis" featured dialogues between Frenchmen and foreigners, who had heavy accents.[52] O'Galop drew a postcard on which a German and a British man approach a French officer on a bike. Both say they are introducing him to a "very French tire." The German says, making all of the supposed errors in French typical for a German, "Che vous brésente oun bneumatique pien vrançais!" The British man counters that "je présentais vô oune tyre tôt à fait française," himself making the presumably classic errors of an English speaker. The French soldier, recognizing the foreignness of these supposedly French tires, opts for a Michelin tire.[53]

Interestingly, even in prewar advertisements in Britain, Germany, and Italy, Michelin did not hide or understate Bibendum's French nationality. A postcard representing a pyramid constructed of Bibendum's racing trophies and Bibendum as the sphinx overlooking Napoleon's troops during the Egyptian campaign appeared in German as well as in French.[54] A drawing for the Italian subsidiary featured Bibendum having a drink with Napoleon himself.[55] Napoleon simply did not bring back moments of German and Italian grandeur, but embodied the grandeur of France at the expense of those two nations. A poster from early 1914 designed for the British market read "Michelin Tyres: A Link with France giving Safety, Economy, Comfort, and Speed" and featured a drawing of a channel tunnel constructed of huge Michelin tires through which automobiles could come from France to England.[56] Despite the entente, it is difficult to imagine potential British clients becoming excited over a tunnel

4. A tunnel built of Michelin tires connects Britain and France, enabling both trains and automobiles to cross the English Channel. Using Michelin's reputation as a French producer of high-quality goods as an asset, Michelin confidently proclaims all of the advantages of closer cooperation with France, apparently convinced that a positive image of French goods would outweigh the fears that a tunnel represented for British nationalists (one that remained strong throughout the twentieth century). The airplane crossing the channel is a further reminder that Britain could not avoid closer proximity to continental Europe. Color poster, 1914.

that allowed French products and French people to invade the British market. Bibendum's assertion of his French origins seems a rather bold proclamation of the place of both Bibendum and France within Europe. International advertising took for granted that Michelin tires could be seen as a superior product, a sort of manufactured French luxury item.

BIBENDUM AS TECHNICAL ADVISOR

In the "Lundis" and other forums, Bibendum became the French technical advisor to automobilists and cyclists. He was an equal to the rich men who owned cars, and over time Bibendum claimed his familiarity as "friendship."[57] In the "Lundis," Bibendum's presence brought a change in tone as the articles became increasingly lighter and more humorous. Even exceedingly technical articles began with a pun, a historical anecdote, or a historical reference. Bibendum made regular appearances in small accompanying drawings or in dialogues between him and "pneu x." After 1908, when Michelin shifted the regular publication of the "Lundis" from the specialized newspaper, *L'Auto,* to *Le Journal,* one of the top four French dailies with a circulation of approximately 750,000 in 1908, Bibendum took on a greater role.[58] The very technical articles found their way into a column in the more specialized newspaper *Les Sports,* a recognition that motoring was including a larger, and thus more segmented group.[59] In *Le Journal,* a small illustration at the top of most "Lundis" showed Bibendum at a desk, presumably writing the "Lundi." Reinforcing the extent to which Bibendum stood for the firm itself, the "Lundis" were usually signed "Michelin," although occasionally "Bibendum." Just before the war, new drawings featured Bibendum in more active poses, often illustrating one of the points made in the individual "Lundi."

As newspaper articles, the "Lundis" gave Michelin space for Bibendum to use several allusions at the same time, to both contemporary politics and literature, before moving on to the issue at hand. For example, a "Lundi" of 1911 explained the company's philosophy regarding its willingness to provide free technical assistance to drivers. It is a typical "Lundi" in many respects, although it also reveals the nuances of Michelin's self-conscious efforts to market tires. The title is "Insistance publique" [Public Insistence] playing on the subject of increasing debates over the place of "Assistance publique" [Public Assistance] in France before the war. The text then begins with Michelin's usual vague citation, "Alexandre Dumas the younger said somewhere that 'Ideas are like nails: the more you hit them the farther in they go.' Our goal, that we have proclaimed a thousand times, is . . . to instruct motorists, to whom we want above all to furnish the means to get the most out of their tires with the fewest hassles. Our industry is directly interested in the continual progress of automobiles, on which it depends: the easier and cheaper that tire use becomes, the usage of automobiles will spread and the more tires we will sell. Drivers will only know how to better serve both their interests and ours by caring properly for their tires."[60]

Even the more purely technical "Lundis" frequently begin with a play on words or an allusion. Several play humorously on European men's assumptions about women and about non-Europeans. One "Lundi" begins with a signed drawing by O'Galop of Bibendum overseeing four young women whose trunks and heads are tires. With reference to finishing school, the title is "Stand up straight, young ladies / Check the alignment of your car wheels [roues]." Playing on the feminine gender of the French word for wheel, roue, the young women are thus being aligned as wheels need to be aligned. The "Lundi" then begins by imagining the driver's complaint after all of the advice from preceding "Lundis": "Check! Again! Oh no, shoot! Is it necessary always to check something else? Tire pressure, application of talcum powder, the tube, the tire, the chain, the grease fittings. What else? The age of the mechanic? And will he have to convert us to Islam, to become Arabs and learn the art of checking dates [l'art de vérifier les Dattes] [italics in the original]?"[61] The pun is thus a word play on dates (dates) (that is, the date of birth of the mechanic) and the fruit called dates (dattes, although with the same pronunciation as dates in French) that Muslim Arabs supposedly spend their time contemplating.

The "Lundi" continues, "Whatever you want our driving friends! As you like it [only the underlined words are in English in the original] (that's how Shakespeare formulated it in his ignorance of French). We just wanted to ask you to please check the alignment of the wheels of your car from time to time! Don't do anything if you don't want to be bothered or if you spend all of your time absorbed by international politics! Let us just point out, however, that in following our advice you will arrange for yourself a pleasant surprise when it is time to do your books." The article then described how one should rotate and align tires and included technical illustrations. The "Lundi" ended, nevertheless, on a lighter note. If one did not follow the instructions, "There will be no alignment and the coupling rod [*la barre d'accouplement*] requires a light correction! There are a lot of pretty women more than happy to oblige!"[62] The "Lundi" reveals an inseparable link among Bibendum as a technical advisor to men, sexual innuendo, and the implicit view of women. On one level, the young women are merely being lined up, as in the illustration and as they would be expected to do in a finishing school. It was an issue of control, even dominance; Bibendum is aligning them. At the same time, the *barre d'accouplement*, a mechanical term, thus needed correction, and the young women would be happy to oblige. The ambiguity of what a "coupling bar" could be in the relationship between a man and woman provides a good part of the humor.

The above "Lundi," like many others far too numerous to recount and analyze here, illustrates Bibendum's somewhat ambiguous relationship with his perceived reader, and thus his role for the Michelin company. Above all, Bibendum was a man addressing other men of the same social standing. He could be quite serious when discussing motoring issues, as one might have expected from an early industrialist explaining the virtues of his products to the French public. At the same time, the articles read like private letters, with jokes and sexual innuendo that seem at times on the edge of inappropriateness in a public forum. Here the existence of Bibendum the young *mondain* son of bourgeois parents was key. Although Bibendum may have gotten his name from Théry's equation of André Michelin and Bibendum, Michelin's early advertising always referred to Bibendum as the son. He was presumably a young man, as an early Lundi refers to his having recently received his *bachot*, the French secondary school degree (*baccalauréat*) degree awarded at eighteen or nineteen.[63] In Arren's *Sa majesté la publicité*, André noted that Bibendum filled him with paternal pride.[64] In several "Lundis," Bibendum refers to "papa Michelin."[65] Yet although André is the assumed father, even that connection is unclear since Bibendum represented the firm as a whole; he could thus also be son to Edouard Michelin, who actually ran the company. In the end, by being an imaginary son rather than a real son, Bibendum had even more license. He could say things that would likely not have ever been attributed to André or Edouard Michelin, at least not in a public newspaper article.[66] Bibendum allowed for a certain distance, enabling a form of humor that could exist in the imaginary more acceptably than anything said by either of the Michelin brothers or their real sons.

Although the "Lundis" include occasional references to Rabelais, Michelin never specifically and consistently associated Bibendum with Pantagruel, another giant and the son of the giant Gargantua.[67] For a well-educated reader, Bibendum bears more than a passing resemblance to Pantagruel. He is strong, extremely active, virile, with a somewhat wild, even ribald sense of humor. His life in some imaginary world is the subject of unbelievable stories, constructed seemingly to make the reader shake the head and smile. Above all, Pantagruel's prodigious appetite for food and particularly drink are reflected in Bibendum's own appetite, especially in the frequent images of Bibendum "drinking the *obstacle*," which was inevitably in a champagne glass. What helped to make Bibendum so successful as a company icon is the extent to which he evoked many cultural images, including that of Pantagruel, without ever being pigeonholed as primarily one of those characters or images to the exclusion of others.

Bibendum even made an appearance as God (seated and smoking a cigar), offering the "Commandments of Bibendum." Here again, it seems doubtful that André or Edouard Michelin would have equated himself with God in the national press, so Bibendum made the advertising possible. Although there were actually thirty-two such Commandments offered in a "Complete Album," only ten appeared in a promotional brochure and one at a time appeared in the "Lundis," where Bibendum explained the technical specifics so that each Commandment might be followed. In the promotional brochure, illustrated by O'Galop, each drawing included a caption. Taken together the captions made a rhyming poem.

"Les Commandements de Bibendum" ["Bibendum's Commandments"]

Un seul pneu tu adopteras	Thou shalt only adopt one [kind of] tire
Michelin naturellement.	Michelin naturally.
Ta voiture tu pèseras	Thou shalt weigh thy car
A tout le moins une fois l'an.	At the very least once a year.
L'humidité éviteras	Thou shalt avoid humidity [bad for tires]
Qui provoque l'éclatement.	Which provokes blow-outs.
Dégonflez tu ne rouleras	Thou shalt not drive without air
Vingt centimètres seulement.	Even twenty centimeters.

have been missed by a prewar French reader—reveals intimacy, an intimacy that may have been acceptable from Bibendum but one that would have been exceedingly forward from an industrialist addressing a potential client.

Each *tableau* of the illustrated theater of the tire used some form of literary reference for an opener, often tying the subject of the play in the supplement to Bibendum's message. The first referred to a nineteenth-century masterpiece by Alfred de Musset, "On ne badine pas avec l'amour" [One does not joke about love], by being entitled "on ne badine pas avec le pneu" [one does not joke about tires], a citation that had no little irony given Bibendum's unceasing jokes about tires. Citing one of Bibendum's Commandments, "Flat thou shalt not drive / even twenty centimeters," this first *tableau* showed a tire massacred by underinflation.[71] Later *tableaux* showed tubes that had not been properly stored, the deleterious effects of the use of chains on tires (when the buyer tried to avoid purchasing nonskid tires), tubes that had not received talcum powder before being placed in tires, the effects on a tire of braking too abruptly, and so on.[72] The last page of the collected *tableaux* included an epilogue. Featuring Bibendum seated as God, more of the Commandments of Bibendum appeared, concluding with the most important: "Thou shalt not demand the impossible / from either the tires or their manufacturer."[73]

The "Théâtre illustré du pneu" shows how well Michelin could juxtapose the purely technical with the theatrical to maintain a reader's attention. The "Théâtre illustré" is a reminder of the assumed market for automobile tires—the only ones pictured in any of the illustrations. The perceived differences between the classes was a fundamental marketing tool in prewar France that Michelin used with aplomb. Clearly, Bibendum was addressing his social equals with *tu;* he was one of them as distinct from working-class men. The only explicit reference to social class in the "Théâtre illustré," however humorous, reminds the reader that he is like a knight in control of his serfs when he uses a tire. "The tire is not at all a 'class-conscious proletarian' and the driver has over him, like a medieval knight over his serfs, the power of life and death."[74] That is, in an age of increasingly organized labor, the reader had real, old-fashioned, unbridled control over his tires, like his automobiles. It was left to his discretion to treat them well, to be a responsible nobleman. As in the case of literary and artistic references and frequent recourse to Latin, Michelin frequently played on class distinctions, reflecting and reinforcing them.

never specifically and consistently associated Bibendum with Pantagruel, another giant and the son of the giant Gargantua.[67] For a well-educated reader, Bibendum bears more than a passing resemblance to Pantagruel. He is strong, extremely active, virile, with a somewhat wild, even ribald sense of humor. His life in some imaginary world is the subject of unbelievable stories, constructed seemingly to make the reader shake the head and smile. Above all, Pantagruel's prodigious appetite for food and particularly drink are reflected in Bibendum's own appetite, especially in the frequent images of Bibendum "drinking the *obstacle*," which was inevitably in a champagne glass. What helped to make Bibendum so successful as a company icon is the extent to which he evoked many cultural images, including that of Pantagruel, without ever being pigeonholed as primarily one of those characters or images to the exclusion of others.

Bibendum even made an appearance as God (seated and smoking a cigar), offering the "Commandments of Bibendum." Here again, it seems doubtful that André or Edouard Michelin would have equated himself with God in the national press, so Bibendum made the advertising possible. Although there were actually thirty-two such Commandments offered in a "Complete Album," only ten appeared in a promotional brochure and one at a time appeared in the "Lundis," where Bibendum explained the technical specifics so that each Commandment might be followed. In the promotional brochure, illustrated by O'Galop, each drawing included a caption. Taken together the captions made a rhyming poem.

"Les Commandements de Bibendum" ["Bibendum's Commandments"]

Un seul pneu tu adopteras	Thou shalt only adopt one [kind of] tire
Michelin naturellement.	Michelin naturally.
Ta voiture tu pèseras	Thou shalt weigh thy car
A tout le moins une fois l'an.	At the very least once a year.
L'humidité éviteras	Thou shalt avoid humidity [bad for tires]
Qui provoque l'éclatement.	Which provokes blow-outs.
Dégonflez tu ne rouleras	Thou shalt not drive without air
Vingt centimètres seulement.	Even twenty centimeters.

Chaque virage aborderas	Thou shalt take each turn [carefully]
A la papa tout simplement.	As would a father quite simply.
Chaque semaine tu liras	Each week thou shalt read
Notre Lundi dévotement.	Our Lundi devotedly.
De gros pneus tu la muniras	With big tires shalt thou equip [your car]
Afin de rouler longuement.	So that thou shalt go for a long time.
Des corps gras tu protégeras	From greasy substances shalt thou protect
Tes caoutchoucs jalousement.	Thy rubbers [tires] jealously.
Lentement tu démarreras	Slowly shalt thou take off
Et freineras très sagement.	And thou shalt brake very sensibly.
Dans notre Guide trouveras	In our Guide shalt thou find
Gîte, garage et monuments.	Shelter, garage, and monuments."[68]

Like any good humor, the Commandments worked on several different levels. First, they were brilliant in capturing the form of the Ten Commandments, with eight syllables rather than the classic French twelve-syllable alexandrine line. By rhyming all of the first lines in the sound *a* and the second lines of each stanza with the nasal *en* or *an,* the Commandments rhyme humorously in French. Moreover, because one's relationship to God is intimate, French versions of the Bible use the familiar *tu,* recaptured in Bibendum's Commandments. The Commandment to read the "Lundi" is not unlike a Protestant injunction to read the Bible.

In addition, the very association of the Ten Commandments with the care of tires involves some humor in mixing otherwise distinct categories of human existence. That is, Michelin places the Scripture in the service of things technical. Bibendum's Commandments exemplify how private companies could use religious traditions to market products, much as religious groups did.[69] The Commandments themselves resemble many of the "Lundis," both before and after the individual Commandments appeared in those articles. The necessity of using the same brand of tires on all wheels, of weighing one's car (to be sure that the tires were large enough to carry the load), of avoiding storage of tires in damp places, of not driving with underinflated tires, of not taking curves too fast (and thus damaging tires), of using tires that were sufficiently big, of avoiding the tire's contact with oil and grease, and of the importance of slowly accelerating and braking to avoid unnecessary wear and tear on tires were all mainstays of Bibendum's technical advice from the earliest "Lundis" in *L'Auto-Vélo*. By combining the religious reference with technical advice, Michelin again used humor to maintain the interest of the motoring public. Even God could be mobilized in the service of Michelin tires.

In addition to the "Lundis," Michelin began in 1911 to take out a one-page advertisement on the back page of the weekly *L'Illustration*'s theater supplement. With its rich photographs, *L'Illustration* catered to likely automobile buyers, the audience to which Michelin appealed to an even greater extent than in *Le Journal*, where Michelin devoted occasional "Lundis" to cycling. The *Illustration théâtrale* featured a new play each week. Michelin dubbed its own space the "Théâtre Illustré du pneu" [Illustrated Theater of the Tire]. *L'Illustration* offered high-resolution photographs of tires so that Michelin could show its readers tires and tubes in various states of disintegration and explain what caused the problems. The advertisements could also be in red as well as the traditional black ink of newspapers.

Eventually, Michelin gathered each of the illustrations into two brochures entitled the "Théâtre illustré du pneu." Every single one of the images featured Bibendum. In the prologue to the first brochure Bibendum explained his purpose. "In reading this Album, driving friend, you [*tu*] will be instructed at the school of others' misfortunes. And the performances [*représentations*] that it will give you will be the performance [*représentation*, thus playing on its meaning as both "performance" and "representation"] of profits, because they will allow you to save some money, without your place in my theater costing you anything. Then, in doing your books, you will quickly see that 'the tire of an informed driver is worth two.' And when you hear the daily refrain 'Life is dear [*La vie est chère*]!' you won't be tempted to also exclaim 'so is the tire!'—Signed Bibendum."[70] Again, the theme of service to the client and the desirability of saving him money are the stated reasons for the publication, along with the implicit desire to promote automobile usage. What distinguishes this prologue from the "Lundis" cited earlier, however, is Bibendum's bold use of the informal *tu* rather than the formal *vous* for the reader. If Bibendum were God speaking to man or vice versa, as in the Commandments, *tu* is the accepted form. Between Bibendum and a motorist, the usage—which could not have been missed by a prewar French reader—reveals intimacy, an intimacy that may have been acceptable from Bibendum but one that would have been exceedingly forward from an industrialist addressing a potential client.

Each *tableau* of the illustrated theater of the tire used some form of literary reference for an opener, often tying the subject of the play in the supplement to Bibendum's message. The first referred to a nineteenth-century masterpiece by Alfred de Musset, "On ne badine pas avec l'amour" [One does not joke about love], by being entitled "on ne badine pas avec le pneu" [one does not joke about tires], a citation that had no little irony given Bibendum's unceasing jokes about tires. Citing one of Bibendum's Commandments, "Flat thou shalt not drive / even twenty centimeters," this first *tableau* showed a tire massacred by underinflation.[71] Later *tableaux* showed tubes that had not been properly stored, the deleterious effects of the use of chains on tires (when the buyer tried to avoid purchasing nonskid tires), tubes that had not received talcum powder before being placed in tires, the effects on a tire of braking too abruptly, and so on.[72] The last page of the collected *tableaux* included an epilogue. Featuring Bibendum seated as God, more of the Commandments of Bibendum appeared, concluding with the most important: "Thou shalt not demand the impossible / from either the tires or their manufacturer."[73]

The "Théâtre illustré du pneu" shows how well Michelin could juxtapose the purely technical with the theatrical to maintain a reader's attention. The "Théâtre illustré" is a reminder of the assumed market for automobile tires— the only ones pictured in any of the illustrations. The perceived differences between the classes was a fundamental marketing tool in prewar France that Michelin used with aplomb. Clearly, Bibendum was addressing his social equals with *tu;* he was one of them as distinct from working-class men. The only explicit reference to social class in the "Théâtre illustré," however humorous, reminds the reader that he is like a knight in control of his serfs when he uses a tire. "The tire is not at all a 'class-conscious proletarian' and the driver has over him, like a medieval knight over his serfs, the power of life and death."[74] That is, in an age of increasingly organized labor, the reader had real, old-fashioned, unbridled control over his tires, like his automobiles. It was left to his discretion to treat them well, to be a responsible nobleman. As in the case of literary and artistic references and frequent recourse to Latin, Michelin frequently played on class distinctions, reflecting and reinforcing them.

Despite the initial gray and eventual carbon black of tires, Bibendum began as and remained a white man, making him a representative of the civilized, progressive West in the face of Africa and the East. In O'Galop's original poster of 1898, Bibendum was white, standing against a black background. The only technical reason for his coloring would be that tires came packaged in white paper; white could deflect sunlight so detrimental to natural rubber.[75] It is noteworthy, nonetheless, that Bibendum remained white with only a few rare exceptions,[76] even when portrayed outside Europe. Bibendum became a white European in a world dominated by European empires.

In the late nineteenth century, companies selling early branded products frequently used racial images of the "Orient" or Africa to advertise their goods. By the early twentieth century, this "commodity racism" was a well-established motif in European advertising, and its existence helps to explain the tenacity of European assumptions of superiority among groups of people who never read even the popular works of Social Darwinists. As recent research has suggested, the notion of racial superiority in Europe was not just constructed and reinforced by racial theorists, politicians, and educators, but also by manufacturers who used those racial images in the attempt to sell their products. Michelin fit squarely into this tradition.[77]

Some of the earliest Michelin advertisements, including the first fully colored advertisement with a certain date, reflected stereotypical views of the "East." André Michelin contacted Pellerin in 1898, the largest producer of *images d'Epinal*, to have an *image* created for the company. Pellerin in turn put Michelin in contact with artist O'Galop (who would do the Bibendum poster that same year). André Michelin's interest reflected the changing nature of advertising in the late nineteenth century. A vibrant market for *images d'Epinal*, named for a town in the Vosges mountains, had existed in nineteenth-century France when itinerant peddlers had sold the images across France. *Images* took a variety of forms, including short illustrated fairy tales and even political propaganda, in the form of representations of political leaders such as Napoleon Bonaparte, Napoleon III, and the General Georges Boulanger on horseback. But in the last third of the nineteenth century, in the face of competition from mass newspapers, the producers began to use the *images* for advertising the products of various businesses, including stove manufacturers, pharmacies, shoe stores, butcher shops, and toy stores. The typical story showed the travails of "a virtuous hero, aided not by magical talismans [or God] but rather advertised goods," who then achieves success.[78]

O'Galop drew an *image* for Michelin called "Le supplice de la roue" [The Ordeal of the Wheel]. With nine scenes on one page, O'Galop told the story of the young poet, "Ali Allo," who dared to look at the sultan's "pearl" as she passed in his cortege. Having thus committed a crime in Turkey, Ali was beaten by janissaries and eunuchs, and sentenced to torture by wheel. In his nighttime reflections, a woman appeared and the voice of the sultana murmured, "apa peurfi ston!" which supposedly meant "Don't be afraid" (*peur* meaning fear in French). The next morning, Ali was placed in a wheel and rolled over a cliff. But alas, at the bottom of the cliff, the unscratched Ali said, "encore." The sultan's assistant proclaimed, "By Allah, I suspected something, it is a Michelin." As the caption reads, "thanks to the sultana the wheel was equipped with a famous Michelin tire, which drinks obstacles." Ali received a pardon, and the sultan equipped all of his carriages with Michelin tires. The final image is that of the sultan surrounded by three black men with stereotypically large lips, men who were presumably slaves. The caricatures of the Ottomans are obvious, as is the presence of Africans, complementing a stereotypical view of the "Orient," where elements of the "other" got all mixed up. Pleased with the results, Michelin ordered one hundred thousand copies for distribution.[79]

A second *image* mixed sub-Saharan Africans with Moroccans. In 1905, O'Galop traced the "Life and Adventures of the Famous Bibendum," beginning with Bibendum's birth. The text under the drawings reads

There was once an illustrious academic who, on a mission in the Sahara, overcome by the heat, fell into a deep sleep. But an ostrich, happening past there, caught sight of his bald head. Taking it for an ostrich egg, she thought she should sit on it; in a few hours <u>naturally</u> it hatched. A bizarre being came out: digesting nails, because he was the son of an ostrich and <u>unpuncturable/tireless</u> [*increvable* means both], because it came forth from an <u>Immortal Being.</u> Because he called out forthwith/incontinently [*incontinent* means both], he was baptised BIBENDUM. Picked up by Moroccans, he was entrusted to the care of a buxom nurse named Drink-your-Nail [Boï-ta-Klou], who was wild about [drinking] nails and crushed glass; the child got a lot out of her. But [she was] soon reduced to a skeleton; Bibendum found himself forced to take . . . the glass of obstacles from [her] hands and gulp it down [underlining and capitals in the original].[80]

Although the *image* then traces the rest of Bibendum's life, including his competitions with other tires, this early section reveals annals about French perceptions of the non-Western world. A place of exoticism, North Africa offered an imagined birthplace for Bibendum. The drawing of the wet nurse is particularly telling in three different ways. First, the *image* offers no explanation for the presence of an obviously sub-Saharan woman of large proportions generally. Her presence, by not being explained, reinforces the image of Morocco as another, distant place where black and North African were presumably indistinguishable. Second, she represents the usual caricature of a large black woman with big lips and huge breasts. Third, despite her earlier size, this black woman is quite literally drained, weakened, and physically exhausted by a growing white boy then man, Bibendum.

In a poster with five scenes from 1913, entitled "A travers les âges," O'Galop again told the life of Bibendum, although this time the academic, the ostrich, and the "Moroccan" black woman were not present. Instead, Bibendum emerged from the clouds to the cry "Noël, Noël, the birth of Bibendum," thus replacing the Old Testament Bibendum of the Ten Commandments with Jesus Christ. Because, once again, Bibendum's appetite was "incontinent [in the French sense of having unbounded appetites], all of the negroes [*nègres*] spread out in hoards in the virgin forests and set about to gather and coagulate precious rubber. The white brothers snatched up, for a king's ransom, these large masses and hurried to carry them into numerous factories."[81] Of course, one might claim that because the portrayals were meant to be humorous, there was no harm done; no one could possibly have believed them, and these images did little to reinforce European images of white and other. It is nevertheless noteworthy that there was not a comparable self-deprecating humor that placed non-Western over Western. Moreover, only a few years after Leopold II of Belgium's abusive control of Congolese rubber gatherers became common knowledge at least among the political elite of Europe, even the reference to blacks profiteering from white manufacturers could not have been funny unless, on some level, the reader assumed without question that Europeans were superior to Africans, minimizing suffering in the Congo.

Even technical innovations could be explained with recourse to non-

Western peoples, here again with a mixing of various people and regions that revealed little knowledge of the world outside Europe. For drivers who found Michelin's introduction of the "lever," a tool for removing tires from rims, a potentially insulting invention given drivers' presumed skill in changing tires, Bibendum responds that one "should think of our overseas clientele, of our friends the colonies, of those who, in Brazil, in the Sudan, in Australia, as in Canada, only want to drive on Michelins. Do you think that these distant friends run into a nice tire dealer every ten kilometers who is ready to give them lessons in mounting tires? Their mechanics are negroes [*nègres*] or coolies." It was for them that the instruments were developed. "Today, they . . . repeat the following, 'Is good, massa Bibendum, is good way to mount big tires without pinching [them] or pinching fingers. No longer get . . . kicks in butt. Now have fun all the time!' [Y en a bon, moussu Bibendum, y en a moyen de monter gros pneus sans pincer, ni pincer doigts. N'y en a plus recevoir . . . coups pied derrière. Même chose rigoler tout le temps!]"[82] Michelin reminds the reader that the fine invention of the levers, supposedly developed for the less capable coolies and blacks, will ease the task for European men as well. In the process, the advertisement strongly reinforced the stereotypical black man's inferiority; he could not even speak proper French. Not surprisingly, the drawing at the top of the "Lundi" shows Bibendum with a colonial hat sitting in a rocking chair. To his right is a black man changing a tire, and to his left an Asian man, both of whom are characteristically smaller than him.

Stereotypical Africans also appeared riding bicycles on a prewar Michelin postcard drawn by "Gaugé." An extremely skinny black man with huge lips and oversized hands, wearing a top hat and a ring in his nose, describes why he has Michelin tires and his wife does not. She, a fat woman with big pants to cover her huge hips, sits passively on tires called "Noyau-de-Pêche." The man says, "Li, rembourrée, monter pneu Noyau-de-Pêche!—Moi, pas rembourré, monter bon pneu Michelin [Her, well-padded, mount Hard-As-Rock / Me, not well-padded, mount good Michelin tire]."[83]

Michelin advertisements consistently portrayed black people, both men and women, with big lips and hands. Important distinctions did exist, however, between images of men and women. The men were consistently skinny and the women were fat. Both wore few clothes. Even the chest of the supposedly Moroccan nurse was bare, as if she were not of, or at least living in, an overwhelmingly Muslim society. One might merely attribute the images to ignorance on the part of advertising people, but that argument does not explain why nonwhites were imagined in this way rather than in some other fashion. The portrayal of blacks in Michelin ads was inextricably tied to cultural notions of superiority and inferiority. For a West that usually associated clothing with civilization and even class status with more as well as better clothing, nakedness represented inferiority.[84] From the point of view of Europeans, whereas the ideal woman needed a small waist—even when big breasts and hips were desirable—the black woman was supposedly huge, despite the paucity of foodstuffs in Africa when compared to those available to the wealthy in Western Europe. Moreover, in the West, a civilization where men were dominant, for women to be huge and men skinny also implied the inferiority of Africans. Male Africans' consistently smaller size further tamed them, as Bibendum towered over them or rode on their backs.

Michelin was not unique in its portrayal of Africans. Turn-of-the-century advertisements frequently offered caricatures of Africans.[85] Advertisements for "La Végétaline," a sort of shortening with coconut oil, featured a poorly dressed black boy with big lips and pants rolled up to his knees, a bandaged leg, and one shoe missing offering up some *végétaline* to a French chef, dressed of course entirely in white, shaped like a pear, and standing with his legs apart, thus in a more dominant pose. The relationship of server to served is unmistakable.[86] Similarly, a poster advertising "Old Jamaïca Goodson Rhum" featured two strangely dressed black men. Although one is dressed better than the other, his color combination signifies his inability to properly dress himself. He is hitting a thinner black man with torn pants over the head with Old Jamaica Rum, so that the man drops his own generic rum. Both have the trademark large mouths and lips.[87] A whole range of European companies peddling food and beauty products, especially soap, used stereotypes of Africans and other non-Western people.[88] In the process, they propagated racial notions made ever more credible by force of constant repetition, reflecting and helping to create European cultures that consistently placed Europeans at the top of a presumed racial hierarchy.

BIBENDUM THE MAN: GENDERING THE AUTOMOBILE AND THE TIRE

Despite decades of acculturation in both Europe and the United States, from a physiological point of view there is nothing fundamentally masculine about driving, car ownership, car care, tires, bicycles, or automobiles, particularly

5. "The Michelin Bicycle Tire: The Best and the Least Expensive." Before the First World War, Michelin advertising often used the expression *le meilleur, le moins cher* although Michelin tires were not generally cheaper than those of its competitors. The poster does illustrate the superiority of the tires; whereas the man in pursuit is losing his tires on the narrow road, Bibendum can get away with the woman by riding over glass. The lesson, from the male Bibendum to the male tire buyer, is obvious: the man who gets Michelins will get the girl. Color poster by Louis Hindre, 1911.

MODANE

Prix du zinc 19.60

6. Particularly before the First World War, tire companies frequently used their success in various races to market their tires. The two races in Milan won by Michelin in 1907 metamorphosed into Italian women whom Bibendum brought back to France. Represented as women, these races (and a host of others) were "taken" by Michelin and its competitors. The company's advertising, like that of auto and other tire makers, thus reflected and encouraged the emergence of automobile consumption as a male domain. Drawing by O'Galop in *L'Auto*, 23 September 1907, p. 8.

Woman, says the Arab proverb, shares our pains, doubles our joys, and triples our expenses. One can say the same thing about the automobile. . . . The automobile, like woman, makes us see the country. . . . To steer her one needs softness, a certain touch, and one recognizes this experience among those who have gotten around [ceux qui ont beaucoup roulé]. You will tell me that the automobile has the superiority of silence. Would you be quiet! A pretty woman and a car, what could be better for expanding one's circle of acquaintances? The owner of a pretty car and the husband of a charming woman never risk having too few friends. . . . And which costs the most, whether you are talking about a woman or a car? Maintenance. . . . I can't pursue this comparison; it could go on into infinity. I can only draw one last conclusion, borrowed from the car and that should give pause to husbands in the mood for marital infidelity. There are mad men who have a pretty limousine and prefer to take a taxi.[96]

After Bibendum's first and only affirmation of the importance of men's fidelity, with a wink of course, he proceeds to suggest a redefinition of feminism as the recognition that the automobile is like a woman: "So am I right to claim that in automotive sports one must be a feminist?"

Parts of automobiles could also resemble women. In addition to the "Lundi" above in which women became wheels, they also metamorphosed into rims (*jantes*). "Like a girlfriend brings you her flowering youth and her tailor's bill, the rim requires a few moments of maintenance."[97] Parisian women in particular supposedly resembled automobile rims. Like a rim, "Parisian women, though the most beautiful women in the world, do not have any fewer of the innumerable qualities and rare faults inherent in their sex."[98] Whereas women were supposedly inconstant and flighty, men were solid yet flexible. *Le pneu,* the French word for tire that was masculine in gender, unlike the solid rubber tire, was ultimately flexible, so much so that he "resembles this ideal spouse . . . not at all jealous, and simultaneously rich and handsome, intelligent and young, dark haired though blond, majestic though slim, of whom all girls dream."[99] A tire, despite the gender of the noun, could also be a woman. As Bibendum encourages men to care for their Michelin tires by keeping them full of air, he asks what they would think of a man who, "having just married an ideal young woman, beautiful, rich, intelligent, and attractive," says, "well that was a good deal," and ignores her. "You would not do that and you should not do that to your tires either."[100] Bibendum also asserts, while implying disapproval, that some women prefer abuse: "It appears that there are women who like to be beaten, just like there are drivers who like to pay for unnecessary repairs by not taking care of their tires."[101]

Michelin advertisements using gender thus served, as in the case of race, to reflect contemporary French assumptions about the social hierarchy of France in the Belle Epoque. Men controlled women just as men controlled automobiles and supposedly inferior peoples. While some images, such as that of Bibendum riding off with another man's mate, might be seen as a spoof on some social conservatives' claims about the deleterious moral effects of the bicycle,[102] the company did not go so far as to feature women actually riding bicycles alone or, with a rare exception, driving a car.[103] In general, advertisements confirmed certain fundamental notions about the "essential" differences between men and women, making what were actually social differences seem like physiological ones, thereby injecting the idea of gender into these new products, reflecting broader cultural notions at the time and reformulating them along the way. The perception of a threat to established gender roles of "modern" or "new" women who cycled or drove after the war (just as they smoked or wore more "masculine" clothing or hairstyles) resulted at least in part from the way that those would-be gender-neutral objects became gendered before the war. Michelin, like other companies, thus took an important, although easily overlooked, role in defining the places of women and men in French society.

BIBENDUM'S FRANCE

Although it is clear that Bibendum was seemingly omnipresent in early-twentieth-century France, one cannot be sure of the reception of Michelin's advertising. The "Lundis" in *Le Journal* had the potential of reaching a mass audience. Postcards, posters, and handbills abounded among those groups most likely to be interested in cycling or motoring, and the *salons* served to make Bibendum known to the same constituency. The phenomenal growth of the Michelin company before the war could be cited as evidence, but sales are not direct proof of the effect of advertising. Michelin's competitors frequently featured a deflated Bibendum in their own advertisements, thus attacking the company's products but, strangely enough, reinforcing the place of Bibendum in the commercial imagination.

Certainly, a section of the political and social elite, centered in Paris, also had some familiarity with Bibendum and some used the phrase *Je bois l'obstacle*. In 1910, Senator Charles Humbert referred to a general in the French army as a "Général Bibendum," although the real Bibendum was quick to correct the senator who had meant by the term "an antique, an impotent senile who reviewed his regiment . . . in his chair with his sword in one hand and his cane in the other."[104] This was far from the image that Bibendum had been supposed to project of the Michelin company, although it suggests how widespread awareness of the figure was. The "Lundis" were quick to mention each time that *je bois l'obstacle* became a frequent expression in certain circles before the war.[105] When Georges Clemenceau at last managed to form a government in 1905, he announced to the press that he was able to "boire l'obstacle." On its cover, the satirical newspaper *Le Rire* then featured the head of anticlerical and nationalist Clemenceau with Bibendum's body, getting ready to drink a huge glass of champagne that included both a Catholic miter and a pointed German helmet.[106] In its anticlerical nature, this was a rather different image of Bibendum than that probably intended by the company. As André Michelin looked back in 1914, he was convinced that "Bibendum has contributed a great deal to the success of our company."[107] These references and André Michelin's assessment are evidence that Bibendum was hardly unnoticed, but there were no doubt a fair number of French people who might well have said about Bibendum, as did the famous peasant about Alfred Dreyfus, "Qui-ça, Dreyfus? [Dreyfus who?]" in the midst of the affair itself. Moreover, it is obvious that the projected image of Bibendum was not always the one received and reprojected by others.

In a sense, good advertising usually represents at least part of a given society, the part for whom the pitch has been designed. It would be tempting to argue simply that Bibendum represented French culture, but such a position would understate the ambiguity inherent in Michelin's advertising or the extent to which any representation of the supposed cultural whole is itself a presumption of what the whole should be. The unspoken omissions reveal as much as the open statements. Although historians have long maintained that the Michelin brothers were anti-Dreyfusard, there is no mention of Dreyfus in their ads; the subject was simply too controversial. Advertisements cannot afford to alienate potential consumers. It is not just what Michelin said about France but what it took for granted, including cultural references. For example, among the thousands of references to French literary figures, La Fontaine, Racine, Corneille, Descartes, Alexandre Dumas, Alphonse Daudet, Guy de Maupassant, and many others abound. In contrast, Voltaire (whose pithy comments were eminently quotable), Diderot, and Rousseau seem to have been largely forgotten. That is, the *philosophes* so admired by republicans by the time of the Dreyfus Affair are mysteriously absent. Similarly, Louis XIV's and Napoleon's pronouncements appear often, while those of revolutionaries and republicans generally do not. Unquestionably, images of Bibendum reflected important and widespread cultural assumptions in prewar France, particularly those held by wealthy French men who controlled that society. To the extent that such men successfully avoided efforts to redefine relations of class, gender, and race in ways less favorable to themselves before the First World War, advertising provided the kind of ideological support that constant repetition can offer.

2

FINDING FRANCE

The Red Guides and Early Automobile Tourism before the War

IN THE FIRST HALF of the twentieth century, Michelin stood out as one of the primary promoters of French tourism. Only the nonprofit association, the Touring Club de France, had a comparable role in defining what people should see and how they should get there. Yet, whereas in the late twentieth century Michelin could be associated with tourism generally, its early promotion began exclusively as the encouragement of automobile tourism, from which Michelin, as France's largest producer of pneumatic tires, stood to reap the rewards. Before World War I, the company left the definition of what one should see to the well-established Guides Joanne published by Hachette, the Baedeker guides, and assorted lesser-known guidebooks, focusing instead on the most pressing problem for early motorists, how to tour by automobile. Along the way, Michelin participated fully in the development, or *aménagement*, as the Touring Club consistently put it, of France.

In several respects, Michelin's early attempts to promote automobile tourism complemented, reflected, and became intertwined with earlier and simultaneous efforts of the Automobile Club de France, the Association Générale Automobile, and particularly the giant Touring Club de France. Established in 1890, founders modeled the Touring Club de France after the English Cyclist Touring Club formed some fifteen years earlier. The TCF, which ultimately advocated forms of tourism from hiking to touring by plane, initially organized cycling trips and other excursions for its members. The organization grew quickly to 65,000 members in 1900 and 104,000 in 1906.[1] The year 1895 marked the founding of the Automobile Club de France, whose membership was smaller and wealthier; both André and Edouard Michelin were founding members of this latter club and also kept themselves well apprised of the activities of the TCF.[2] The Association Générale Automobile, founded in 1895 with a membership that overlapped with the other two groups, also worked to advance the cause of automobile touring, particularly by developing early road signs; this group maintained the closest relations with Michelin, adopting Michelin's red guide as its own *carnet de route*.[3] After 1895, the Touring Club de France increasingly shifted its focus to the automobile, advocating the interests of automobilists as it had cyclists.[4] Although not supplanting the ACF or the AGA, the TCF remained the largest associational advocate of the interests of motorists as well as cyclists until after the Second World War.

Although ostensibly politically neutral, the TCF's political philosophy as revealed in the many articles of its monthly review, *La revue mensuelle du Touring Club,* was solidarist. A set of ideas most identified with the politician Léon Bourgeois, solidarism outlined a middle road between an absolute free market (with no state intervention) and socialism (with too much state intervention, in their solidarist eyes), claiming to avoid the extremes of both. Solidarism matched nicely the voluntarist, associational spirit of the group, although it is obvious that the group's sense of solidarity included fellow group members but not society as a whole.[5] The association became an outspoken advocate for better roads, the protection of historic and natural sites, and the banning of billboards, while offering its members the opportunity to share their touristic experiences in the pages of the review so that others might learn from their mistakes. In short, the TCF was an extremely active, even vibrant association of members who worked together for the common good of the group. Early tourist guides such as the Guides Joanne and the Baedeker guides had solicited corrections from their readers, but the TCF took such involvement much further by publishing the contributions of its members. Early bicycle and automobile touring thus instituted a level of participation in a larger process to promote touring, involving individuals to a greater extent than train travel had.

In 1900, Michelin tapped into the spirit of early touring as defined by the TCF. Michelin's free touring guides, its free itineraries from its own tourist office, its donations of road signs, its organization of a petition to have French roads numbered, and its cheap maps were all billed as service to the client, *service au client*, a veritable leitmotif of Michelin's publications both before and after World War I. Michelin made profits selling tires and absorbed the cost of these efforts, claiming merely to want to help its past and future clients. Above all, the company encouraged, at times even demanded, people to write in with corrections for the guidebooks, the itineraries, and the maps. As in the "Lundis," the company cast itself as the friendly assistant of motorists, an expert willing to fix problems and explain the use of its products. Like the TCF, Michelin referred frequently to one big family of tourists who needed to work together for the good of France. In developing France for touring, Michelin closely associated its name with the nation-state in an age of growing nationalist sentiment, much as the TCF did. In promoting tourism in France, Michelin could gently remind the client that Michelin tires were not only supposedly better tires, representing the quality of French artisanal production, an idea well established among the nineteenth-century bourgeoisie, but that they were French tires, as opposed to the tires sold by its primary competitors. Like articles in the TCF's review, Michelin claimed to be helping tourists to discover France, to be assisting in the economic growth that would result from greater tourism. In the process, the company contributed as much as any institution, and a good deal more than the French state, to the creation of a culture of touring in modern France.

THE GUIDES: REPAIRING AND SLEEPING BEFORE WORLD WAR I

In 1900, Michelin published the first *Guide Michelin* to France. Containing 399 thin pages with small print and a red cover, and measuring approximately 3 1/4 by 6 inches, the guide was designed to slip easily into a tourist's pocket. The preface noted that "this work desires to give all information that can be useful to a driver traveling in France, to supply [the needs of] his automobile, to repair it, and to permit him to find a place to stay and eat, and to correspond by mail, telegraph, or telephone." Although the guide itself focused on these

7. "The rail vanquished by Michelin tires." French, English, and Italian versions all portrayed men driving so fast that they outran the train. Until the middle of the first decade of the twentieth century, when likely automobile and tire buyers consisted of racing enthusiasts at least as much as bourgeois tourists, companies did not hesitate to use speed in order to appeal to drivers. In fact, the original French version pictured a peasant to the left, nearly run over and obviously frightened in the wake of the automobile. Soon after this poster appeared, both Michelin and the Touring Club appealed to motorists to drive sensibly. Color poster by E. Montaut, 1904.

simple facts, the company optimistically articulated its larger vision. "This work appears with the century; it will last as long. Automobiles have just been born; they will develop each year and the pneumatic tire will develop with them, because the tire is the essential organ without which the automobile cannot go."[6] By offering the guide free of charge, the company recognized that by encouraging automobile travel it fostered the consumption of tires.

Michelin announced in this first guide that an updated version would be published each year, allowing the guide to evolve with the needs of motorists. In this first edition, the fueling and repairing of patently unreliable early automobiles were the most pressing needs, so the contents were overwhelmingly technical, more so than in any subsequent editions. The guide of 1900 included three parts. In the first section (pages 17–50), the company described in excruciating detail how to use tires, how to inflate them, how to change the tire tube, how to reinstall the tube, how to change the tire, how to request that Michelin do repairs at the factory, and how to do one's own repairs, of both automobile and bicycle tires. Abundant drawings in black and white illustrated the various parts of the tire with the appropriate terminology. Early tires, and the valve stem apparatus in particular, were technically quite complicated, and the tubes and tires were fragile enough that mishandling was a serious concern. The first section also included a comprehensive list of all Michelin dealers, or *stockistes,* all of the small businesses that had contracted with Michelin to carry a full line of Michelin products (*les stocks*), divided into summer (May 1–October 1) and winter (October 2–April 30) depots. In all, forty-eight French cities had dealers in the summer, while there were only six that remained open all year. Not surprisingly, all of the latter (Biarritz, Bordeaux, Lille, Marseille, Nice, and Pau), with the exception of Lille, were in the south, where a portion of the upper bourgeoisie and aristocracy went for the winter social season. Although it would quickly expand, the list of necessary stocks to be considered a Michelin dealer included only a range of different sizes of tire tubes.[7]

The second section (pages 54–280) listed French cities and towns alphabetically. Here, too, the focus was mostly technical, with hotels reduced to one among many necessities while traveling; restaurants separate from hotels received no mention at all. The only two criteria for a town's inclusion in the list was whether it possessed a mechanic or a place to buy gasoline. In the early days of the automobile, before the installation of actual gasoline stations, motorists purchased gasoline in 2, 5, or 10 liter containers in small groceries (*épiceries*) and hardware stores (*quincailleries*) as frequently as in repair or bicycle shops. Michelin listed such establishments as well as the brand of gasoline sold, be it Automobiline, Stelline, Motonaphta, or others.[8] Above all, the list noted the address and phone number of the Michelin dealers and mechanics. For the thirteen provincial cities with rudimentary maps, the locations of hotels, mechanics, Michelin dealers, the railroad station, and places to buy gas all appeared on the maps. The third section (pages 281–344) included practical information about the rules of the road, taxes assessed on automobiles, information about maps, and a bibliography of tourist guides. This last section also included advertisements from various French automobile and auto part manufacturers with long descriptions of how to use and install their products. At the end of the guide, the company included a calendar so that the motorist might keep track of days of the week and holidays, sunrises, sunsets, the moon, distances traveled, the consumption of gasoline and oil, as well as how much the car owner spent on these.

Although historical accounts have focused on the vision and ingenuity of Michelin in producing the first guide, the company's primary contribution was in building a better mousetrap (at least from the perspective of people traveling by automobile) and rebuilding an ever better one with each successive edition of the guide. Completely overlooked by everyone who has even briefly praised Michelin's first red guide is the extent to which Michelin borrowed both its format and much of its information from the *annuaires*, or directories, published by the Touring Club de France in the 1890s. While Michelin did introduce the inclusion of its dealers, places to buy gas, and city maps, the TCF pioneered the comprehensive list of mechanics as well as hotels.[9] Michelin's genius was in adapting the TCF's approach and in quickly altering the red guides to changing conditions of automobile travel.

The Touring Club published its first *annuaire* in 1891, listing its leaders and regular members with their addresses. The *annuaire* also included a short list of towns with hotels that had offered the TCF guaranteed prices for meals and a room as well as a short list of towns with mechanics. As the membership grew, the TCF could not list all members, shifting its focus to practical tourist information. By 1899, the year before the first Michelin guide, the TCF's *annuaire* (gray in color, of approximately the same dimensions as the first red guide; the TCF's special *annuaire* for motorists sported a red cover) featured a list of French cities and towns with the number of inhabitants, the department in which each was located, whether it was the seat of a canton, subprefecture, or prefecture, and whether it had a train station, a post office, a telegraph office, each of which was represented with a symbol to save space in the list. The first Michelin guide duplicated both the items and the symbols. Moreover, the Michelin guide used distinctions established by the TCF in evaluating mechanics. Claiming that Michelin's close relationship with mechanics, who often made decisions about what kind of tires to stock and to install on automobiles, precluded the firm's objectivity, the Michelin red guide merely listed whether a given mechanic had met the TCF's certification procedure for minor or major repair jobs.[10]

Initially, Michelin's list of hotels differed markedly from that of the Touring Club. Although Michelin listed essentially the same hotels as the Touring Club and Automobile Club *annuaires*, Michelin did not—unlike the TCF—guarantee exact prices in 1900. Instead, Michelin had three categories of hotels with suggested price ranges; and Michelin did not vouch for the quality of the accommodations. Hotels marked with three asterisks in the guide were those where an average room, candle (for lighting before widespread use of electricity), and three meals including wine cost 13 or more francs daily. Hotels with two asterisks offered the same items for 10 to 13 francs, and those with one asterisk charged less than 10 francs.[11] After a hotel's listing, the notation ACF indicated that it was a place recommended by the Automobile Club de France, whereas the symbol TCF meant that it was a hotel where TCF members received the 10 percent discount negotiated by that organization on behalf of its members.

Michelin did not hide its debt to "similar works," pointing out instead the advantages of its own guide. Namely, it was free (the TCF *annuaire* cost a franc), regrouped information otherwise dispersed in three or four different works, and offered information about how to care for tires. More significantly, Michelin assured the reader that more and better changes would be introduced as early as 1901, making the red guide even better. "Next year it will give an indication that . . . will be much appreciated by readers; to facilitate the driving through a large city, it will use conventional symbols on a map of each important metropolitan area to show hotels, mechanics, recommended dealers, the placement of Michelin stocks, tramway lines, train stations, post offices, and telegraph and telephones that can reach Paris or nearby cities where one might find a replacement part at one of our dealers. Finally, the map will give the directions for going to nearby towns with their distances in kilometers." The small, barely legible maps in the 1900 guide were thus only a beginning. In addition, the company promised that in 1901, "We will indicate, in as many departments as possible the picturesque roads [*les jolies routes*]. We admit, in fact, that at the moment a good brand of automobile can make it through on all of the listed roads. But very often the *routes nationales* [national roads] are boring, whereas by taking a secondary road [*un chemin d'intérêt commun ou de grande communication*], one can take a more picturesque journey by prolonging the trip a bit." For this, the red guide was necessary because "if you ask the people from the area [des gens du pays] which is the prettiest way, most of the time they will tell you the one that is the most direct and the widest, which is often the least picturesque."[12] That is, the company already promised to enlarge the guide to include more touristic, and proportionately less technical, information. Moreover, the 1900 edition of the guide already captured a fundamental irony of twentieth-century tourism: often guidebooks, designed for people traveling by means of modern cars and trains, point out the supposedly timeless, beautiful countryside undisturbed by such modern means of transport. In this case, the locals think, with an appreciation for the modern, that a larger, paved road is better, whereas the tourist at the wheel of a car is looking for the picturesque road defined in part by the extent to which it is less traveled, less well maintained, and thus less modern.

Michelin's red guide most resembled the work of the Touring Club in its appeal for readers to assist in correcting and improving the guide, making a plea reminiscent of the Touring Club de France's constant calls for joining hands (se serrer les coudes) in order to work together. As Michelin put it, "The present edition . . . will inevitably be considered very imperfect, but the work will improve each year; it will be perfect as quickly as drivers respond carefully and in the largest number to the questionnaire we are asking them to fill out. . . . Without them, we are capable of *nothing;* with them, we can do *anything* [emphasis in the original]."[13] Like the TCF, Michelin also began to establish a certain control over mechanics and especially hotel owners by working with tourists. Already in 1899, the TCF had worked out an elaborate system for ensuring that its members were not taken by mechanics and hotels. TCF mechanics accepted a price list of maximum prices for routine automobile and tire repair; the TCF then purged those mechanics overcharging from the list. In return for a listing in the TCF *annuaire*, hotel owners made even more concessions. Hotel owners promised to offer a 10 percent discount to all TCF members. Moreover, the TCF made hotel owners guarantee the published price of room and board, including wine, for the duration of the year in which the *annuaire* appeared.

Michelin solicited readers' help in several domains, asking them to send in comments. First, readers were supposed to check Michelin's calculations of the distances between towns, which always appeared in the guides. Second, readers were to report any absences from the stock of Michelin products that dealers had committed to carry. Third, any mechanics listed who were not good or who were good but not listed needed to be reported to the company. Fourth, Michelin asked readers to report whether the sellers of gas actually carried gas. Fifth, Michelin wanted readers to write with any information about mechanics who could charge electric cars. Finally, Michelin wished to exercise oversight of hotel owners by ensuring that readers did not pay significantly more than the averages reported by the hotel owners. Michelin asked readers to provide specific details, "particularly whether there are bedbugs" in a given hotel. Michelin, like the TCF, assured readers that their corrections would be acted upon with rigor. "We promise to purge without pity any hotels that [drivers] report as having inadequate meals, rooms, toilets or service; and poorly stocked gas dealers."[14] Along with corrections, the company asked drivers to supply information about themselves, their car, and their tire brand, allowing for early market research as well as improvements to the guide. The company then promised that anyone writing in with corrections would have the guidebook mailed directly to their homes in 1901, signaling the importance of driver participation in the improvement of the Michelin guides.

Drivers' participation remained a constant theme. In 1901, Michelin's "Lundi" reminded clients what would happen if they avoided sending in corrections for the guide. If no one reports back, "Monsieur" will find himself entrusting his car to a mechanic who damages it instead of fixing it. "Madame" will exit the hotel covered with "little brown marks as disagreeable in odor as cannibalistic." Similarly, the 1902 guide reminded the reader that Michelin would remove mention in the guide of any poorly kept hotel, having bedbugs or making people with the Michelin guide pay to park their car, a cardinal sin for a company preoccupied with encouraging automobile tourism (both Michelin and the Automobile Club de France demanded free parking of all listed hotels).[15] Michelin quite articulately appealed to drivers' sense of belonging to a larger, although exclusive group that needed to stick together against predatory hotel owners.

Michelin appealed to men, whom it assumed planned the trip, drove the car, and used the guidebook, thus reinforcing turn-of-the-century gender roles regarding the automobile. Michelin addressed men's sense of patriarchy in this patriarchal society. "Yes you like [the guide] very much . . . like those weak parents who do not correct their children because the sight of tears gives them an attack of nerves. . . . *Drivers, the Michelin Guide must be your work. Please don't be easy on it for the sake of the Guide.*" It simply could not be left to others: "You are all part of the same big family. More than any other, [this family] needs its members to practice the motto *All for one and one for all.* In particular, you *fathers* give [the guide] to *your sons* during your vacation. Have them go through the details. Let those young brains so taken with novelties come up with an original idea [for improvement of the guides]. *Thus you will have well served the cause of motoring* [emphasis in the original]." The Michelin guide of 1905 even included a postcard to make the process easier, so that the driver might "serve all of his *brothers in motoring* [in italics in the original]."[16] The language was unmistakable; men were not only in control of their families as well as their cars, but motoring itself was a brotherhood, a fraternity of equals for which sons, but not daughters, might be prepared.

In 1908, bemoaning that "most motorists who drive across France have other things to do besides point out the deficiencies of the guide" so they never bothered to write in, Michelin set up a competition to see who could find the most corrections. The company devoted 3,500 francs, with the first-place winner getting 1,500 francs and the next seven finalists amounts that declined with one's placement. A committee of "eminent geographers and important personalities" in the world of tourism and motoring chose the winner. An elaborate system awarded varying numbers of points for different corrections. Whereas, for example, an error in the tabulation of kilometers from one town to another earned one point, the correction of the number of a *route nationale* earned 3 points. In the spring of the following year, after Michelin received all of the entries (and incorporated the corrections into the 1909 edition of the guide), the company announced the winners. The limited number of entries, 207, despite the awards at stake, is a fair indication of the divergence between Michelin's appeals to solidarity and reality.[17]

Over time, the Michelin guide gave increasing importance to accommodations over technical matters. In addition to more maps, more dealers, more mechanics, and more information generally, hotels became a focal point for improvement. Beginning in 1902, the guide included a questionnaire that a hotel owner wishing to be listed needed to fill out and send back to Michelin in Paris. The questions reveal much about early urban bourgeois tourists in the French countryside as well as about Michelin's efforts to make hotel owners more accountable, an effort that complemented the TCF's own rigorous work. The Michelin questionnaire reads:

1. Is your hotel open all year?
2. How much should an automobile tourist expect to spend daily at your hotel?
   —for an average room including service and lighting?
   —for breakfast in the morning?
   —for lunch?
   —for dinner?
   (Specify if wine is *compris*)
3. Do you offer TCF members a reduction of 10%?
4. Do you have hygienic rooms [*chambres hygiéniques*] in the TCF style?
5. Does your establishment have a bathroom?
6. Do you have advanced WCs [*des WCs perfectionnés*]? (We call advanced WCs those that are equipped with water flush mechanisms with mobile seat, and of which the walls are covered in tile or earthenware, kept in extreme cleanliness and always equipped with toilet paper).
7. Do you have a dark room for photography?
   —Does it have a red light, basins, and water?
8. Do you have in the hotel itself a covered garage?
   —Is it completely closed/locked [*fermé*]?
9. Do you make people pay for the garage? How much?
   —Do you agree to let people with the Michelin guide park for free?
   (All hotels listed in our guide agree to this condition)
10. Do you have a repair pit *in the hotel?*
11. Do you have a stock of gas *in the hotel?*
12. Do you have *in the hotel* a source of electric energy that would permit

motorists to recharge an electric car? Or only batteries for lights [*accumulateurs d'allumage*]? Do you know of any in your town? What is the address? What are the prices?

13. Do you have an intercity telephone? What is the number?

14. Do you have a telegraph number? What is it?

15. What are the sights [*curiosités*] to be seen in your town? (attach a page)

16. What are the interesting excursions to be made nearby? (attach a page)

In order to assure the honesty of the hotel owner, two members of the Automobile Club de France, or if there were no local ACF members, two members of the TCF, needed to attest to the accuracy of the hotel's responses.[18]

The questionnaire reveals the centrality of care of the automobile for early automobile tourists. Electric cars needed to be recharged and drivers of gasoline-powered cars needed to ensure the supply of gasoline. Before many automobiles were enclosed, a covered garage was useful, and given the value of cars, a locked garage a reassurance. Early cars were unreliable enough that a repair pit, a hole in the floor that allowed one to get underneath the car, might also prove handy.

The other most pressing questions concerned the level of accommodations. Michelin needed to verify the average prices of rooms it continued to list. At a time when a bathtub was not standard equipment in hotels, the company needed a specific statement that one was present. Tourists' growing expectations for hotels centered, however, on the WC. Although specific information about the WCs in a hotel had not been part of the first Michelin guide or the early TCF *annuaires*, WCs were important enough to merit detailed questions by 1902. Here the red guide reflects the preoccupations of an urban French bourgeoisie which was increasingly adopting new hygienic standards and bemoaning the lack of them among hotel owners in the provinces.

In the late 1890s, the regular articles in the *Revue mensuelle* of the TCF focused above all on WCs. In 1896, an article complained that either the installation or the maintenance of WCs was inadequate, even when they were present. In 1897, Emile Gautier called for "a crusade" in the pages of the review. Implicitly equating provincials without WCs with natives in the colonies, Gautier claimed that "cleanliness is an indication of progress, a sign of civilization. All savages are dirty," whereas "all civilized people are clean." Europeans were supposed to know better. "How many individuals, how many cities, in the heart of our European societies, so proud of their prodigious flowering that have not yet picked themselves up from the apathy of barbarian races!" He was convinced that the English provided the model for improvement, "one knows that the English people, at least those of the cultivated elite, are the cleanest race in the world: it is noteworthy to observe that it is also one of the most powerful, one whose influence is simultaneously the most widespread, the most profound, and the most solid." There was, however, cause for hope. The French needed to go beyond English standards of cleanliness; to do so, France required a "national league," and that league was the TCF, who would lead the crusade for WCs for the sake of the French nation. "It is a crusade to undertake. To succeed, one need only want it. The Touring-Club wants it. It will thus have deserved [praise not only] from the nation [*patrie*] but from humanity. So be it!"[19] Regularly thereafter, the TCF reported on the progress of equipping hotels with WCs. In 1899, it provided drawings of WCs that it offered for sale to hotel owners, according them a 25 percent reduction from the retail price, with the difference covered by the TCF.[20] The TCF similarly subsidized purchase of toilet paper to encourage its adoption. The TCF even showed its model toilets at the Universal Exposition held in Paris in 1900 along with an entire "hygienic [hotel] room."[21]

The WC could even guarantee the economic progress of France. In 1901, an article in the TCF *Revue* entitled, "The Defense of National Interest," reminded readers that "France is both better gifted [in touristic treasures] and worse served than most of her neighbors."[22] WCs were key to the salvation of France because they helped to make French hotels competitive with those of other countries. As H. Berthe wrote in the *Revue*, "We should thus consider the hotel industry not as a private enterprise, but as an essentially national work, destined to raise up in large measure the intellectual level of diverse social classes [he does not indicate how] and to contribute powerfully to the financial prosperity of the country [*pays*]."[23] Thus, clean toilets in hotels could improve overall hygienic standards of lower classes while bringing in money that would enrich France as a whole. Hotels, and the WCs in them, were thus key to the economic future of France. André Michelin, so attuned to trends in the world of tourism and arguments for it, no doubt realized the centrality of

WCs for rich French tourists in the first decade of the twentieth century, hence the preoccupation of the red guides.

In 1908, Michelin completely reorganized its presentation of hotels, eliminating some of the uncertainty of the earlier rating system. Rather than divide the hotels by the average price of daily room and board as had been done since 1900, the guide began to place hotels in one of five categories from "the most sumptuous palace to the good village inn." Then, as the Touring Club had done since the late 1890s, Michelin engaged listed hotels to set a minimum price for each room and each meal, prices that anyone possessing the Michelin guide would be charged. Thus the tourist knew in advance how much room and board should cost in a given hotel and could ask Michelin to intervene should the hotel not honor the published prices. The action removed the uncertainty for the tourist and continued to provide free advertising for the hotel, while reinforcing tourists' collective control over the latter.[24]

The 1908 guide also eliminated potential conflicts of interest by refusing advertisements from hotels. The guides had, since 1900, been littered with advertising that reflected the perceived desires of rich motorists: automobiles, auto parts, bicycles, furs, various other luxury goods had pride of place and so did hotels.[25] The company continued to take advertisements for goods or services besides hotels until after the First World War.

In the years before the war, the Michelin red guides became a major marketing device for the company. Print runs climbed from 35,000 for the 1900 edition, to 52,815 in 1901, to 70,000 in 1911, to 86,000 in 1912. The 1900 edition contained 400 pages with 13 city maps, the 1901 edition 600 pages with 80 city maps, the 1912 edition with 757 pages and approximately 600 city maps.[26] The guide received a hard cover, so that it would hold up better, and it took on a larger format, reaching by 1912 the rough dimensions of late-twentieth-century red guides. Although tourist sights, or *curiosités* as both the Touring Club and Michelin consistently called them, received one-line entries under the nearest city, the red guides did not provide any significant details as did other guides. Michelin did, however, expand the approach of the red guides beyond France just as it had expanded its tire production and sales outside France. By 1912, Michelin published guides to the British Isles, the Alps and Rhineland (northern Italy, Switzerland, Tyrol, Bavaria, southern Württemberg, the Rhineland, Belgium, Holland, and Luxembourg), *Les pays du soleil*

(Algeria, Tunisia, Egypt, southern Italy, Corsica, and the Riviera), Spain and Portugal, and Germany in several different languages. In 1912 Michelin boasted that the combined international and French distribution of guides had totaled 1,286,375 between 1900 and 1912. The staff had the responsibility for more than 1,300 city maps.[27]

The red guide complemented Bibendum in representing the firm in France (where the guide was published in both French and English). As in the case of Bibendum, the red guide reflected not only changing expectations of early automobile tourists, but also the perceived familial relationships among those tourists. Men presumably cared about their cars and women about how bed-bugs might hurt their appearance. Men too busy to keep track of errors in the red guide needed to delegate that task to their sons, so supposedly interested in novelties and things technical; thus advertising for the red guide reinforced the notion that boys were naturally outward bound and technical while girls and women worried about their appearance. Advertisements for the red guides also played on the idea that men, the providers, needed to supply women with a comfortable place to stay.

In one telling "Lundi," Michelin recounted the story of newlyweds traveling without a red guide. After the chauffeur informed them that a mechanical breakdown would leave them stranded overnight, the Viscount René de la Ribaudière (a name suggesting bawdiness) and Giselle, his new wife (the text notes that "she was not yet [really] the countess"), got a room in a hotel that was, according to the owner, "the best in the region." After retiring to their room, they found a bat, and

> it took a quarter of an hour and all of the eloquence that M. de la Ribaudière had in order to calm down Giselle. However, the little viscount did not waste any time, and he quickly addressed his very imminent wife [*sa très prochaine femme*] the most legitimate compliments on the beauty of her legs and the finesse of her ankles [*attaches*], when suddenly he cried out in distress. "Ah! my God, what is the matter?" Giselle asked him. [He replied,] "my darling, where did you get this bit of red on your shoulder which was so white a moment ago?" The same exclamation came out of both of their mouths, "Bed bugs." They killed 10, then 100, then 577; they could not have fought off the yellow invasion with more ardor. Finally, overtaken by sleep, Giselle resigned herself to stretching out on her un-

comfortable and hard bed. And the viscount wanted to begin the conversation again. "Oh, no, my dear," she told him . . . When the sun rose, Giselle was still not yet Madame de la Ribaudière, though she looked like cream with strawberries [that is, her cream-colored skin had red marks resembling strawberries].[28]

By playing on the notion of a legal consummation of the marriage, Michelin could politely make the point that the viscount, however desperately he may have tried, did not get to have sex with his new wife because he had not ordered a copy of the red guide, so he did not realize there was a fine hotel nearby. The idea, no more unfamiliar to an early-twentieth-century reader than to a late-twentieth-century one, that men wanted sex and delicate women were more reluctant, was thus confirmed. Having not fulfilled his role as good provider, the viscount could not fulfill his role as a man in the act of sex. Thus, the red guide—which began ostensibly as a list of mechanics and places to buy gas—could assert certain assumptions about the appropriate behavior of men and women in French society: men were supposed to take care of the practical details while traveling, by buying a red guide and handling the chauffeur, and women were to worry about their appearance.

The red guides also allowed Michelin to associate itself ever more closely with French national identity. The TCF appealed to a sense of patriotism, broadening the concerns of individual tourists into veritable national issues. The group desperately wanted clean toilets in provincial hotels so that their own relatively recent expectations of proper hygiene might be maintained while traveling. On some level, at least some members of the TCF believed that more foreign tourists really would visit France if the country could maintain the reputedly high standards of hygiene in British, German, and Swiss hotels. Some may even have believed that hotel toilets might serve as a model of cleanliness for provincials, thus elevating them to a higher "intellectual level." In the process of associating WC installation in hotels with the national interest, however, the TCF pursued its constant theme: that tourism was good for the country. It was, in the words of the *Revue*, an "important economic factor."[29]

While we cannot evaluate whether such a thing was true, it is obvious that touring France and making France tourist-friendly became closely associated by the first decade of the twentieth century with French patriotism, at least

among tourists. Michelin, by advocating French tourism in a language so reminiscent of that of the TCF, thus also subtly reinforced a certain intellectual connection between Michelin and France. Like the TCF, Michelin claimed that it was opening France up for both French people and foreigners. According to Michelin, the automobile made it possible for the French to see "the prettiest country in the world, ... to admire in detail the innumerable aspects of the most diverse country in the world, their own." Meanwhile, the red guide made it possible for them to do so with convenience; the "Guide Michelin has revealed France for the French."[30] "The auto has finally permitted the French to discover their country and our guide shows them the ways to profit from such a discovery, which is worth every bit as much as the discovery of America [presumably by French tourists as well as Columbus]."[31]

Michelin's ability to associate its company with the French nation became even more obvious in its legal and promotional struggles with the German tire company Continental just before World War I, much as it had battled Dunlop at the turn of the century.[32] Since the 1890s, Michelin and Continental had been competitors in France and across Europe. But given the early size of the French automobile market compared to that of Germany, Continental's success rested on penetration of the French market as well as the German, and the company used strategies quite similar to Michelin's in order to help tourists associate Continental with France more than with Germany.[33] In 1904, Continental's own guide to France first appeared, looking remarkably like Michelin's in format and in listings, and it made the same appeals to motorists to help each other by sending in corrections. In 1910, playing on Continental's role in making rubberized cloth for French dirigibles and airplanes and thus tying the firm to French aviation, Continental began to call its publication a "Road and Air Guide," although it included very little information specifically for airborne tourists.[34]

Michelin claimed publicly and in court as early as 1910 that Continental had plagiarized the Michelin guide. In April 1913, the Tribunal de Commerce de la Seine ruled that Continental's 1910 guide did indeed plagiarize the Michelin guide. Authorities seized 32,000 copies of the guide, and Continental paid 10,000 francs in damages. In July 1914, the Parisian court of appeal upheld the ruling. Michelin's legal victory provided superb ammunition for its battle against Continental as a foreign firm posing as a French one, thus reinforcing the Frenchness of the Michelin guide and of the Michelin company.[35] Conti-

nental had publicly maintained in the automotive press that the company actually produced automobile tires at its plant in Clichy, but in the course of legal battles, Continental admitted that, while other rubber articles were produced in France, tires were not.

After 1913, in the "Lundis" and in pamphlets that showed stereotypical Germans with funny hats and plaid clothes cutting up the Michelin guide and sneaking in their foreign tires, Michelin stridently pointed out that Continental was a foreign firm that had attempted to "naturalize" and "Frenchify" (*franciser*) what were in fact products made in its Hanover factory while simultaneously "abusing French hospitality" by copying the Michelin guide. Although in the "Lundis" Michelin attempted to soften the tone by only referring to Continental as "X," anyone who knew anything about tires was aware of the case and Michelin had referred to Continental as "X" at least since O'Galop's first Bibendum poster in 1898.[36] On the eve of the First World War, Michelin artfully used its own "French nationality." In announcing the appearance of the 1913 red guide, the company noted that the guide, in the service of France, took "a continued effort on the part of an elite personnel. Two statistics will convince even the most skeptical: we have so far paid more than a half million just for the salary of our employees charged with the Michelin guide and spent more than 200,000 francs gathering original documentation, and we are being plagiarized." Michelin then made the nationalist charge. "One must be very bold and not very respectful of the laws where one receives hospitality to not only 'denationalize' one's products and pass them off as French but also to brazenly plagiarize a French work [*une oeuvre française*]."[37] Thus, Continental not only was a bad foreigner by passing off its German products as French, but it also went so far as to plagiarize a "French work." Michelin, confirming its patriotic image, employing a term usually used for literature rather than for a list of hotels and mechanics, thus implied that the red guide itself was a sort of cultural expression, a French literary work (*une oeuvre française*).

INDIVIDUALIZING THE ITINERARY: MICHELIN'S TOURIST OFFICE

Since the nineteenth-century expansion of the railroads, travelers in France had the itineraries of their trips set by the existence of train lines. In its train station bookshops, Hachette sold hundreds of editions of the leading French guidebooks, the Guides Joanne, named for their founder Adolphe Joanne. The

Joanne, like its competitors, organized guidebooks according to train routes. Several Guides Joanne followed a single railroad, such as the one from Paris to Lyon, which listed the seventy-two stations en route with the exact distance of each from Paris and Lyon. Even the title, the *General Itinerary of France: The Paris to Lyon to Mediterranean Line* (the Paris-Lyon-Méditerranée was the major line to the southeast), revealed the predominance of the rail line. But even within the guidebooks to a particular historic region, such as Normandy or Brittany, the guides were organized by the railway from Paris to get to the region (the assumption was always that the tourist was a Parisian seeing the provinces) and also by railway lines within the regions.[38] Similarly, the guides published in French by Karl Baedeker's firm, the well-known German producer of guides, divided France into four regions: the northeast, northwest, southeast, and southwest. Within each region, however, Baedeker organized the guidebook into various possible itineraries, which were by definition limited in number because they too took the train lines for granted.[39] In the 1890s the Touring Club de France also published possible itineraries for excursions essentially for bicyclists, who often took the train to get to an area for excursions, went by bicycle, and returned by train, or met a train that carried their bags to the destination. Even in this case the railroad determined the trajectory of the trip.[40]

By the first decade of the twentieth century, motorists began to make intercity trips. Because they were traveling by roads, automobile drivers could not easily recognize where they were headed. Articles published in the TCF's *Revue mensuelle* describing people's travels included not only what they saw but how far they traveled and often what route they took. The TCF also sponsored the publication of a series of guides to roads, concentrating on the road conditions and grades of hills so cyclists and motorists knew what to expect.[41] The Association Générale de l'Automobile edited a series of itineraries by individuals belonging to it, the Touring Club, and the Automobile Club, which read much like directions, to aid automobile tourists otherwise wandering between cities, dependent on the local populace and a few maps that had not been designed for motorists.[42] There were few road signs except the occasional *borne*, or milestone, which only occasionally included the number of the road. There were few signs at crossroads indicating which road went where. In short, if tourists were to reach their destinations, they needed detailed directions, not just within cities but between cities.

Early Michelin guides, like the TCF *annuaires,* recognized the problem and provided short notations of the distance and eventually the road that one would take to get from one town to nearby towns. Within the alphabetical list, a tourist could thus read in the 1900 Michelin guide under La Rochelle, "Paris 439 kil[ometers]." By the end of the decade, however, the company began to supply ever more information, including itineraries for excursions from a principal city to surrounding villages in 1909. By 1912, the guide included a special section of more than fifty pages that outlined possible automobile excursions.[43] Although no doubt helpful to tourists given the lack of information on the roads themselves, the guides allowed for little variation for individual drivers not interested in taking the usual routes—whether it was one selected for the condition of the road or its picturesque quality—from one city to another.

Michelin, like other auto makers and tire makers, had a vested interest in helping tourists find their way. Given that the automotive world remained small, elite, and based in Paris, with its concentration of automobile owners, tourist offices were a viable alternative for helping individual clients. Like Continental, which had established a tourist office in its Parisian headquarters, Michelin established a tourist office at its address on Boulevard Pereire in Paris. In 1907, in a "Lundi" devoted to the appearance of the 1907 guide, Michelin stated that "we are happy to announce that our Guide Service in Paris is at the disposal of all motorists to supply them with all useful information: studies of itineraries, documentation on the best route to take, indications for the purchase of maps, the delivery of brochures from the syndicats [d'Initiative] or the regional Automobile Clubs, etc."[44] The tourist office served each individual motorist as he "established a practical and reasonable itinerary" for a trip. In "Lundis" and in its guides, Michelin repeatedly told tourists they would find "a large well-lit room with comfortable easy chairs, a complete collection of guides for all countries, detailed maps, stereoscopic photographs that permit them to see in advance the route they want to take, and especially the *oral and detailed* [italics in the original] information that nothing can replace."[45] The individual tourist could thus receive free, personalized service that encouraged him to travel and, hopefully, his sense of brand loyalty.

Michelin claimed special expertise so that it could tell a tourist which route would be the shortest with the fewest *obstacles,* which was boring or in poor condition, and finally "the one that you should take if you have a little time."

Those who could not come personally could send a letter tracing the projected voyage and in a few days they would receive a complete and detailed typewritten itinerary. The company assured tourists not to fear to use the service intensely, noting that they had received a letter requesting an itinerary including "Paris, the Rhine valley, the Black Forest, Munich, Innsbruck, Tyrol . . . the Italian lakes, return to Paris via Mont-Cenis, Savoy." Michelin wrote that the next day the client received an itinerary of ninety pages. The service, even by mail, emphasized personal contact with the firm. "Thus will we maintain useful and constant relations with our clients and friends, for their own good and for the greater progress of tourism, whose methodical and reasonable organization needs to be here as in Switzerland one of the components of the national fortune."[46] As usual, the company asserted that it wanted to develop tourism, thus automobiles and ultimately tires, using the same language of friendship, solidarity, and patriotism found among the associations, notably the Touring Club.

Michelin also used other cultural images, such as assumptions about men's marital duty in planning trips, in order to encourage use of the tourist office and reliance on Michelin. Just after the First World War, in its weekly article in *L'Illustration*, Michelin provided three drawings of men. In the first is a man who did not prepare for his trip. He has become completely lost, and his elegantly dressed wife looks furious. In the second, a content man in a bathrobe leans back in his chair, smokes a cigarette, and says "allô Michelin" into the phone as he orders his itinerary. In the third, the man has taken to preparing his own trip. His surroundings are more modest than that of man number two, indicating that he simply does not understand the keys to success in earning any more than in touring. He sits frustrated at a messy desk. His wife stands across the room next to the bedroom, looking tired and ready for bed.[47] Clearly, as in the case of the Michelin guides, the successful man who can keep his mate and himself happy relies not on a do-it-yourself definition of masculinity but on one that allows Michelin to help him.

MAPPING FRANCE FOR THE AUTOMOBILE TOURIST

Ideally, tourists used maps along with the detailed itineraries in order to visualize their destination. In fact, a good map had the potential of nearly displacing the itineraries, showing the best routes, both in terms of the size and condition of roads and of their "picturesque" or "boring" views. In the 1890s,

France had excellent maps just as it was so often reputed to have the best, most dense network of roads in the world. However, existing maps drawn for the French state were of little use to cyclists and motorists. As historical geographers have recently pointed out, maps have always been drawn for specific purposes and need to be read not as transparent reflections of reality but as partial constructions of that reality.[48] Nineteenth-century maps thus reveal only part of reality but much about the makers and the administrative division of the French state, an important backdrop for the later development of tourist maps.

The most obvious reason for the French state to commission maps was their military function. In the course of the nineteenth century, French Army Staff Headquarters (the *Etat- Major*) undertook the mapping of all French roads. Between 1832 and 1882, beginning in the north and finishing in southeastern France, engineers in the employ of staff headquarters developed a map of France in 267 separate pages with a scale of 1/80,000e, or one centimeter for every 0.8 kilometers.[49] Given that the intended user was the French army, the topographical information necessary for troop movements and planning for battles was the primary focus. The roads did not include their official numbers (except in the case of the *routes nationales* [national roads]), because the army was not responsible for road maintenance and needed only to know of their existence and how passable they were. Knowledge of whether a road was atop or at the bottom of embankments had obvious military uses as well; of course the army did not necessarily expect an enemy ambush deep in the French interior, but European armies in the nineteenth century were kept for internal as well as external threats, hence the necessity of knowing where a unit might be ambushed.

The Ministry of Public Works and the Ministry of the Interior had maps of French roads because they oversaw maintenance and new construction. Answering to the Ministry of Public Works, the Corps des Ponts et Chaussées, the elite corps of engineers created by Louis XV, had as its primary responsibility the *routes nationales* (national roads), the roads organized, numbered, and formerly named the *routes impériales* under Napoleon and the *routes royales* during the Restoration. Combined with canals, these roads had been the arteries of the French state before the advent of the railroad. Because these were the major roads, the corps did not need maps as detailed as those of the *Etat-Major*, so the maps were drawn to the scale of 1/200,000e, or 1 centimeter for every 2 kilometers. Done in 3 colors with 141 different pages, only 100 of which were complete by 1900, this map distinguished different kinds of roads (that is, the *routes nationales* from secondary roads) and the "agricultural, industrial and administrative physical circumstances that had an influence on traffic."[50] The corps could thus correlate maintenance with the level of use of various sections of the *routes nationales*.

Since the 1830s, *agents-voyers* (road surveyors/inspectors) reporting to the Ministry of the Interior supervised work on local roads, the *routes départementales*, the *chemins de grande communication*, and the *chemins d'intérêt commun*.[51] Completed between 1879 and 1893 with the information from the inspectors, the Ministry of the Interior's map, which relied on information from the *agents*, distinguished the roads for which the inspectors were responsible with five different colors.[52] Modeled on the *Etat-Major* map but in the scale of 1/100,000e, or one centimeter for each kilometer, this map had the advantage of allowing tourists to measure their distances more easily, a fact necessary for calculating speed as well as distance traveled because these secondary roads often lacked milestones (*bornes*).

None of the maps worked well for early cyclists or motorists. Most were too big to be easily handled. None provided both the detailed topographical information, such as the grades of hills and the quality of roads, with numbers and a scale of distance that made it easy to calculate the number of kilometers between two points. After marketing maps of the *Etat-Major* in the early 1890s, in 1897 the Touring Club began to make its own maps for cyclists to be completed in time for the Universal Exposition of 1900, in an effort overseen by Henry Barrère, a trained cartographer. Given the Touring Club's close connection with government engineers, including those in the elite Ponts et Chaussées, the participation of the membership in helping to supply Barrère with information made the TCF's assessments of the quality of given roads a vast improvement over earlier maps. The Touring Club's maps were drawn in the scale ideal for cyclists at 1/50,000e (1 centimeter per 0.5 kilometers) for the area around Paris and at 1/400,000e (1 centimeter per 4 kilometers) for the rest of France, the latter divided into fifteen separate pages.[53] The automobile manufacturer Dion-Bouton edited tourist maps in four colors that also distinguished *routes nationales* and *départementales* from the lesser *chemins de grande communication* and *chemins d'intérêt commun*. They listed the number of kilometers between cities and noted the more difficult grades, making it, according to the Michelin guide of 1900, "specially suited to the owners of fast automobiles." Its scale, 1/800.000e (1 centimeter to 8 kilometers), made it considerably less detailed than the others.[54]

André Michelin's earlier work as a cartographer in the Ministry of the Interior allowed him to combine his understanding of what automobile tourists needed from a map with the knowledge of what was technically possible.[55] Even the first Michelin guide of 1900 reveals an appreciation for maps, in its early inclusion of thirteen city maps, in its promise eventually to include maps of all French cities, and in its two-page description of various maps that might be useful for tourists. In 1905, Michelin edited a map for the "Circuit d'Auvergne," the French elimination race for the Coupe Gordon-Bennett competition held near Clermont-Ferrand. The map showed not only the race course but also the other regional roads that tourists interested in seeing the race or Auvergne might need. The map resembled that of the Ministry of the Interior in distinguishing carefully between *routes nationales, départementales* or *grande communication,* other roads accessible to automobiles, and those lanes and walking paths not accessible by automobile, although it also included information about hills and sharp curves. It did not include the numbers of the various pictured roads.[56]

The company continued to develop and refine a map that would be useful for automobile tourists. In 1906 the red guide to Belgium, Holland, and the Rhine valley began to include a map on the scale of 1/1,000,000 within the guide itself. In 1907 a map of France, again on the scale of 1/1,000,000, first appeared in the red guide. Although Michelin assured its users that this was only the beginning, the maps represented several innovations. By using a large scale and breaking the map into seventy-two pages within the guide, Michelin enabled a motorist to open the guide and see 160 x 160 kilometers without opening up any bigger map. According to Michelin's calculations, even the fastest of drivers would not need to turn the page more frequently than every fifteen minutes.[57] The new Michelin map distinguished between *routes nationales,* where drivers could depend on a good road, and secondary roads. However, the map's distinctions among secondary roads depended not on their classifications into *routes départementales, chemins de grande communication,* and so on, so important for administrators, but rather on the quality of the road itself, which was what mattered to the motorist.[58] Within the 1907 guide, Michelin admitted that the map was very definitely a work in progress.[59]

In 1908 Michelin offered a complete map of France both in the guide and for purchase in four big sheets designed for planning trips, for which the map of France divided into seventy-two pages was unworkable.[60] In 1907, recognizing that motorists might prefer another map to the one inserted in the guide, Michelin began selling small leather map holders with a celluloid window so that motorists could fold the map, then protect it from the elements as they drove.[61]

Between 1905 and 1909, André Michelin undertook a series of steps toward perfecting a tourist map for France. Using the Ministry of the Interior maps as a base, he sent an extensive, detailed survey to the engineers in Ponts et Chaussées and to the *agents-voyers* in order to get accurate information not just about what kind of roads existed or vague notions of their viability but also very detailed information about each stretch of each road, so that the quality of the stones (*pavés*) or macadam (compacted crushed stone), the grades, the turns, the railway and tramway crossings could be inserted.[62] This early work culminated in a detailed map of Auvergne included in the red guide in 1907, which reappeared free of charge as the first part of an entire new map of France in 1909.[63]

In 1910 Michelin began to sell its new maps at one franc for the paper version, 2 francs for the cloth. Done in the scale of 1/200,000 (1 centimeter for 2 kilometers), the map divided France into forty-seven individual maps. As a tourist map, it was a clear breakthrough. The maps were the first to unite information about the size of the road, about the condition of the road at any given point, about its surface, about whether a road was picturesque or very picturesque, and about whether railway crossings were above the road, below the road, or at the same level (the bane of motorists who hated to wait for the attendants at the *passages à niveau* to open the passage). Moreover, the maps clearly distinguished the size and administrative importance of towns, as well as the locations of churches, châteaux, ruins, and other sights (*curiosités*). The maps showed with small markings the grades of different hills. Most important, the maps were the first tourist maps to include the number of each road and the distance in kilometers between intersections, so that motorists could use milestones to be sure they were on the desired road and to check their progress (useful before the widespread use of odometers). Finally, Michelin resolved the problem of portability. Unlike earlier tourist and government maps, the Michelin map folded accordion-style, so that one could move from one fold to another without opening the entire map and refolding it. It was even small enough to fit into a large pocket.[64]

In January 1910, Michelin began with the publication of the maps covering the Riviera in time for winter touring by the wealthy spending the winter months in the south. It then published the maps for Paris, those covering the primary route between Paris and the Riviera, via Lyon, and finally those for the rest of France.[65] By 1913, the task was complete, although the map for Alsace-Lorraine would be added after the First World War. Like earlier tourist and government maps, the actual information on the maps—so new in 1913— quickly became dated. By the 1930s Michelin had committed to biannual reeditions of the guides in order to keep pace with the tarring of roads and other improvements necessary as automobile use became more widespread.[66]

LABELING AND NUMBERING PROVINCIAL FRANCE

Maps, however well-drawn, provide an abstract overview of an area toured. They are obviously not photographs and do not look like what tourists see on the ground. French towns did not have their names posted at the town limits, because earlier travelers either knew the roads and towns en route, or they stopped to ask. Early motorists, who were both highly conscious of their social position (and somewhat reluctant to rely on locals who might resent their expensive contraptions) and preoccupied with speed, could not easily and quickly distinguish where they were. In a sense, the touristic development, or *aménagement*, of the French territory for use by tourists was a process in which the resolution of a "problem" frequently led to another "problem." That is, Michelin's tourist maps of 1910–13 solved a fundamental *obstacle* for early tourists by including notations of the precise road they would be taking so that they might match the road on their map with the road they were on, as Michelin so often instructed its "Lundi" readers to do. But milestones were not always present because they had had only limited use for anyone besides government employees or the army; local users certainly did not need the markers as they hauled goods to nearby railway stations. Even in cases where the markers had been well maintained, they had not been designed for use by motorists, who could not read the stones without getting out of their cars and examining them carefully. Thus the availability of maps with names led to the perceived necessity of providing towns with signs noting entry and exit as well as to the renovation of milestones for motorists' use.

In 1910, at the same time that the company began to publish tourist maps with road numbers, Michelin began offering free signs to all of the cities, towns, and villages of France, complementing earlier efforts of the TCF to

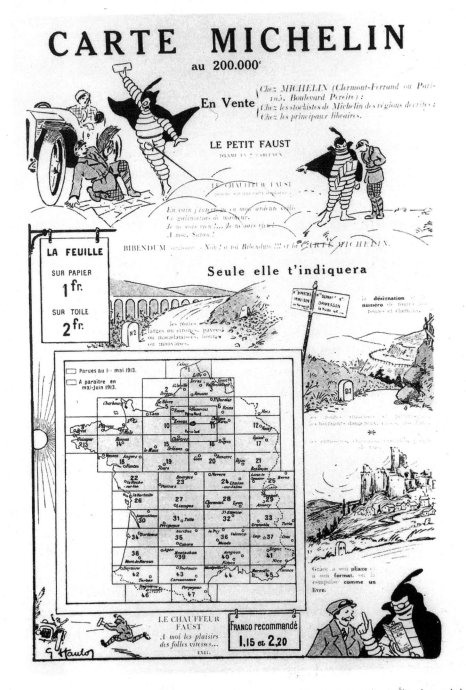

8. Before the First World War, Michelin marketed its tires to the wealthy, including classically educated the-atergoers. In a "Little Faust in 7 scenes," Faust was lost, unable to use his inadequate map, crying out at last "à moi Satan [come to me, Satan]!" Bibendum, as would Mephistopheles, corrects him, "No, you mean, come to me Bibendum and the Michelin map." For the motorist, the truth could thus be obtained with a good Michelin map, apparently without the driver having to sell his soul. From "Théâtre illustré du pneu," which grouped together ads from *Illustration*'s theater supplement, c. 1913.

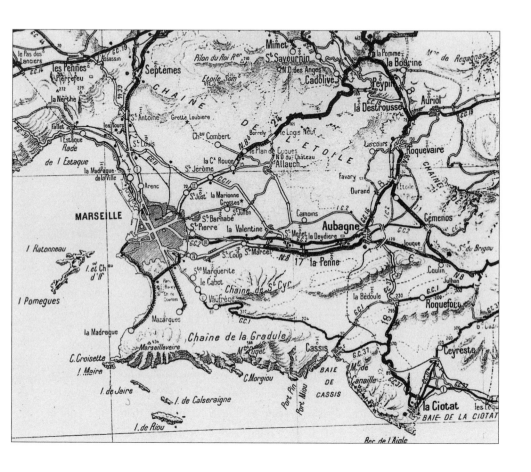

9. Motorists with early Michelin maps needed to be trained to read them. Here a cutout from the map for the region around Marseille instructed potential map buyers. Because each marking meant something (thus allowing for greater clarity), Michelin offered a complicated key for reading them. Eventually the colors, widths, and topographical signs became accepted as the norm, much as late-twentieth-century Americans would assume that blue lines are interstates and green lines earlier turnpikes as Rand McNally featured them. From the brochure "Ce que Michelin a fait pour le tourisme," c. 1912.

place road signs across France. Michelin's signs were 1 meter wide and 0.60 meters long and hung from a horizontal shaft attached to a building at the edge of town. Upon entering from one end of town (thus on one side of the sign) one read "Donated by Michelin / Please slow down / [the name of the town] / Watch for children" and then the name of the road, such as Route Nationale No. 24. On the back of the sign, as one exited, one could read "Donated by Michelin / Thank-you / Route Nationale 24."[67] Michelin gave its signs to mayors just as the TCF had, gaining the company some positive publicity both among current motorists and anyone else eventually buying tires. It offered the notation to slow down and look out for children as an effort both to tame motorists and to avoid more stringent speed limits on the part of municipalities. Joking about contemporary fears of depopulation, Michelin explained its intentions.

> Slow down / Look out for children. Rest assured. This is not a Malthusian [meaning birth control in French] appeal that we are issuing to young households. On the contrary, we have no other desire than to serve the national cause of repopulation in reminding motorists that there are cases where speeding is not only a danger for them. However little you drive on the roads of France you must know that nothing bothers a driver more than all of these disorderly reminders of order that are placed in front of him at the edges of villages: *Prescribed speed: 2 kilometers per hour! Slow down or face a fine! No faster than the pace of a man!* . . . Even God contented himself with ten commandments. . . . The Frenchman, a born rebel, does not like to be bothered like that; he considers such instructions to be provocations, and he doesn't hesitate to thumb his nose at them. It is only a question of listening to each other, and all of these questions can be resolved with mutual courtesy [emphasis in the original].[68]

In short, the signs avoided more rigorous speed limits and supposedly improved relations between motorists and mayors at the same time that TCF members were adopting a more moderate approach to the issue of speed limits.[69] Interestingly, Michelin did not ask motorists to slow down for the sake of individual children, whom the TCF *Revue* often portrayed as unpredictable urchins who needed to learn to stay out of the street, but for the sake of French repopulation; the welfare of individual children was humorously poured through the sieve of the patriotic discourse of early-twentieth-century French advocates of tourism and pronatalism. Whatever mayors thought of Michelin's approach, the towns had a financial incentive to take the company up on the offer of a free sign and free installation. Of course, the biggest technical innovation of the Michelin signs was for tourists; the signs included the road number, so that tourists would know when entering or leaving town whether they had taken the right route. In December 1911, the company congratulated the mayors of ten thousand French towns who had ordered Michelin's signs. Thirty thousand signs were in place by 1914.[70]

Michelin also had an important influence on the effort to number French roads. The Touring Club had presented a model sign that included road numbers at the first International Congress of the Road held in Paris in 1908.[71] In April 1912, the Office National du Tourisme, in cooperation with the TCF and the ACF, undertook the installation of a whole array of new metal road signs, including signs at each crossroads indicating the direction and distance of nearby towns on the entire route from Paris to the Norman resort town of Trouville. The estimated cost was 51 francs per kilometer.[72] Michelin argued that the ONT was wasting taxpayer money and proposed instead the painted numbering of all French roads.[73] Thus, having provided French tourists with maps, Michelin now advocated that all roads have their names and numbers on milestones so that tourists could easily and quickly find their way by combining the abstract overview of the map with the specific markings on milestones.

When the head of the ONT criticized "diverse persons [who] published articles in newspapers" advocating a numbering system that would make people dependent on "a special map," he noted that the ONT experiment actually better served all users of the roads.[74] In the face of what it considered bureaucratic opposition, Michelin launched a drive for a petition in favor of its proposal, unveiling the petition in its booth at the yearly aviation salon at the Grand Palais in early November.[75] André Michelin presented a huge volume resembling a *livre d'or,* or visitors' book, to the crowd. Government officials always made an appearance at these occasions, and in this case the president of the Republic, Armand Fallières, was the first to sign the petition, giving Michelin useful grist for the mill. The company promoted the petition in the "Lundis," in other advertisements, and indirectly by means of various newspapers reporting the campaign. Michelin distributed 80,000 sheets to people willing to gather signatures and send them back fully signed.[76] National newspapers had a field day, either penning their own articles or reprinting a prepared one by André Michelin.[77] By early December, 200,000 people had signed the petition (at a time when there were about 125,000 cars in operation in the country),[78] and Michelin delivered copies to the Ministry of Public Works and the Ministry of the Interior.[79]

In its advocacy of the numbering of roads, Michelin used the same patriotic language in favor of developing tourism that the company and the TCF had consistently employed. Numbering French roads became a matter of pride and of economic growth. According to Michelin, it received a letter from a certain "Lord Jimmy" from Britain who pleaded that the French adopt Michelin's system; thus British tourists would spend their money in France rather than in Switzerland or Italy. In an advertisement in *Le Plein Air*, Michelin asked, "Why do rich foreign tourists hesitate to visit our country that they love and admire? Because it is very difficult for them to get their bearings on our roads that are so poorly marked. It would be different if they were numbered. France would then see the development of automobile tourism take off, and she would become *Switzerland*, rich and prosperous as a result of motoring [emphasis in the original]." The illustration above and below the text featured a cowboy and a woman (riding in back), a stereotypical male American Indian and a woman (who is slightly behind him even though they are in a coupe), a black man with huge lips and a top hat driving with a woman and a baby (in back, barefooted, facing backward), an Englishman and woman (again in back), a Russian man and woman (in back), and a Chinese man and woman (in back). Playing on all of the stereotypes, particularly of those of non-Europeans, Michelin made the point that if France had numbered roads, every people in the world would come to tour.[80]

In pushing for the numbering of roads, Michelin also played on the social distinctions between urban French bourgeoisie and stupid peasants with bad French that were made by early motorists out in the provinces. When milestones were not numbered, according to Michelin, one gets lost and then "at last there is a native [*un indigène*]." "Hey, my friend! The route to Fouilly, please," says the driver. The "native" responds, "Oh! mon pauv' mossieu, vous n'y êtes point! A l'avant-dernière à gauche, faillait toruner à vot' drete. Mais vous vous retrouvez facilement. Continuez cor un brin, puis prenez le chemin qui tourne d'vant le cabaret du grand Colas, à eune bonne petite lieue d'ici. Ensuite à drete manquez pas d'passer sul' pont. Puis la deuxième à gauche et ça vous mènera directement!" That is, the "native" gives him complicated directions with such a heavy brogue that the driver could not hope to understand him. Michelin then comments that "our driver will be a lucky devil if the directions from the native don't lead him into new errors and lamentable wanderings on impassable lanes." Moreover, *indigène* implicitly compares him to a "native" of the colonies, a frequent comparison at the time.[81] In an illustration in the "Théâtre illustré du pneu," a motorist asks a local Norman peasant in sabots how to find his way in the absence of numbered milestones. The dialogue has the request in perfect standard French, "Say there, my friend? We would like to go to Bretteville?" The peasant, too stupid to even understand or too stubborn to offer directions to Bretteville, responds with a heavy accent, "Eh ben! j'vous en empêche-t-y, moué? Allez-y, quiens. J'vous r'tiens point!"—in short, roughly translated, "go ahead, I'm not holding you back, am I?"[82] Clearly, the advantage of numbered roads would be that drivers would no longer have to depend on locals for help; they could see the sights and appreciate the place, that is, France, without having to deal with people, particularly rural Frenchmen.

Michelin won its campaign. In March 1913, the minister of public works ordered the painting of milestones essentially in accordance with Michelin's proposal. On the side facing the road, the stone would have the abbreviation *N* (for *routes nationales*) and the number of the road as well as a notation of which kilometer it was. On the side that a motorist first saw would be towns, the next one and the next major one after that. A month later, the minister of the interior followed suit, ordering that stones have the abbreviations *D*, *GC*, or *IC* (for *routes départementales, chemins de grande communication,* or *chemins d'intérêt commun*) with the road number on the side facing the road. On the side that motorists first saw would be nearby towns, provided that the stone was large enough for the additional information. Michelin rejoiced that its stereotypical bureaucrat, "Mr. Office" (Monsieur le Bureau), was finally doing his job.[83]

DEVELOPING FRANCE

By 1914, Michelin had implemented a veritable tourist system in France. It supplied guidebooks, prepared individual itineraries, and designed maps to be used in concert. By helping tourists plan their trip, find their way, and get a

10. Attempting to persuade the French government and driving public of the necessity of well-marked roads, Michelin compared the motorist to a pedestrian in Paris. A pedestrian near the Arc de Triomphe could use street signs and a map of the city to take the shortest route; so too, Michelin argued, a driver ought to be able to find the name of a road and visualize where it went. The class difference between motorists and others is taken for granted; while a peasant in sabots and patched trousers paints a milestone, the map of Paris features one of the most exclusive neighborhoods of the city, presumably one with which the reader was familiar. From the "Théâtre illustré du pneu," c. 1913.

good hotel, all while escaping the clutches of hotel owners and peasants, the company legitimately claimed to have reduced the place of "risk" for travelers, even to have "suppressed the unexpected."[84] In short, a whole apparatus of touring could allow automobile tourists to have the "liberty" that automobile travel supposedly entailed, but to have it packaged so that "they would not be exposed to the dangers of lost children in the forest of the ogre."[85]

Michelin participated fully in the development of automobile tourism, working alongside and taking cues from the associations that advocated tourism so fervently in prewar France. Catherine Bertho Lavenir has pointed out that the French government, despite its reputation for centralized control, generally responded to associational lobbying rather than taking the lead in touristic development in the early twentieth century. The Office National de Tourisme had, to be sure, a role of coordination that increased over time, but the TCF and the related automobile associations with their high-ranking members, including presidents of the Republic, ministers, and government engineers, exerted crucial pressure at several junctures. Michelin similarly worked not only to make France accessible to tourists but also, as in the case of the petition for the numbering of roads, to force the government's hand.

In promoting automobile tourism, Michelin obviously stood to increase its sales. A bigger market meant more opportunities for the tire maker. But the company's actions went further in that Michelin placed the company in the service of the client, proclaiming the wonders of visiting France and the importance of making changes that might entice foreigners to do the same. The company's adoption of the TCF's patriotic and solidarist language helped to associate the Michelin name not only with tourism but with France itself. The company praised the way that the automobile made it possible for the French to admire the "most diverse and the most varied nation in the world: theirs."[86] With the proper numbering of roads, according to Michelin, "foreigners will leave their wealth . . . in the country that they all prefer the most, France."[87] Moreover, Michelin's use of social distinctions, between wealthy car owners and rural peasants, or between men and women, may have reflected widespread bourgeois assumptions in the Belle Epoque, but it also served to strengthen them by force of repetition. Above all, Michelin was not a foreign firm like Continental or Dunlop, but a French firm providing the kind of service that bourgeois men might expect in the form of "service to the client." The company thus reinforced both French national identity as well as the collective urban, bourgeois, and male identities of individual prewar tourists. In the end, Michelin's experience reveals that a private company—like government, schools, newspapers, political groups, and other actors—had a role in defining and disseminating a form of national consciousness among the French bourgeoisie.

3

TOURING THE TRENCHES

Michelin Guides to World War I Battlefields

NOT SURPRISINGLY, World War I brought an intensification of various manifestations of French nationalism, penetrating aspects of everyday life previously closed to such patriotic frenzy. Advocates for tourism, who had used a patriotic language before the war but at the same time encouraged international travel and even use of Esperanto, quickly adopted a virulent anti-German rhetoric.[1] Everything, including appeals by companies wanting to change wartime governmental regulations regarding their business, or to get permission to export goods, came clothed in the same patriotic garb. Although companies limited their advertising during the war, it focused on what companies and their employees had done to save the nation and civilization itself from the German "barbarians."

In a practical sense, the Great War presented Michelin with fundamental challenges. Although the company had through its promotion of tourism associated its own name with the French nation before the war, the very limits of its product offerings and the relative absence of publicity early in the war made it difficult for Michelin simply to continue the marketing strategy of the prewar years. Above all, despite Michelin's advocacy of national interests, it produced pneumatic tires, not the solid rubber ones necessary for the war effort. Military leaders eventually realized just how useful motorized transportation could be, a realization embodied by tales of the role of Parisian taxicabs in the battle of the Marne and of the *voie sacrée,* the road that supplied Verdun in 1916.[2] But the motorized vehicles most central to the war effort, trucks and buses, had solid rubber tires capable of crossing rough terrain without concern for blow-outs. Some military vehicles and bicycles did have pneumatic tires, supplied principally by Michelin, but they provided neither a large market for the company nor an important symbol of its contribution to the war.

In an effort to preserve its workforce and to continue its business, Michelin undertook a multitude of initiatives during the war. Initially, the company created a hospital, instituted some of the most generous benefits for widows and orphans in France, and produced assorted goods to supply the army. Pursuing its interest in aviation, Michelin also began producing Breguet airplanes for the war effort. In the area of marketing and public service, maintaining its central role as a promoter of French tourism, Michelin undertook the largest interwar effort to produce guidebooks to the World War I battlefields, the first of which appeared in September 1917, more than a year before the armistice. In the next four years, the company produced a total of twenty-nine such guidebooks, covering the entirety of the western front. Including translations and later reeditions, at least fifty guidebooks ultimately appeared.

By and large, Michelin guidebooks to the battlefields used a relatively moderate, objective, and outwardly apolitical tone. Designed above all for a well-off tourist traveling by car, the guidebooks helped to create and solidify an interpretation of World War I as a defensive war in which the fatalities and casualties of ordinary *poilus,* the French nickname for footsoldiers (meaning "hairy," thus strong and virile), were subordinate to the victories of armies led by larger-than-life generals. In this interpretation, suffering resulted solely from German aggression, while French leadership and valor ultimately prevailed. Marketing of the guidebooks also reinforced an idea—dear to the French Right but largely shared across the political spectrum in the interwar years—that stagnant population growth weakened France before and after the war, making "depopulation" a more important cause of the war than the diplomatic entanglements of the Belle Epoque. Michelin continued in yet another way to associate its interests with French national interests, helping to define the French experience by writing one of the earliest and most widely disseminated versions of its immediate wartime history.

MANUFACTURING AND PATRIOTISM DURING THE WAR

The outbreak of the war forced Michelin to scramble in order to maintain and then improve its market position. The company immediately faced the loss of most of its male skilled labor force when the war began in August 1914. At the outbreak of the war, Michelin did not have the right to maintain any conscripted employees at their usual posts because the company's production was not deemed essential for the war effort. In contrast, Bergougnan, Michelin's competitor in Clermont-Ferrand that produced solid rubber tires, had the right to maintain its skilled laborers at their jobs. Moreover, although the army was willing to buy Michelin stocks for its own depots at prices considerably higher than those paid by auto makers to Michelin, the army had limited needs for pneumatic tires, and these needs were largely satisfied by existing stocks. Meanwhile, the company could not sell either to private individuals in France or to foreign clients. Both Michelin and Bergougnan successfully argued for the right to sell any production in excess of the army's needs (but there was virtually no internal civilian market), and the government even lifted the early ban on exports in September 1914, although multiple authorizations continued to close the export market to both firms.[3]

Annie Moulin-Bourret has convincingly argued that Michelin's "spectacular patriotic initiatives" begun between August and October 1914, like those of its competitor Bergougnan, must be seen at least in part as an attempt to influence the government and public opinion as the company aggressively appealed for the right to sell its goods, and particularly to export them to allied and neutral countries. In August, the company transformed a warehouse into a state-of-the-art military hospital to receive war wounded. Michelin began to produce rubber bags for horses' feed, gloves, tents, and sleeping bags in order to contribute to the war effort. On 6 August 1914, André Michelin announced in an open letter to President Raymond Poincaré that Michelin would provide a million francs to Poincaré to be distributed to any pilots undertaking spectacular actions during the war. In October, the company began to offer an award of 5,000 francs for every French pilot who shot down an enemy plane. As early as 20 August 1914, the company even announced that it would give a hundred airplanes to the government so that a bomber squadron might be developed. The offer turned out to be merely for the bodies of planes designed by Breguet to be produced by Michelin with Renault engines supplied by the government, meaning that Michelin covered about one-fifth the cost of the squadron while imposing the model of airplane and the creation of the squadron itself.[4]

Airplane production ultimately helped to keep Michelin from closing its doors and laying off those employees not drafted. Closure would have meant not only hardship for workers, no doubt a concern in a rather paternalistic family firm, but also the loss of remaining skilled workers, including many wives of drafted men. Production of the Breguet airplanes along with bombs and the devices for dropping them, to a greater extent than the technologically simple goods, allowed Michelin to argue for scarce resources during the war. Although the early versions of the Breguet airplane promoted by Michelin were not well-received by pilots, later designs proved more effective. By 1918, Michelin was producing 6 percent of the airplanes built in France.[5]

By the end of the war, Michelin had fully succeeded in preserving its labor force. In 1917 and 1918, demand for Michelin tires grew as the more heavily mechanized American units arrived in France, requiring pneumatic tires. Despite struggles at the outset of the war, Michelin ultimately profited handsomely, causing no little tension between Edouard Michelin and a French state that attempted after the war to impose a hefty tax on war profits. Michelin's workforce actually grew during the war years by 57 percent to almost seven thousand employees. In the end, the war years allowed Michelin further to outpace Bergougnan (whose workforce grew by a third, to 2,015 employees), its only potential internal rival by the end of the war. Furthermore, Michelin's initiatives, including its hospital, its substantial aid to large families as well as to widows and orphans, and its airplane production allowed Michelin to claim during and after the war that its long-term interests were essentially patriotic. By 1916, well before the company's or France's eventual wartime success was obvious and as the company kept an eye open for ways in which to associate its name with French national interests, André Michelin had the staff that normally produced the red guides and maps begin work on the guides to the World War I battlefields.[6]

In launching the guides to the battlefields, Michelin managed to tap into a vein of language about the First World War used by both the Touring Club de France and the Office National du Tourisme during the war. Initially, tourism itself ground to a halt in 1914 and advocates for tourism, expecting a short war, suspended their efforts. The *Revue mensuelle* of the Touring Club de France did not appear between August 1914 and April 1915, and then it appeared only every other month. When the Touring Club did emerge from silence, its language was far more virulently patriotic than before the war. Gone was its old, somewhat nationalistic cosmopolitanism. Germans became "barbarians" and *Boches,* and Germany *Bochie* (roughly, Krauts and Krautland). The group organized the sending of packages to soldiers and the equipping of army units with wagons to sanitize water supplies paid for with patriotic fund-raising such as the "day of the 75," celebrating the French 75 millimeter gun. At the same time, however, the TCF did not merely use its strong association for such typical wartime relief. Rather, it began in 1915 to do what it had always done so well, work as a sort of lobbyist and cheerleader for tourism. The resulting problem was obvious: as *poilus* died by the tens of thousands, the TCF was advocating tourism for pleasure. There were few civilian-military divides as extreme as the difference between the son of a peasant or worker dying in the trenches while wealthy tourists enjoyed the seaside or took a cure in a posh spa. At least in part, the outpouring of quite overt flag-waving of the TCF legitimized (no doubt within their own minds since their patriotic entreaties seem quite sincere) their simultaneous promotion of touring during the war.

Although the Touring Club had long argued that tourism allowed French and foreigners to "discover" France and to aid in its economic development by spending money, the fusion of tourism and patriotism became a leitmotif in the *Revue mensuelle* during the war. In August 1915, Henry Defert, vice president of the TCF, argued that Baedeker guides, so valued by the *snobbisme* of some French tourists, would have to be banned from France after the war, along with so many other German products; he neglected to mention, as regular readers no doubt would recall, that the TCF had frequently run advertisements for Baedeker guides to France before the war. Throughout the war, the Baedeker guides were a *bête noire* for the TCF, to the extent that the organization signed on as a sponsor of Hachette's attempt to replace the Baedeker guides with a new series of guides, distinct from its earlier Guides Joanne, called the Guides Bleus. By doing battle against Baedeker, the TCF fought the enemy by proxy. Tourists were supposed to do their duty by touring in France (sans Baedeker), just as soldiers did theirs by fighting and sometimes dying. In 1916, before the United States joined the war, the president of the TCF, Abel Ballif, argued that while soldiers pursued their *revanche* (against Germany for the Franco-Prussian War and for the annexation of Alsace-Lorraine), civilians needed to undertake the *revanche* of commerce and industry, preparing for postwar tourism of the battlefields, including an influx of Americans interested in making a "pilgrimage" (*pèlerinage*) to see the battlefields.[7]

Each summer for the duration of the war, the TCF exhorted Frenchmen of means to see France for the sake of the country. In September 1916, Léon Auscher, another vice president of the association, reported that summer tourism had seen a considerable improvement over 1915. He was especially pleased that well-heeled French people, who before the war had traveled outside France, now had to remain within the country. He thought that the greatest obstacle to yet further tourism, now as before the war, was the dearth of hotels. Above all, French spas and hotels needed to reopen. Both spas and many hotels had been requisitioned by the government at the outbreak of the war and remained in government hands as housing and recuperation areas for soldiers. Auscher attacked the French "administration" for not having, like the British, built "clean, well-ventilated, quickly built barracks" that would cost the government less than damages to the hotels. Above all, he wrote that the "stubbornness" of government bureaucrats jeopardized postwar tourism and the "millions in foreign gold" that postwar tourism could bring.[8]

This surreptitious attack on the right of *poilus* to stay in nice hotels is not evidence that the Touring Club was unpatriotic. Rather, it shows that their notion of patriotism could not be separated from their assumption that society was supposed to have fundamental and unalterable distinctions. The TCF argued constantly that France was a prettier, better country for tourism and that the best French hotels and spas rivaled those of Germany and Switzerland. At the same time, however, these tourist destinations were inappropriately luxurious for common footsoldiers who belonged in hastily built barracks. It is also noteworthy that the TCF could see itself as better representing national interests, in the form of advocating development of the hotel industry, than the government and its bureaucrats. This was not a new idea, particularly not on the French Right, where the army and the Church had been held up as better manifestations of the "true France" than politicians and bureaucrats paid by the French Republic.[9] Clearly, the TCF leaders' vocal patriotism may have been heartfelt, but it did nevertheless also mask the fact that the TCF was advocating touring while young Frenchmen died in the trenches.

Like Michelin and the Touring Club, the Office National du Tourisme, part of the Ministry of Public Works, spent much of the war planning postwar tourism. In particular, early in the war the Office commissioned Pierre Chabert to travel to the United States and Canada to study the tourist infrastructure with an eye to making France a preferred destination for North American tourists. Completed by March 1916 (although not published until 1918), more than a year before American entry into the war, the report circulated widely in the Ministry and among advocates of tourism; Abel Ballif cited it heavily in the Touring Club's *Revue* as early as May 1916.[10] After devoting ninety pages to the expectations of Americans and the organization of American tourism, Chabert noted that four hundred thousand Americans had come to France between April 1913 and April 1914, and he estimated that six hundred thousand to seven hundred thousand Americans would visit France in the first year after the war. Above all, he—like members of the TCF—believed that the primary attraction would be the battlefields. "Visits to the battlefields are the decisive factor that will cause a throng of Americans to come to France." But unless France wanted to lose the tourists to Germany, which would try to "steal" them in visits to the Rhineland or spas, "pilgrimages [to the battlefields] must be organized *without delay* [his italics]."[11] Interestingly, Chabert took for granted not only that tourists would be coming to France in droves but also that the prewar boundary between France and Germany would remain unchanged.

Americans and their money would ultimately help to rebuild France and redress France's imbalance of payments with the Americans, provided that reconstruction did not remove the traces too quickly. "This rush of Americans has major consequences for France. We have an immense fortune in our ruins . . . on the condition that we do not rebuild too quickly and that all is well organized." Chabert was insistent that Americans liked their touring to be well-organized by tourist agencies and railroad companies with smooth transitions from one form of transportation to another. Above all, he claimed that they loved to plan everything, hence the necessity of guidebooks. In order to be ready as soon as the war ended, Chabert claimed that France needed to decide immediately "under what conditions the battlefields will be visited." The guides, "in English, will tell the history of the battle and indicate the major movements [of troops] and point out certain spots in the terrain."[12]

Of course, as was true in Michelin's own efforts to promote battlefield tourism, there existed for Chabert a fear of sacrilege of the secular shrines to the nation that the battlefields represented. For him, as for so many other promoters of tourism, touring the battlefields was supposed to be a form of pilgrimage. He noted that in the case of Americans there were fears that "they will come to see our battlefields with the same curiosity that makes millions of people go to see a boxing match or [a game] of that big national sport, 'le base-ball.'" He attested on the basis of his travels and knowledge of Americans, however, that Americans would remove their hats and act respectfully when visiting the battlefields.[13] Like the leaders of the TCF, Chabert noted in conclusion that Americans drawn to the battlefields could be attracted to other sights in France, spending their entire vacations and their dollars in France rather than in Germany or Switzerland.

Wartime advocates for postwar tourism clearly viewed the battlefields as a tourist attraction unique to France. André Michelin thus undertook the design and publication of the Michelin guidebooks at a time when other advocates of tourism believed that the World War I battlefields provided France with an opportunity to increase the number of domestic and foreign tourists, whose spending would help to rebuild the devastated areas and fill French coffers generally.

Between September 1917, when the first guidebook to the Marne battlefield of Ourcq appeared, and April 1921, when the creation of new guidebooks abruptly ended, Michelin's Parisian tourist service produced twenty-nine different guidebooks, each one considering an important battle site or locale associated with the war. These years coincide closely with the rush to build war memorials.[14] Not including translations and reeditions, the Michelin guides included approximately 3,500 pages, 4,500 photographs, and 1,000 maps. This number of twenty-nine does not include translations or later variants, not to mention two later French editions of the Verdun guidebook in 1926 and 1936. Moreover, Michelin translated nineteen of the twenty-nine guidebooks into English and one—that on Verdun—into German (although not until 1929). The firm's advertising between 1919 and 1921 consisted almost entirely of producing and attempting to sell the World War I guidebooks.

André Michelin decided as early as 1916 to publish guides to the battlefields, claiming that the army's propaganda service had requested that Michelin undertake such a project.[15] Because it had been rumored that Baedeker was preparing guides to the battlefields in France, preparation of a Michelin guide became an ideal forum for reinforcing Michelin's association with a French national identity. André Michelin's closest collaborator, Fernand Gillet, who had run the Paris tourist service before 1914, spent the war in the geographical and propaganda services of the French army. He claimed after the war that Michelin had been informed that the German army had already charged Baedeker with producing a guide to the devastated areas of France. According to Gillet, "Our enemies [were] thinking about a work of propaganda. By encouraging such works, the German government [thought] it [could] plead the case of German innocence in the conflict and deny the horror that they have caused with their devastation and cruelties."[16]

Internal notes of ideas for publicizing the guides were even more direct. In plans for a prospectus that never seems to have appeared, the text proclaims that not a single ally of the entente should have a *guide boche Baedeker* and that no Baedeker should escape *la Bochie*.[17] Michelin's publisher, Berger-Levrault, similarly claimed the guides were necessary because "our enemies are already preparing guides to the war that will deceitfully plead the case of German innocence." Henry Defert, announcing the agreement giving TCF members a 20 percent discount in return for Michelin's right to declare the TCF's patronage, wrote that the Michelin guide would compete against "all of the kraut guides that Germany is preparing to dump on the world with their ingeniously biased presentation of the facts and this boldness in lying to which she is accustomed. Our comrade Michelin has understood that. It will be to his honor to have established a guide that is clear, precise, done with honesty and clarity, a truly French guide."[18] This notion that any German guide would by definition lack objectivity and that a French guide by Michelin would embody it, although hardly surprising given the intensity of propaganda at the time, was one of the founding myths of the guides.

Michelin could thus serve the national cause simply by producing the guides. By announcing, however, that profits from the venture would be paid to Dr. Jacques Bertillon's Alliance Nationale pour l'Accroissement de la Population Française, the best-known pronatalist group during or after the war, Michelin could further be seen to be placing the interests of France above its own. Michelin thus shielded itself from any potential charges that it would be making a profit from the sales of guides to the battlefields, a fact that the company was quick to point out in advertising the guides.[19] The strategy was brilliant; Michelin could be an utterly disinterested patriot, covering its costs, and donating the profits to an organization working for France.

Appearing in September 1917, Michelin's first guidebook, *L'Ourcq (Meaux, Senlis, Chantilly)*, was the first of three guides devoted to the battle of the Marne. Fernand Gillet claimed that l'Ourcq had been chosen for its proximity to Paris, making it easier to verify information about the current state of travel conditions.[20] There was no little symbolism. The battle of the Marne represented France's ability to halt and roll back, at least partially, the German onslaught. The guidebook appeared in time for the third anniversary of a battle that had already attained mythological proportions. The anniversary made a public initiation of the guides more appropriate, and the Office National du Tourisme and the Touring Club de France (whose members received a discount on the guides in return for free advertising in the pages of the review)[21] sponsored a "tour of the Ourcq" on 27 September 1917, inviting both the French and the foreign press.

The tour offered participants the possibility of tracing the itinerary laid out in the Ourcq guidebook so they might see just how useful the guide could be. A banquet followed at the castle of Chantilly, which was both a center of the learned Institut de France and had been Joffre's headquarters and residence until he was appointed Marshal of France, not to mention a regular meeting place for Allied war planning.[22] At the banquet, members of the Institut (including historian Ernest Lavisse), the president and vice presidents of the Office National du Tourisme and the Touring Club de France, André Michelin, and Dr. Jacques Bertillon all addressed the assembled crowd, praising the guidebooks and instructing people in how they should be considered and used. André Michelin again expressed that the guides were not only to facilitate touring but also to combat presumed German lies.

The enemy is preparing special guides destined for the so-called "visit to the battlefields," but the secret goal is especially to prove that the Germans did not commit the atrocities that they have been accused of or at least that

MARKETING MICHELIN

98

it is always the French, Belgian, or English who started it, who set things on fire, who raped [*violé*], who made reprisals indispensable. . . . We have to struggle energetically to annihilate such efforts; each time that the opportunity presents itself we have **to prove without pretty phrasing and without literary embellishment,** by briefly recounting the facts, **that the Germans are waging an abominable war** a war that surpasses in horror all of the atrocities committed by their ancestors, the Huns. . . . Not only must **we fight against such infamous propaganda,** but we must **beat it.** The whole world must one day be convinced that the Krauts have **scientifically organized the abominations of the current war** [quotation marks and bold in the original].[23]

On the one hand, Michelin maintained that its own approach was honest and objective, whereas any German guides would be misleading denials of German actions. Although the tone of the Michelin guidebooks was actually less militant than Michelin's comments, André Michelin, like his contemporaries, equated objectivity very closely with the French version of events.

Repeatedly, speakers at Chantilly asserted that trips to the battlefields were no ordinary tourism. They were pilgrimages. Henri Welschinger proclaimed that such tourism was not a question of "vain curiosity" but instead a "veritable pilgrimage," not a bunch of "profane curiosities," but "a patriotic work." Ernest Lavisse said that "it is not without apprehension that I think of these tours. They cannot resemble pleasure trips by train. The battlefields must not resemble fairgrounds." Louis Baudry de Saunier, that indefatigable advocate for automobile tourism, implied that the Michelin guides were one possible form of "sacred advertising." Later, the Touring Club de France referred to the "patriotic piety that must inspire all organization of visits to battlefields that have been drenched in so much generous blood."[24] Michelin guides and advertisements for them joined the refrain; battlefield tours were pilgrimages.

In a sense, the idea of battlefield pilgrimage was an extension of the "secularization" of formerly religious language during the First World War. With cause, historians have taken very different positions about whether there was a wholesale secularization and a sort of a "nationalization" of religion or whether use of religious terminology was in fact evidence of a more simple adaptation of religious faith during and after the war.[25] In any event, no one denies that, for at least much of the population, there was a certain fusion between religious faith and nationalism within society at large, making it difficult to generalize about where religion ended and the nation began.

Historians of World War I have tended to make precisely the same distinction between pilgrimage and "normal" tourism that contemporaries did.[26] Here again the difference was highly arbitrary. The distinction between mere tourism and the religious experience of pilgrimage had existed at least since the mid-nineteenth century, when contemporaries applied it to religious pilgrimage sites, especially Lourdes. Just as recent scholarship has pointed out how arbitrary this distinction was at Lourdes, so too was it arbitrary when applied to the World War I battlefields.[27] In fact, it may have been the very ambiguity of encouraging rich people from France and abroad to spend their money on expensive hotels and food (penny-pinching was not going to contribute much to the rebuilding of devastated areas or the French economy) while visiting the battlefields that engendered such forceful distinctions between the "sacred" and the "profane" forms of travel. It is even possible that an individual traveler's insistence that a trip was a pilgrimage, not a tour, could be a means of easing a grief-ridden conscience that felt strangely guilty for surviving the war or an effort to set the "pilgrim" apart from the others, the tourists.[28]

In practice, one visitor's place of silence was another's picnic area. The same visitor could visit a gravesite, gather souvenirs, and have fun looking around nearby cities. The very structure of the Michelin guidebooks—with information about tourist sites in nearby cities and towns as well as about the battlefields themselves—reveals how arbitrary the distinction could be, even for the individual tourist. Moreover, touring the battlefields for pleasure and pilgrimage was hardly a new phenomenon, except to the extent that spectators actually had to wait until the front had moved before going to nose around; in the American Civil War and during the Franco-Prussian War, tourists had been able to watch fighting and bombardments while they were actually occurring.[29] The battlefields from both of those wars, like Waterloo before them, also became later tourist sites, and Gettysburg in particular became a place of pilgrimage as well as tourism.[30]

Immediately after the war, Michelin's advertising reflected the ambiguity of the notion of pilgrimage. On 10 May 1919, Bibendum reported on a tour he had made in April to Reims and Soissons. He noted that "the roads were better than I dared hope. One can get around without encumbrances. The battlefield has remained intact and sublime, with its decor of trenches, communication trenches, barbed wire, shell holes, and abandoned tanks. In the hollow of a thicket I even had the tragic vision of a German soldier whose hands were still holding a grenade and a rifle." Thus, Bibendum is concerned about the condition of the roads before admiring the supposed authenticity of the battle-field, including the "decor" of the trenches, even playfully imagining an enemy soldier. In the same "Samedi," Bibendum then states his real problem: where to eat and sleep. He suggests a box lunch and a bottle of wine. Accommodations presented a real problem; there was nothing in Soissons or Reims. In Verdun, men and women had to sleep separately in the dormitory of a secondary school, "while waiting for the sexes to become equal."[31] In short, accommodations were in a state of crisis, and Michelin asked readers to send in their ideas to solve the housing crunch. Michelin thus gave away what it thought the priorities of its guide users might be: they would want to experience the "reality" of the trenches without forsaking the comfort of eating and sleeping well. Visitors may have seen their trips as pilgrimages and no doubt generally felt sympathy for the men who had fought and died, but Michelin suggested that ease of travel and the availability of acceptable, even excellent places to eat and sleep were also foremost concerns.

TOURING LIKE A GENERAL: A GUIDE, A PANORAMA, A HISTORY

From the outset, Michelin billed the guides as three things in one: "Un Guide, Un Panorama, Une Histoire" was the slogan printed under the title on the French versions of the guidebooks. Although advertisements held out at least the possibility that the guides might be used by others, including school-children, the obvious market consisted of those bourgeois men and women wealthy enough to have both leisure to tour and money for an automobile. One could of course sit at home and look at the pictures (as Michelin suggested for the winter months when one could not tour), but these photographs were smaller and thus somewhat less impressive than those in the weekly *Illustration* during and after the war. Michelin designed the structure and contents of the guides specifically for tourists in their own cars. Moreover, intentionally or not, the battlefield experience laid out by the Michelin guides, including their "guide, panorama, and history," was the perspective of French officers, particularly generals, men of essentially the same social position as the tourists following in their footsteps and tire tracks.

André Michelin clearly knew his audience. At the end of the Great War, automobile ownership was still an elite phenomenon, and the guidebooks to the battlefields addressed that elite. In writing for automobile owners, however, Michelin related a version of the war that differed markedly from that of individual combatants, not just in terms of their own experience but also in their own narratives of what had happened on the battlefields. In the 1920s, hundreds of former French soldiers published novels and memoirs of their years at the front. In 1929, Jean Norton Cru, a former soldier and a professor of French literature at Williams College, listed and commented on some three hundred of these texts, judging how well each offered what he considered to be a reliable account of soldiers' lives in the trenches. There could not be a greater divergence between Norton Cru's narratives and the traditional account offered in the Michelin guides. The contrast seems even more marked in light of recent historiography focusing on the experiences of ordinary footsoldiers and their own attempts to commemorate the events of the war.[32] The Michelin guides to the battlefields, designed for wealthy Frenchmen, helped to construct a narrative of the war, one building on the official accounts of army officers, and evolving into what became a traditional top-down historiographical portrayal of the war. Although valiant *poilus* certainly appeared in the Michelin guides, the discourse from the trenches, that other narrative of the war, found little place in the guides to the battlefields.

In Michelin's tripartite "guide, panorama, and history," the "guide" or itinerary, was of obvious necessity for a tourist given the lack of road signs or any indications of direction in areas still filled with shells, destroyed roads, and other dangers. It is noteworthy, nonetheless, that during the war the only men who had needed a guide for automobiles, presumably a human one in the form of a driver, were the top brass. Michelin realized the similarity between French generals and leaders who rode on Michelins during the war and tourists, who were supposed to ride on them during their trips to the battlefields. After the war, Michelin advertised its solid steel wheels as the wheels that "contributed in beating the Krauts, by rolling on the cars of the army staff . . . The wheel is an old friend of the front. And the tourist . . . will save himself a lot of troubles and breakdowns (even small ones!) by equipping his car with good, solid Michelin wheels." The advertisement included an illustration of men riding in a car equipped with such wheels. The caption read "practical and solid, the Michelin wheel never left the Tiger stranded as he rolled over the Boches, among shell holes and rut-filled trails of the battlefields."[33] The tourist on Michelin wheels thus traveled like the general staff, even like Prime Minister Georges Clemenceau himself.

Just as tourists could drive on the same wheels as officers, they could also sleep in the same rooms and in the same beds. To alleviate the housing crisis for early postwar tourists, Michelin began to advocate renting from villagers the rooms that the army had formerly used as officers' quarters. As the company put it in its weekly "Samedi,"

> No doubt along the entire front, numerous hotels are being founded or reopened; but they will be inevitably bursting with the peaceful invasion. It goes without saying that a resourceful hotel owner knows how to find rooms somewhere in town for clients who arrive at the last minute. That's good for a traveler getting off of a train. . . . [But] the motorist's plight is complicated in general by a car that cannot easily sleep under the stars all night. But the cities of France are not to my knowledge built in the middle of deserts. There are villages nearby, and they are never very far. During the war, these villages were organized in billets; the best rooms were reserved for officers, and the state paid one franc daily to the owner. The officer was often accompanied by a military car . . . and his chauffeur, a resourceful soldier [*poilu débrouillard*] always knew where to dig up a stable or barn to shelter his car. . . . [And] the room, often quite agreeable, clean, and cheery, furnished with a good, comfortable bed, had not been filched by the Boches. . . . There you will find a warm welcome, good lodging, fresh eggs, and a smile, all for a good price.[34]

Complimenting the valiant *poilu* for his resourcefulness in doing his duty for the officer, the advertisement also reminded readers of the divide between an officer and his driver, a difference not unlike that between a tourist and his own chauffeur.

Michelin devoted several "Samedis" to listing the places where such rooms might be found, and charged its tourist office with coordinating offers and demands, so that a tourist might write or call Michelin to learn of the address of villagers willing to rent former officers' quarters.[35] For the tourist, such rooms offered parking, still an issue given the value of cars, and an assurance that the rooms must be decent, both because officers had slept in them and because Michelin would receive bad reports about rooms it had suggested that were not deemed acceptable by tourists. In essence, Michelin—while refusing to organize tours, which would have been counterproductive to automobile travel—did facilitate the location of housing. In the process, the company further re-created the wartime experience for tourists, although it was the experience of the officers, not that of *poilus*.

More than mere "guides," the Michelin guidebooks also offered "panoramas" in the form of photographs of the former battlefields. These were generally wide-angle shots from a high point and included an overview of the terrain of the battlefield. To a certain extent, the "panoramas" resembled the photographs of various French tourist sights offered each month in the TCF's *Revue mensuelle,* except that they provided less height and much more width. From the perspective of the troops during the war, of course, few *poilus* ever saw a panorama of a battlefield; theirs was an immediate, circumscribed existence. By contrast, officers mounted on horseback used such overviews of the territory to command their troops.[36] Michelin's publisher, Berger-Levrault, made the parallel between perceived generals' views of the war and that of postwar tourists when it described the panoramic photographs as "taken at the exact spots where leaders [*chefs*] followed the progress of the troops."[37] Similarly, Michelin's superb maps of battlefields in the guide reflected a reality experienced by officers, not by most *poilus*.

11. "No vacancies? Then to the [officers'] quarters!" Bibendum explains how tourists could rent a room from local peasants near the former battlefields. During the war, in the illustration on the left, the officer talks to a woman alone with her child. After the war, the man has returned (although she does the negotiating), and there are signs of normalcy. There are several chickens, the boy has now grown into boys' clothes and out of the gown/dresses both peasant boys and girls wore, and smoke rises from the chimney. The billeting of tourists appears to allow tourists to see the battlefields and locals to increase their incomes. Drawing by E. L. Cousyn, in "Le Samedi de Michelin," *L'Illustration,* 19 July 1919.

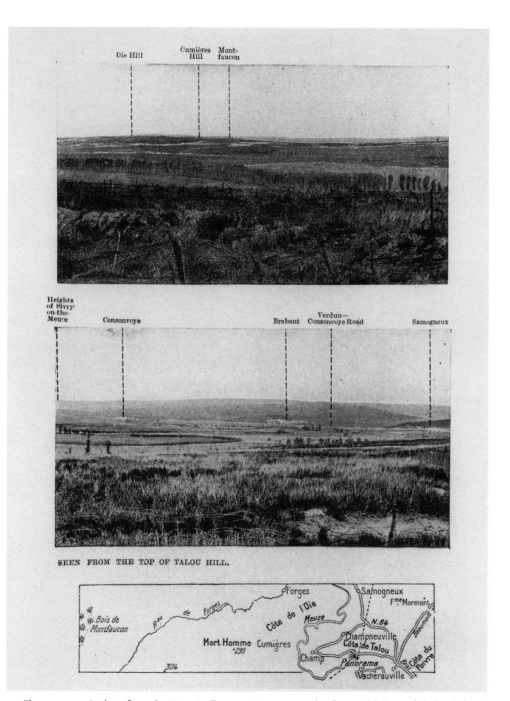

Die Hill   Cumières   Mont-
            Hill   faucon

Heights
of Sivry-
on-the-
Meuse   Consonvoye   Verdun—
                                     Brabant   Consonvoye Road   Samogneux

SEEN FROM THE TOP OF TALOU HILL.

12. These panoramic shots from the Meuse Valley, reminiscent not only of generals' views of the battlefield but also of the Touring Club's panoramic photos of tourist sights, show two large swathes of the battlefield. The photographer appears to have taken both photographs from the "Panorama" featured on the diagram in the lower right. The diagram itself, like the legends above the photos, helped to orient the tourist bearing a Michelin guide. *The Battle of Verdun*, 1919, pp. 86–87.

Advertisements for the guides frequently described the photographs as revealing the "reality" of the war. They were supposed to provide a sense of authenticity. Of course, Michelin photographs came primarily from the army propaganda service, where Michelin protégé Fernand Gillet had used photographs and films for wartime propaganda, and from individuals whom Michelin called on to send any pictures of the front that they had. The photographs provided a photographic record of the war, but they were no more authentic than prose and could be quite misleading. Given the early rhetoric that the guides would show the devastation done by the Germans and the utter lack of any broader context in the guides, the photographs portrayed what the Germans had supposedly done to France as if there were not two armies and as if there were no larger diplomatic and military context that in any way implicated the French government as well as the German.

Michelin also claimed that its guides provided the "total history" of the battlefields. Here the notion of history not only focused on generals during the war but also took for granted the idea that the tale of the battles, and the men who led them, was the whole history of the war. According to the company's description, the guides provided "a precise and brief description of the military operations, which permits the reader's imagination to reconstruct the battle's movements, to sort of see the immense horde of gray uniforms [feldgrau] advance and then retreat, settle in, then retreat again and finally lose its footing and succumb before our allies and before the big French army in blue [grande Armée bleu horizon]. . . . And just as the Michelin will describe all of the battlefields from the sea to the Vosges and from 1914–1918, the collection will contain all of the history of the Great War [toute l'Histoire de la grande Guerre] [their underlining]." Then, unable to resist, Michelin continued: "After that, if you still prefer the Baedeker there is at least a chance that you will find some deficiencies."[38] So whereas the Baedeker provided incomplete coverage, the Michelin guide would tell "all of the history" of the war by describing "all of the battles."

The guides consistently offered a sort of historical perspective oriented around great men by tracing the history of the area that became the battlefield, focusing on the kings, princes, and saints since the Romans. Inevitably, the position of the Gauls against the Romans and particularly the battles between Gallo-Romans and the "German barbarians," including the "Alboches" and "Huns," received mention. The modern period received little attention except

MARKETING MICHELIN

106

for destruction of local religious structures during the French Revolution and of important battles under Napoleon and during the Franco-Prussian War. The guides' treatment of World War I took a similar tack. The guidebooks feature abundant photographs of generals, and they never question generals', even Nivelle's, decision-making. The history is that of regiments and their movements on the field of battle led by these great men. Given how historians and teachers taught history in turn-of-the-century primary and secondary schools, where well-off French tourists had generally received their education, this was a familiar view of history. It was a simple, comprehensible story of French victory, without much detail about French losses, and without the complications of understanding the political, economic, or social context of the war.

NARRATING THE WAR FROM THE BATTLEFIELDS: REIMS, VERDUN, AND ALSACE

Michelin designed each of the French-language guidebooks to trace the history of an area or city and nearby battles. Focusing on the movement of tourists through the space of the battlefield without getting bogged down in lengthy introductions about the war as a whole, the Michelin guides provided little context for understanding the causes or conduct of the war. The diplomatic alliances and the arms race received no mention, nor is there a place for a description of international brinksmanship in the summer of 1914. The guides, like those of Michelin's competitors, took for granted that the German invasion of Belgium and France caused the war. The eastern front appeared only in the few cases where it very directly affected fighting in the west, such as the signing of the Treaty of Brest-Litovsk, and the movement of German troops from the east to the west. Battles in the Balkans and the Middle East were absent.

In recounting battles on the western front and describing nearby sights, Michelin guidebooks offered somewhat more context. Nevertheless, the divergence between Michelin's interpretation and that of historians of the First World War serves as a reminder of the extent to which Michelin reflected contemporary versions, disseminating an understanding of the war historians would today consider incomplete at best. The guidebook to Reims, entitled *Rheims and the Battles for Its Possession*, is one of several guides that used a single city for orientation, tracing nearby battles throughout the four years of the war. The guidebook began with a short history of Reims, including its Gallo-Roman past, the barbarian invasions, and the city's place in the kingdoms of both the Merovingians and Carolingians (1 1/2 pages). It then considered Reims since the High Middle Ages, focusing on its importance in the coronation of French kings and on battles during the Hundred Years' War (4 1/2 pages). With very brief mention of the intervening years, the guidebook then recounted the history of the battles of World War I (8 pages). Michelin then proceeded to the primary topic, the destruction of Reims and especially of the cathedral (11 pages).[39]

The guidebook's account of the shelling and resulting fire in the cathedral at Reims typifies Michelin's approach.

> To protect the Cathedral, which the Germans had fitted up on the 12th for the reception of their wounded [when the Germans briefly occupied the city in September 1914], some seventy to eighty German wounded were accommodated on straw in the nave. [After the German retreat] the Red Cross flag was displayed on each tower, and notice given to the enemy. On the 18th . . . the Cathedral, in spite of the Red Cross flag, was struck by 8-in. shells, which damaged the outside sculptures of the lower windows of the main transept, smashing the 13th and 14th century stained-glass. . . . On the 19th, . . . towards noon, incendiary shells were rained on the centre of the town. . . . A shell fired [sic] the wooden scaffolding round the northwest tower which had been under repair since 1913. The fire spread quickly to the roof, the molten lead from which set fire to the straw in the nave. In spite of a rescue party, who risked their lives in getting out the wounded, a dozen of the German wounded perished in the flames. The conflagration spread to the Archbishop's Palace, from which it was impossible to remove the tapestries or the prehistoric Roman and Gothic collections [Michelin's translation].[40]

The tone is obviously quite descriptive, consonant with Michelin's billing of the guides as clear and objective, contrary to supposed German lies. But there are oversights and much innuendo. Most glaring for any historian of World War I, the guidebook neglects to mention that the French used the cathedral tower for observation of nearby battlefields. Although prewar agreements had been designed to protect cultural edifices such as the cathedral, the agreements had specified that the structures could not be used by either warring party. In addition, rather than focus on the fact that the French had hoped to use German wounded to protect the cathedral, the guidebook implied German insensitivity not only to cultural treasures and Red Cross flags but also to their own men, proving the assertion in advertising for the guides that the Germans were barbaric.

The remainder of this 170-page guide consisted of carefully planned itineraries for visits to the city and nearby battlefields, richly illustrated with photographs of monuments and buildings before the arrival of the Germans and of the rubble after their departure. These before and after photographs reinforced a fundamental notion underlying the guides: the war had been caused by the Germans, whose much-vaunted *Kultur* (as Michelin advertisements and commentaries in the TCF *Revue,* like wartime propaganda, pointed out) was a facade for unparalleled brutality and destruction. Guides to Amiens, Lille, and Soissons repeated this structure and stressed the theme.[41] Photographs, billed as objective evidence of German misdeeds, provided superb propaganda, allowing Michelin to employ a neutral, objective tone so that the guides do not seem to be typical wartime propaganda. The best propaganda, like the best advertising, may well be that which can maintain an image of objective treatment of facts while subtly tipping the scales.

Other guidebooks used battles for orientation, as was the case in the Verdun guide, which described the infamous months-long battle of 1916. In the guide to Verdun, after a very brief history of the area since Attila the Hun (2 pages) and the description of the battle (28 pages), Michelin described the visit to the city (with a focus on city hall, the cathedral, and the bishop's residence as well as the Citadelle), and then a visit to the battlefield followed. The battlefield itinerary had two parts, the visit to the right bank of the Meuse and another to the left bank. In the section on the itineraries, photographs not only of things to be seen but also of traffic intersections were supposed to aid tourists in a terrain without signs or often even any remaining buildings for orientation.[42]

In the Verdun guide, most illustrations of the battlefields were, as in army photographs generally, devoid of men. As in *L'Illustration*'s coverage of the battle of Verdun in 1916,[43] there were few wounded, certainly none who were French, and French men were mysteriously absent in the photographs. There were a few pictures of troops before they went into battle, in addition to the

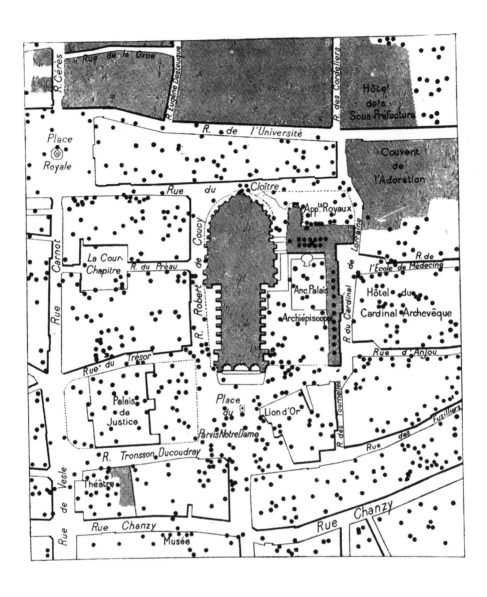

13. In an effort to illustrate that German destruction of the cathedral of Reims, a longtime symbol of France that had served as the place of coronation of French kings, was no accident, Michelin showed each place that a shell fell. The diagram's legend notes that "the shells which struck the Cathedral were far too numerous to allow all of them to be shown on the above plan." *Rheims and the Battles for Its Possession*, 1920, p. 20.

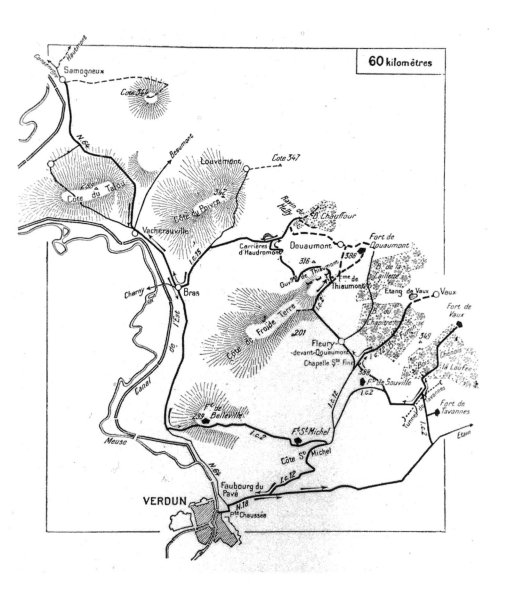

14. Michelin guides to the battlefields featured carefully drawn maps of the itineraries described. Tourists could visualize both the roads they were taking (including road numbers) and the overall topography of the front before following all or part of the itinerary to various battlefields. *The Battle of Verdun*, p. 57.

multitude of individual photographs of generals. The presence of devastated lands without dead bodies points out the horrible destruction caused by the Germans without including casualties incurred by the French, except for those already buried. Of course, everyone knew that the war wounded and killed millions of French men, but their absence from the photographs nevertheless minimized the suffering, placing the battles more squarely into a narrative of German destruction of France, not of individual men. Germany assaulted France, its monuments, its buildings, its infrastructure, its land, and in a general sense its men. The guidebooks were quite specific about how many acres and buildings and which cultural edifices the Germans destroyed, but the numbers of French soldiers taken prisoner, wounded, or killed received no mention; during the war, censorship ruled out inclusion of such numbers but guides appearing after the war, even in the late 1920s and 1930s, still omitted the dismal figures.

Not surprisingly, given Michelin's prewar advocacy of the importance of

15. This photograph of the Vaux Pond near the Fort of Vaux illustrates the makeshift nature of battlefield tourism by automobile. The guidebook advises that the motorist needs to "turn the car around 100 yards from the pond, at the place where a narrow-gauge rail-track formerly ran" then shows that spot along the road. Tourists could verify that no trace remained of the village that stood on this site. *The Battle of Verdun*, p. 69.

roads and automobile travel, Michelin included an extensive description and a nicely detailed map of the *voie sacrée*, the road connecting Bar-le-Duc to Verdun via Souilly. The Verdun guide duly noted that because the Germans had fourteen different railways supplying their forces and every French rail link except the narrow gauge track running along the Meuse had been destroyed, only a single road with constant truck traffic kept the French armies supplied. With repairs done as some 1,700 trucks rolled past each way each day, the "motor service along the 'Sacred Way' was organised to such a pitch that it was able to ensure the transport of the troops, the evacuation of the wounded, and the revictualing of 250,000 combatants."[44]

The guidebooks also managed to repeat and lend credibility to several legends about the war, and particularly of those surrounding the battle of Verdun. The most telling example is of the *tranchée des baïonnettes* (the bayonet trench), a case in which Jean Norton Cru claimed as early as 1929 that tourists themselves had created the legend. It is true that in June 1916, under heavy German fire, two sections of the French 137th suffered heavy losses. After the battle, onlookers saw a trench where rifles remained standing with the bayonets poking out of the ground. Either the rifles had been buried by nearby explosions of shells, or as likely, the bayonets had been stuck in the ground to mark a makeshift collective grave, a practice known elsewhere on the front. But a legend, capturing the popular imagery of the bayonet, soon emerged: the men of the French 137th had supposedly been buried alive by German bombardment after they attached their bayonets and held their rifles upright.[45]

Although the first Michelin guide to Verdun never mentioned the incident, the later edition of 1919, as well as all subsequent editions, told of how "our brave Vendéens and Bretons" had been buried alive and how "the trench kept the appearance which it was found still to bear at the Armistice." The Michelin account featured a photograph of the monument donated by "the American citizen, George F. Rand" in order to protect the bayonet trench from the elements.[46] The bayonet trench was of course a hoax, revealing more about tourists' and others' ideas of trench warfare than the reality. As Norton Cru emphatically stated and Antoine Prost has more recently argued, the incident could not have happened. Given that they were being bombarded, and were not on the offensive, there was no reason that the soldiers would have attached their bayonets. And most obviously, shelling did not quickly bury men alive by filling in trenches. Even shells landing just before the trench could not fill in an entire trench evenly, leaving only the bayonets sticking out of the ground; shelling could only cause trench walls to crumble.[47] The trench of the bayonets became, however, through the efforts of Michelin and so many others, an important legend of the Verdun battlefield.

The *Battle of Verdun* and other guides followed a similar narrative structure. The Germans initially advanced, met stiff French resistance, and ultimately retreated under French pressure. Even at Verdun, a site of mutual bloodletting in the course of 1916, Germans advanced but were forced to retreat. Here as elsewhere, German losses dominated the narrative. As the guide's summary of the battle put it: "The Battle of Verdun in 1916 was not merely a severe local setback for the Germans; by using up their best troops it had also very important strategical consequences. Their successes were few, temporary, and dearly bought. Advancing painfully, each step forward was marked by a mountain of corpses. Up to the end of the War, even after the Battles of the Somme and Aisne in 1916 and 1917, and after the Battle of Champagne in 1918, Verdun remained a hideous spectre for the German people, while their soldiers surnamed [nicknamed] it 'the *slaughterhouse* of Germany' [Michelin's translation]."[48] Verdun was a French national victory, and a slaughterhouse of dead Germans, not a slaughterhouse for both sides.

Not surprisingly, General Philippe Pétain emerged in the guidebook to Verdun as the heroic leader, capable of leading his troops to victory. At Verdun, "General Pétain managed to imbue all under his command with his energy, activity, and faith, and the enemy's drive was stopped."[49] As interesting, however, was the level of respect and dispensation from criticism that the guides accorded all of the other generals, who often fared less well in wartime or postwar assessments. General Robert-Georges Nivelle received no criticism for the disastrous 1917 offensive at the Chemin des Dames, even though Pétain got better coverage; in fact, the guidebooks did not portray the 1917 offensive as a disaster. Not surprisingly, the guidebook to the Chemin des Dames made no reference whatsoever to the widespread mutinies that ended in Nivelle being relieved of his command, although indirect reference to the mutinies did appear in the introduction to the guidebook on the battle of Picardy: "During the year 1917, after the allied offensives of the spring a wave of lassitude had darkened the morale of certain units of the French army, but General Pétain brought the remedy. In March 1918, the French army had never had a morale that was higher or stronger."[50] The guidebooks with their pictures of generals and words of praise projected an image of the war as an inevitable struggle against the German aggressor that French generals led with great skill. Traditional authority remained intact by featuring generals' successes and ignoring their errors.

All but two of the twenty-nine guides considered actual battlefields of World War I or cities at least partly touched by the destruction of the war. The two exceptions were the guides to Strasbourg and to Colmar-Mulhouse-Schlestadt (sic), designed to acquaint potential French tourists (Michelin never translated the two into English) with the reannexed departments of Alsace. In neither case did Michelin describe actual battles. Strasbourg, for example, never suffered from shelling or even aerial bombardment. Instead, the guide to Strasbourg encouraged French tourists to see the reconquered city, and repeated French myths about the fate of Alsace and Lorraine "under the [German] boot," as the guides put it. Volumes on the Alsatian cities, like those devoted to the battles in Alsace, repeated the myth of a loyal Alsace, protesting the German annexation and desiring nothing more than to rejoin France. The German "occupation" of 1870–1918 was a mere interlude for these fundamentally French areas. In Strasbourg, the cathedral and a walking tour of the old city were the primary high points, taking up forty-six pages of description. The entire new section of the city, which the Germans built to house administrative offices and the university, occupied just over one page. In volumes considering Alsace, abundant photographs grace the pages as do drawings by pro-French cartoonists Hansi and Zislin. The perspective in both books is that France freed Alsace-Lorraine from the German yoke; the volumes on Alsatian cities, like those on the war, offered readers a very specific, politically loaded interpretation of the recent past.[51]

In the Alsatian guides, the presentation of "facts" about Alsace is so misleading as to pass occasionally into pure error. In a section entitled "Ethnic differences," the Michelin guide asserted that the Alsatians, who looked so much like Germans, were really French. As was typical in France, the guide put Alsatians into a distinct category, one of many regional peoples in France and fundamentally different from the Germans: "The Alsatian race, despite the multiple invasions that the region [*pays*] has been subjected to through the ages, has remained basically Celtic or at least close to the Celtic type: a wide but shallow skull, dark eyes and hair. . . . Even the blond Alsatian, characterized by a clear complexion and thick, ashen blond hair cannot be confused with the pure Germanic type, with its bigger build, more colored complexion, and finer, more reddish hair. The same observations apply to the women, who have a more svelte form than German women; they more closely resemble French women of the Nord region."[52] Alsatians were thus Celtic, just like the rest of the "French race" (used in that turn-of-the-century sense essentially meaning ethnicity) however much many Alsatians seem to have had the blond hair and blue eyes of so many Germans.

Similarly, the guides repeated a frequent French assumption that Alsatian was not essentially an Alemannic dialect, closer to any German dialect than to French. In the guide to the Alsatian battlefields, one would think that Alsatian is a dialect of French: "Alsatian is blended with French words [L'alsacien est panaché de mots français]. A very simple language with a rudimentary vocabulary, it is from French that it borrowed in order to express abstract ideas and to designate new objects. Naturally since the schools became German . . . German words have been substituted for a certain number of French words. Often, instead of *mie parapluie* [my umbrella], they have for some years been saying *mie schirm;* instead of *an d'Gâr* (à la gare) [to the train station; their parentheses], *an der Banhof.* In its vocabulary, syntax, and intimate nature [*nature intime*], Alsatian is mixed, between the Latin and Germanic languages."[53] By restraining discussion to vocabulary (into which French words had entered at least since the French took possession of Alsace in 1681) as well as syntax, the guide could claim that the dialect was at least as much French as German and that the place of German resulted from the annexation of 1870, not from the origins of the dialect itself. Of course, the origins of Alsatian were Germanic, as the guide inadvertently indicates: even the *an d'Gâr* uses the Germanic prepositional construction rather than the French *à la.* But by denying that the Alsatian "race" and the Alsatian "dialect" were Germanic, the guidebook implicitly asserted that Alsace was rightfully part of France, forming a region like any other French region.

That said, the overall tone of the guides, including even those on Alsace, is exceedingly informative and, at least given the context of 1917–21, reasonably moderate. Although it is fair to term the guidebooks anti-German, as Rudy Koshar has recently done, the guidebooks also reveal a fair degree of restraint. Compared with portrayals of Germans offered in French primary school textbooks, for example, the Michelin guidebooks at least maintained the pretense of objectivity.[54] The term *Boches* was not used, and I have found only one use of the term *Bochie* in the guides (although advertisements for the guides used both terms repeatedly, while asserting that the guides themselves were "objective"). At the same time, the guidebooks did refer to "the enemy" far more than to "the Germans," depersonalizing Germans as much as possible. Far more troubling for anyone who knows the causes of the First World War and the collective responsibility for its outbreak, the guidebooks are placed squarely within the context of the individual battles of the western front, which means that Germany can be seen as a sole aggressor against France and Belgium, and France is cast as valiant defender of its own national sovereignty. The overall causes of the war, the interlocking alliances that exacerbated the events of 1914, and the eastern front are all absent, eliminating the crucial context for understanding the war. The absence of the overall context may in part be inherent in the format of a guidebook, focused on place rather than more abstract political decisions, but the guides nevertheless offered a perspective of the war far from the advertised "total history," however much it may have resonated with well-off automobile tourists after the war.

The single long-term cause of the war that Michelin suggested was the one so frequently cited by Bertillon's Alliance Nationale: Germans had more babies. In the front and back matter of the guides and in its advertising, Michelin emphasized that the profits to the guides went to Bertillon's organization. At the "tour of Ourcq" launching of the publication of the Michelin guides, Bertillon himself laid out the single most important cause of the war; it is a position that the organization repeated ad nauseum after the war, at the same time that Michelin bankrolled the association. Bertillon wrote that "if you doubt that the depopulation of France was the cause of the present war, open the Yellow Book published by the French government at the beginning of the war. As early as the third page, you will find a report sent on 15 March 1913 by our military attaché at the embassy in Berlin, Colonel Serret, in which he announced the coming war. . . . 'The Germans believe that for our 40 million inhabitants we have too great a place in the sun.' . . . And it is true: sixty years ago their population equalled ours; today there are 65 million of them."[55] Michelin further argued that demographics alone could bring about war. In a weekly "Samedi" signed by Bibendum, the company noted that "No barrier, not even the Rhine, ever kept the Germans from overflowing into Gaul. And no scrap of paper will ever stop the invasion of a more prolific people into the deserted field of its neighbor. There is no humanitarian theory that can hold against these inevitable laws of Nature. . . . **Depopulation [Dépeuplement] is the suicide of France** [bold in the original]."[56] That is, World War I could have been, and any future wars with Germany might be, averted, provided that France had a birthrate that allowed it to keep pace with the Germans. The war was thus part of a larger social Darwinistic struggle among peoples, in which the most fit were those able to reproduce in greater numbers.

GUIDEBOOKS FOR THE FOREIGNERS

From the beginning, Michelin wanted foreigners as well as Frenchmen to buy the guidebooks. At the "tour of the Ourcq," André Michelin announced that the guidebooks featured "for the first time, the grouping of all elements to encourage the coming, to welcome, to transport, to lodge, to guide, to interest, and to retain all foreigners in our country."[57] Although there were plans for Spanish and Portuguese translations, and Michelin produced a couple of guides in Italian to Italian battlefields, the targeted market was English-speaking. The British, Australians, Canadians, New Zealanders, and Americans would presumably want to see the battlefields of France. Nineteen of the early editions of the guides appeared in English soon after their publication in French.

After American entry into the war, the Americans came to represent the primary postwar foreign tourists. In one "Samedi," Michelin reproduced an illustration from *Life* magazine showing a veritable horde of American tourists getting off a huge ship and heading up from the coast. They follow a sign that reads "To the historic places of recent European history." In the text Bibendum expresses his satisfaction, "I couldn't wish for a more vivid 'illustration' [a play on the weekly's name, *L'Illustration*] of my predictions of the impending visit of our friends from the United States. Here they are, throwing themselves in hurried lines toward France, Kodaks in hand."[58]

Michelin did not wait for postwar tourists coming from the United States. At some point late in the war or just after the war, the company published a small tourist guide to Clermont-Ferrand for American pilots stationed at the bombing school located at the local airport of Aulnat, where American pilots learned to use Michelin bombing equipment. The company sent employees to other American bases to find out what American soldiers read and what they did with their time. André Michelin even approached U.S. commander General John Pershing to get permission to offer its guides to American soldiers on their bases.[59]

Recognizing that the early guides to the battlefields appealed most directly to a French or British audience with notations of French and British units and photographs of their leaders, Michelin published a special series of guides to all battles where Americans were present, so that Americans could find the actions of American troops in three volumes, the *Americans in the Great War* series. While providing background to the relevant sectors before the arrival of the Americans, the *Americans in the Great War* focused on the American contribution, laying out the various American divisions and providing photographs of American generals. Photographs uniting French and American generals, particularly Marshal Foch and General Pershing, stressed the importance of American assistance in the war. Michelin thus hoped to lure American tourists to France and to keep them in France, an agenda it shared with other interwar advocates of tourism in France. For Michelin, there were potential dividends in the North American market as well. Michelin had worked to break into the American tire market for some time, to the point of establishing a factory at Milltown, New Jersey, in 1908 to produce tires. In fact, Michelin listed Milltown as a place of publication of the *Americans in the Great War* series. Advertisements for the Touring Club and for Michelin tires in the front and back matter are in English, specifically for Americans.[60] Clearly, Michelin attempted to appeal to as large an audience for the guides as possible.

SELLING THE GUIDES, COMPETING WITH RIVALS

Whether Michelin ever reached the large audience it sought is a question that depends on one's perspective. By the company's own measure, the guidebooks were a mixed success. Although Michelin had sold 855,000 guides in just the last 7 months of 1919, the pace slowed thereafter. In 1921, despite announcements that English translations of French guides were soon to appear, Michelin abruptly ended production. Michelin eventually sold up to 2 million guidebooks, including 300,000 guidebooks to Verdun.[61] Sales of the guides covered costs but did not generate profits.[62] Having promised to contribute the proceeds of sales to Bertillon's Alliance Nationale, Michelin had clearly been optimistic that there would be some kind of profit. In the end, Michelin offered the Alliance Nationale direct grants, including a "Michelin prize" for the best pamphlet describing the demographic predicament faced by the French and the necessity of ever more reproduction.

The mixed success of the guidebooks may well have resulted from Michelin's own ambition. Michelin shared fully in the high expectations that French advocates had for the arrival of foreign, particularly American, tourists after the First World War. Although there were battlefield tourists, the expected numbers simply did not arrive immediately after the war. In the summer of 1919, "international shipping and Continental transportation were still too disrupted for much American tourism." Widespread international advertising organized by the Office National du Tourisme billed 1920 as the year in which the American invasion would occur, although the ONT had already revised Pierre Chabert's optimistic wartime estimate of 600,000 to 700,000 Americans down to 200,000. The deluge of Americans did not come. "Financial instability in America, and [press] stories of overcrowded trains, inadequate accommodations, and price-gouging" discouraged trips to the battlefields. "Fewer than 100,000 Americans visited all of Europe in that year," and many of these were immigrants who had not been able to return to Europe during the war (and who were very rarely French).[63] By April 1921, Michelin had halted publication of the guides. There does seem to have been a weak link between the number of tourists and the number of guides sold; it is noteworthy that the 1926 French and 1929 German editions of the Verdun guide roughly coincided with increased levels of international tourism in the late 1920s, with the late 1920s "war books boom," and with the 1927 and 1928 meetings and trips to the battlefields of the American Legion and British Legion.[64]

The cost of production must have been relatively high in comparison with the prices of the guidebooks. The guidebooks themselves reveal that they were by no means cheap to manufacture. The collection of details they include about the war from regimental histories, army files, and other sources so occupied the Michelin tourist service that it never produced the 1921 red guide to France. The guides, unlike most interwar publications, were not produced on that thin, fragile paper currently disintegrating in libraries; the pages are thick, glossy ones. The reproduction of numerous photographs, often two per page, entailed substantial costs. At the same time, Michelin's effort to make the guidebooks' use as widespread as possible caused the company to charge relatively modest prices. In 1919, the guides ranged in price from 1 franc 50 for the thin ones (with as few as 64 pages) and 3 francs 50 for the thick ones (between 100 and 200 pages) at a time when the red guide, admittedly much longer but made with lightweight paper and without any photographs, cost 7 francs.[65]

Although the largest producer of battlefield guides, Michelin nevertheless faced numerous competitors. André Michelin had worried about the possibility of competitors, particularly the Touring Club de France. The agreement between Michelin and the Touring Club, whereby TCF members received a 20 percent discount on the Michelin guides in return for Michelin's right to claim the TCF's patronage had helped, in the words of one company memorandum, to "suppress the only competition that could really be dangerous."[66] And while the TCF never embarked on a project like Michelin's, it did work with the Automobile Club de France and the Office National du Tourisme to publish a small, less detailed edition of possible itineraries for the Verdun battlefields. Nevertheless, the seventeen-page guide featuring itineraries and maps (but no details about the battles or photographs) established trips for the motorist in a hurry to see only "what one must see on the battlefields of Verdun."[67]

The Office National du Tourisme itself issued a 290-page "official guide to the army zone" establishing thirty-one different itineraries of the battlefields from the Channel to the Vosges. The itineraries provided few details besides the order of towns that one would pass through. It offered short notations of the battles and information about the towns, including the number of inhabitants, the name of hotels, and notation of local specialties. The guide thus featured both less practical information than the Michelin guides and much less information about the war; its primary advantage, like that of most competing guides, is that it grouped all sectors of the war into a single volume for motorists.[68]

One of Michelin's primary interwar competitors, BF Goodrich, attempted to copy Michelin's approach, fashioning its own guidebook to the battlefields. Grouping together itineraries of all of the major battlefields into one volume, the Goodrich guide offered much less detail as well as fewer maps and smaller photographs of the generals. Guides appeared in French and English. Goodrich, unlike Michelin, used the first twenty-nine pages to offer information about Goodrich tires that amounted to little more than advertising. For example, with photographs of its French and American plants in Colombes and in Akron, Ohio, the firm noted that it was "one of the most important industrial enterprises" in the world with the "biggest factory in the world [in Akron]," playing on French stereotypes of the size and technological advances of American business.

Goodrich, like Bergougnan, could also use its role as a supplier of solid rubber tires for the French military to its advantage as in an advertisement from the guide announcing that "le bandage plein Goodrich broie l'Obstacle [the solid rubber Goodrich tire crushes the obstacle]," a clear poke at Michelin, whose pneumatic tires had not been used for trucks by the French army. Goodrich employed a strategy not unlike Michelin's; Goodrich's account of the war does not even use the terms *France* and *the French*, opting instead for *we* or *us* and *our*, thus identifying Goodrich the firm with France rather than with the United States. Moreover, Goodrich's account of the *voie sacrée* further asserted Goodrich's place as a guarantor of French national interests. Describing the convoys of trucks that supplied Verdun, Goodrich noted that "without pause, the teams . . . repaired the roads that wore out quickly under the excellent tires on our heavy trucks. Here too 'Goodrichs' served very well. Among the various kinds of tires used, they distinguished themselves by their resistance and their ease in adaptation to all different kinds of terrain. There was a lot of skidding along the roads of Verdun but never by those vehicles equipped with 'Goodrichs.' . . . The road from Bar-le-Duc to Verdun received the name 'voie sacrée'; it witnessed the passing of thousands and thousands of heroes going to the battle with an invincible ardor or coming back bruised but covered with immortal glory."[69] Thus, like Michelin, Goodrich attempted to tie its name to the war effort and to France by showcasing its own patriotic contribution. It did so to the extent of using the "sacred way" and the "immortal glory" of French soldiers to sell its tires.

Michelin, accustomed to dominating automobile tourism, faced its greatest guidebook competition from publishers and tour operators who assumed that tourists would generally take mass transportation. Although people in buses and trains could of course opt for a Michelin guidebook, it had not been designed for their use, omitting such basic information as schedules. Hachette, the publisher of the Guides Joanne designed specifically for tourists traveling by rail, offered the Guides Diamant to the World War I battlefields, some of which were translated into English. An entire series used geographical areas of the front and near the front for the organizational framework. The guides featured many fewer illustrations than the Michelin guides and less historical detail. Above all, the Guides Diamant were quite practical guides, focusing on the timetables of trains and buses as well as the prices of hotel rooms and meals. About half of each guide consisted of advertising, of which there were only a few pages at the front and the back of the Michelin guides.[70]

French railway companies even published guides to the battlefields. Having established special open-air trains that ran on Sundays, the Chemin du Nord published guides to provide additional information about the "martyred towns" (*villes martyres*), destroyed by "des Huns," along the tracks connecting Albert, Arras, and Lens. The guide provided commentary on cities en route, such as Amiens, and then used the rail line for the itinerary, offering information about both the towns and the history of individual battles. Because tourists were on day-trips, there was no practical information; tourists needed to pack their lunches. At 2 francs each for only forty-seven pages, they were more expensive than the Michelin guides.[71] In addition, the Chemin de Fer du Nord and the Chemin de Fer de l'Est produced a volume entitled *Battlefields of France* in both French and English editions. This guide was not tied to a particular train itinerary but opted instead to describe five zones of the battlefields and surrounding areas: Alsace, Vosges, et Lorraine; Hauts-de-Meuse et Argonne; Champagne et Soissonnais; Noyonnais et Picardie; and Artois et Flandre. Each zone received a topographical survey and then a chronology of the major military engagements in the area. The eighty-page guide closed with a list of the battlefields in the zone with the nearest train station, leaving it to pedestrians and cyclists to determine what they wanted to see and how to get there. The guide provided no other practical information.[72] For British visitors, both the Cook company and the St. Barnabas Society offered small guidebooks, comparable to the French railway guides, to accompany their guided tours to the former battlefields. The British travel agent Pickfords also offered battlefield guides to its clients.[73]

Michelin was thus not the only producer of guides designed to accompany the visitor to the battlefields, and it is unclear whether returning veterans and families who visited the gravesites of loved ones bothered with any guide.[74] Most French "pilgrims" to the gravesite of a son or father, many of whom needed subsidies from the French state to go at all, used public transport and had little need for the itineraries or the descriptions of nearby cities, which they had neither the time nor the money to tour. Of course, one could buy a Michelin guide without ever touring the battlefields, even though the itineraries were for motorists. Michelin certainly hoped to attract an audience of people curious about the war and desirous of reading Michelin's illustrated history of it. As early as December 1919, aware that the tourist market may not be the only one available, Michelin advertised other uses for the guidebooks, including their appropriateness as *étrennes* (traditional French New Year's gifts) for children.[75] Michelin also initially offered free copies of the guidebooks for the awards made at school graduation ceremonies in hopes of later sales, and the last six of the published guides bore the statement that the guides were published under the auspices of the Ministry of Public Instruction as well as the Ministry of Foreign Affairs, a sign of their pedagogical potential.[76]

MICHELIN'S FRANCE

Although Michelin did not make any profits for the Alliance Nationale, as marketing tools the Michelin guides were hardly a disaster. The guidebooks associated the company name with the French nation more closely than ever, at a time when companies regularly pointed out their contribution to the national war effort. Moreover, the publication of the guidebooks broke even; most advertising campaigns actually cost companies money. While one cannot directly tie the production of the battlefield guides to increased tire sales after the Great War, there is no question that the guides offered Michelin at least the opportunity to place its name in the service of France. Given the limitations of pneumatic tires during the war and the advertising boost that producers of solid rubber tires such as Goodrich received, Michelin's production of guidebooks, like its production of airplanes, positioned the company for the postwar market.

The guidebooks allowed Michelin again to assert its service to motorists. By taking no profits and by putting its tourist office in the position of offering help to motorists with battlefield lodging, Michelin continued its role as a linchpin for automobile touring. The guidebooks themselves offered many of the trademarks of twentieth-century tourism. Like other guidebooks, they promised to make the otherwise inaccessible accessible, allowing motorists to see the sights on their own time, distinguishing these lucky few from their contemporaries who took the train tours or the bus tours, or had to stay at home.

The guidebooks also guaranteed a certain authenticity. Prewar touring had long promised entry of tourists into "authentic" provincial or colonial settings, settings that themselves became in some respects "inauthentic" or at least differently "authentic" by the very presence of tourists seeking authenticity. Tourists equipped with Michelin guides had "realistic photos" for planning their trip and the company's panoramic shots and "total history" further promised that the war could be discovered as it had really happened. Of course, preoccupations with decent dining, a nice room, a garage for the car, and the ease of movement by automobile meant that the "authentic" experience more closely resembled the experience of tourists' likely social equals, the generals and other high-ranking officers, more than that of the *poilus*. Yet, tourists could not really experience the war as generals had either; only the more introspective of tourists might have recognized their collective responsibility for the deaths of hundreds of thousands, whereas generals lived with the reality that men's lives depended directly on their decisions. Nevertheless, the guidebooks promised the truth about the war, packaged a fabricated "real" experience of it, and maintained the notion that such a tour was authentic, thus expanding prewar notions of possible tourist sights and experiences.

Like other advocates of tourism during and just after the war, Michelin used strong patriotic language in promoting its guides to the battlefields and French tourism more generally. Michelin's guidebooks and promotion of them reflected the strong imprint of a preexisting yet evolving nationalism, which came to dominate French public discourse during the early twentieth century, but the company also contributed to the formulation of French nationalism after the First World War. In its guidebooks, some of the earliest efforts to tell the history of the war, Michelin interpreted the war as the result of German aggression, made inevitable by Germany's larger population, not of a broader diplomatic, social, or economic context. By promising to donate the proceeds of sales of the guides to France's best-known pronatalist organization, Michelin reinforced the importance of the French birthrate for the French nation, thus tapping into and helping to create a widespread consensus that the rate of population growth would largely determine the future of France.

4

Saving the French Nation

Pronatalism and Paternalism

AFTER THE MID-NINETEENTH CENTURY, all major European nation-states eventually experienced a decline in birthrates. That decline, often attributed to the effect of industrialization or "modernization," occurred first in France, which actually industrialized more slowly than either Britain or Germany. Although the French population continued to grow in the nineteenth and early twentieth centuries, except during several individual years when deaths exceeded births, the rate of population growth fell in France before it declined in Britain, Germany, and other large European states. Between 1800 and 1900 the population of France increased only by 43 percent while that of Britain (including Ireland) 164 percent, and Europe as a whole 113 percent.[1] The cause of the French decline was the willingness of French couples to practice voluntary birth control, which generally took the form of abstinence and coitus interruptus because of the high cost of condoms and other early methods of contraception before the mass production of rubber in the early twentieth century. As a result of these "Malthusian" measures, the population of France, described as the most populous country in Europe in 1700, was surpassed demographically during the nineteenth century by those of Britain and the newly united German Empire. On the eve of the First World War, there were 41 million French people, 64.9 million Germans, and 45.5 million British.[2]

Many French observers worried about the decline, casting it as a crisis.[3] Even the terminology used, including the word *decline,* connoted trouble. Just as late-twentieth-century population experts with an international perspective preoccupied with overpopulation would find a low rate of population increase salutary, late-nineteenth and early-twentieth-century French demographers feared the "decline" because they understood it in a cultural, economic, and political context of assumed conflict between individual nation-states and their peoples, or "races," as they so often put it. Moreover, the phenomenon emerged at a time when bourgeois men dominated public discourse. The very language of early demographers, like the notion of population decline, was from the outset gendered. By the latter half of the nineteenth century, Louis-Adolphe Bertillon, a statistician for the city of Paris, and other early demographers began to define new ways of collecting and analyzing French population statistics and viewed their findings with alarm. They asserted, with the assistance of statistics that seemed to represent objective reality, that the French were not having enough babies to compete economically or militarily with other major powers. Moreover, in assuming that women should have as many babies as biologically possible, they increasingly proclaimed that the problem was simply that French women were not fulfilling their responsibility in having enough babies to keep France strong.[4] French men's complicity, so patently obvious when the primary methods of birth control were abstinence and coitus interruptus, received somewhat less attention. Early demographers thus understood the indisputable fact of a declining birthrate, itself a larger abstraction, as a specific social "problem" jeopardizing French power and arising largely from the behavior of individual French women.

Although France lost the Franco-Prussian War for a host of reasons, not one of which had anything to do with the size of its population, after 1870 French observers increasingly viewed their country's "stagnant" population as a sign of weakness vis-à-vis the Prussian-dominated German Empire, particularly after the number of French deaths exceeded births in 1890, 1891, 1892, and 1895. In 1896, Dr. Jacques Bertillon, son of Louis-Adolphe, founded what became the most influential among many pronatalist groups, the Alliance Nationale pour l'Accroissement de la Population Française (the National Alliance for the Growth of the French Population). He used the organization to lobby authorities to undertake measures that would supposedly increase the French birthrate. Prodded by Bertillon and perceiving interest among the electorate, politicians increasingly turned their attention to the population "problem." In 1901, Minister of the Interior René Waldeck-Rousseau established a commission of demographers to examine the population issue.[5] Despite enlivened discussion in meetings of the extraparliamentary commission and in parliament itself, the most important pronatalist measures to make their way into law before World War I were laws in 1913 that mandated family allowances (*allocations familiales*) for large families that could prove need and maternity leaves for working women.[6] The passage of even these two, rather limited measures occurred in the context of heightened international tension after the Moroccan crisis and after deputies had examined carefully German and French population numbers when deciding to extend the draft from two to three years (France's strategy for fielding as many troops as the more populous Germany).[7]

It was precisely in 1913, when pronatalism began to receive more public attention in France, that Michelin began publicly to support the Alliance Nationale. Contributions by André and Edouard Michelin would quite literally bankroll the Alliance Nationale, as the two became the association's single largest public donors in the 1920s; as one specialist of French pronatalism put it, "there is no doubt that the Michelin connection played a major part in ensuring the healthy state of the Alliance Nationale's finances."[8] Michelin's advertising began just after the war to feature the company's association with the group as well as its own pronatalist measures. Michelin's advocacy for the pronatalist cause—an approach that would seem quite odd in the context of American tire firms at the time—reflected widespread cultural fears in early-twentieth-century France that France was being endangered by a "stagnant" birthrate, and the company helped to shape those fears as well as possible solutions.[9] While not marketing in the traditional sense, Michelin's very public efforts on behalf of the pronatalist cause, at a time when that cause had managed successfully to portray itself as the salvation of France in the eyes of a large number of French people, placed the Michelin name once again in the service of France.

Jacques Bertillon founded the Alliance Nationale in order to alert the French, and particularly French politicians, to the danger of "depopulation" (*dénatalité*). Its monthly *Bulletin* (and later *Revue*) provided statistics on the French birthrate, articles on depopulation, and accounts of the parliamentary debates affecting the Alliance's program. Before World War I, the group focused its efforts on providing statistics and proposing legislation to deputies. The French parliament needed, according to the Alliance, to provide aid to pregnant women and a host of benefits to fathers of large families: a fast-track to promotion in the civil service, tax breaks, reduction in the length of military service, and family allowances (*allocations familiales*). Although some of its proposals, such as aid to pregnant women and maternity leaves, could potentially fit into a solidarist program to help French women (whether married or not) out of mutual social commitment, most of the association's program could be better called what Karen Offen has termed "patriarchal paternalism," for attempting to increase the birthrate by reinforcing the authority of men over their wives and children.[10] Even before the war but particularly after it, the Alliance wanted to increase the French birthrate by keeping women at home and supplementing fathers' salaries. The Alliance also promoted unsuccessful prewar legislation to limit the availability of artificial methods of contraception and to punish with greater severity those caught performing abortions. Finally, the Alliance advocated the "family vote," a proposal that would have given men with families not only their own ballot but also ballots for dependents, including their wives (French women did not get the right to vote until after the Second World War). Although the organization was adamant about its secularism and political neutrality, the Alliance's program was in some respects rather close to social Catholicism, particularly in its rejection of economic, social, and political liberalism, not to mention in its advocacy of unrestrained reproduction.

On the eve of the war, the Alliance changed its tack. While maintaining its lobbying efforts, the organization increasingly focused on influencing potential voters as a way of forcing members of parliament to accept its ideas. The Alliance, which had only 230 members in June 1913, attempted to build its membership and undertook "propaganda" (its term, which did not have the same negative connotation that it does today), creating a subcommittee for propaganda headed by Fernand Boverat.[11] Boverat, a pronatalist writer and a donor himself, spearheaded the Alliance's attempts to reach a broader audience and to raise more funds for the effort. Producing pamphlets and posters for voters as well as for candidates was extremely costly, requiring sums that the membership fees of two hundred members could never support. In October 1913, André Michelin stepped into the breach, offering the Alliance Nationale a check for 1,000 francs.[12] The French state officially recognized the group as an *association d'utilité publique* in 1913, placing it under the patronage of the president of the Republic and giving it an annual subsidy. The group's propaganda quickly expanded as did the membership rolls; there were 1,321 members by January 1914. In April 1914, Edouard Michelin donated 10,000 francs specifically for the production and distribution of brochures and posters for the upcoming legislative elections.[13] Michelin's support enabled the group to distribute 1.5 million copies of an electoral "poster depicting two French soldiers being bayoneted by five Germans. The caption explained that each time two future soldiers were born in France, five were born in Germany." A second poster warned that France would be abandoned by her allies unless the birthrate increased. In the end, "over a third of the candidates who published a political program either referred directly to depopulation as a major national menace or proposed to introduce pronatalist measures in the next parliament."[14]

Although the Alliance did less to promote its cause during the war, to many French the war itself seemed to lend legitimacy to the pronatalist program. The Alliance suspended most of its propagandistic activities during the war, focusing instead on supplementing the small family allowances paid to women with large families whose husbands had been drafted. For this effort, Edouard Michelin repeated his earlier "patriotic gesture," to use the words of the Alliance, and gave another 10,000 francs in 1915.[15] French state and private propaganda during the war, including widely diffused postcards with pronatalist themes, used the arguments that the Alliance had frequently made before the war: what France needed, above all, was more babies.[16] During the war there emerged the idea that if France had had more people and thus more men in arms in 1914, Germany would never have resorted to war; French depopulation thus became the primary cause of the war. In 1919, Premier Georges Clemenceau reminded French deputies that "the treaty [of Versailles] does not say that France must undertake to have children, but it is the first clause which should have been included in it. For if France turns her back on large families, one can put all the clauses one wants in a treaty, one can take all the guns of Germany, one can do whatever one likes, [but] France will be lost because there will be no more Frenchmen."[17]

Although there were by 1922 at least eighty pronatalist groups in France, the Alliance Nationale's influence was unparalleled after World War I.[18] In the heavily gendered discussions about the future of France and the respective places of women and men in that future, actual legislation largely mirrored the Alliance Nationale's proposed reinforcement of patriarchy. The conservative postwar majority of the *chambre bleu horizon*, named for the sky blue uniforms still worn by veterans sitting in parliament, imposed a supplemental tax on single men and women. Although widows were exempt, divorced women and women who never married were not. The tax hit single women proportionately harder, both because they earned less than men and because the demographic reality, imposed by the approximately 1.5 million men killed during the war, meant that there were approximately 2 million more single women than men after the war. The Alliance was also a major proponent of the French state's adoption in 1920 of "medals of the French family" to be given to French women who bore several children; mothers of five children received a bronze medal, those of eight a silver one, and those of ten a gilded one. Moreover, in 1920 the chamber made any "neo-Malthusian" advocacy for or information about contraception or abortion illegal and imposed harsher penalties on those getting or practicing abortions, a measure that many historians have rightly seen as a symbolic attempt to control women's reproductive lives while not tampering with men's; the exemption of condoms from the law for the control of venereal disease not only made the law less effective in limiting the birth control practiced by couples, but also implicitly accepted men's sexual liberty with women other than their wives.[19]

Even after these early victories, the Alliance Nationale remained in the interwar years close to the halls of power. After the war, the board of directors (*conseil d'administration*) of the Alliance Nationale comprised a host of former generals and government officials, as well as private entrepreneurs, including André Michelin. When French Premier Alexandre Millerand, himself a member of the Alliance, created a new Ministry of Public Health in 1920, he placed Jules-Louis Breton, another member of the group, at its head. Breton's first official act was the creation of a permanent advisory body of thirty people to coordinate government efforts to promote natalism in France. The new Conseil Supérieur de la Natalité included Jacques Bertillon, Fernand Boverat, and three other members of the Alliance Nationale's board, including André Michelin, who served until his death in 1931.[20] Throughout the interwar years, premiers and ministers on the Left as well as the Right regularly met with representatives of the Alliance to get electoral support. A further sign of the organization's growing influence after the war was its expanding membership, which reached a peak of nearly forty thousand in 1930.[21] Although the Alliance never reached the numbers attained by groups with a similar agenda but lower membership fees, such as the Ligue Populaire des Pères et des Mères de Familles Nombreuses, which was reputed to have had hundreds of thousands of members, the Alliance had an influence out of proportion to its numbers. Along with other pronatalist groups, the Alliance ultimately claimed credit for the national family allowance legislation of 1932 and 1938, as well as for Edouard Daladier's *Code de la famille* in 1938–39.

ADVERTISING THE GUIDES TO THE BATTLEFIELDS

Michelin's decision in 1917 to give the profits from sales of the guides to the battlefields to the Alliance Nationale served two purposes. Michelin could shield itself from charges of profiting from the war by selling guidebooks to the battlefields; this not only distinguished the Michelin guides from those offered by Hachette and a host of other publishers but further reminded potential tire buyers of the Michelin brothers' own patriotism. At the same time, the Michelin brothers could offer financial support to the Alliance, clearly an interest even before the war.

The earliest advertising for the guides focused on the Alliance Nationale. The guides themselves always included notation that the profits went to the Alliance Nationale. Moreover, Berger-Levrault's handbill, the first advertising for the guides, featured Jacques Bertillon's oration at the "tour of the Ourcq," the Marne battlefield, in September 1917. At Ourcq, the ailing Bertillon delivered a long oration in which he placed the blame for the war quite squarely on French depopulation, particularly the French predilection for families with only one or two children. Not only could the French place fewer men in the field than Germany in 1914, but prewar economic growth had supposedly suffered because France had a smaller labor force and fewer consumers. Bertillon repeated a frequent theme in Alliance propaganda, addressing French men to the exclusion of women: "every man has the duty to contribute to the perpetuation of his country [by fathering children], just as he has the duty to defend it." The anticlerical Bertillon declared that the two-child family was "blasphemy" against which "the tombs that cover the battlefields of the Marne protest." It was thus the "duty" of French men to repopulate France; women, who had raised children and done paid work in the absence of their husbands, received no mention. Bertillon also suggested the theme implicit in several wartime postcards, in which making war and love were equated. The well-known postcard, "Un bon coup de baïonette," in which babies hung from a vertically held bayonet that suggested the erect penis of a soldier on leave, was only the most graphic of many portrayals of French men's duty to father children during and after Great War.[22]

In Michelin's own advertisements for the guides to the battlefields, the Alliance Nationale received pride of place, and its arguments appeared in short-ened form. Echoing themes in Bertillon's speech as well as in the pages of the *Bulletin de l'Alliance Nationale,* Bibendum proclaimed in 1920 that the task before France was "to confirm and perpetuate the victory. [France] can only be fertile if we ourselves are. The children of today will assure the peace of tomorrow, but only on condition that they are numerous and strong. France will only maintain its position, first in the world, through production, first of all by the prettiest [production] of them all, that of babies. I thought it proper that the war itself contribute to this work of life [*oeuvre de vie*]. That is why all of the profits that I might have made will be given to . . . the Alliance Nationale."[23] With the exception of the Taylorist language of production that equated the production of babies with that of widgets, this is rhetoric from the pronatalist camp: the only assurance of peace and French power was a larger population.

In its "Samedi," the company's weekly advertising section in *L'Illustration,* Michelin described the Alliance Nationale and justified the transfer of profits from the battlefield guidebooks to the group. Using the apocalyptic language that the Alliance and Boverat in particular had used before the war and that became a mainstay of Alliance propaganda after the war, Bibendum briefly summarized the work of the organization. "What is this Alliance Nationale for the growth of the French population. . . . Of what does it consist? *It consists, quite simply, of the very existence of the country* [emphasis in the original]."[24] The Alliance, like Michelin in its support for the Alliance, attempted to mobilize support for the pronatalist cause by exaggerating reality; given that France had just won the war with a smaller population, it took such rhetorical sleights of hand to make the argument that France needed more babies than French couples wanted, or could afford, to have.

Like the Alliance, Michelin used presumably objective statistics to assert a presumably self-evident reality. As Bibendum put it,

It suffices, alas, to let the numbers speak for themselves. Their eloquence is frightening:

NUMBER OF BIRTHS

	In France	In Germany
In 1881	937,000	1,682,000
In 1911	712,000	1,871,000

In 1918, France (not including the invaded departments) [had] 400,000 births and 800,000 deaths, not including the soldiers dying in the army! Two times more coffins than cribs! All commentary would be superfluous. If this march toward the abyss continues we will soon be in no position to impose on Germany the execution of the Treaty of Versailles, in no position to resist her attempts at revenge: there are no soldiers, however heroic they may be, who can fight and win when it is one against three, and there are no faithful allies for peoples who give up [s'abandonnent].[25]

As Bertillon had claimed from the earliest days of the Alliance Nationale, the presumably neutral statistics would speak for themselves. But as recent works by Joshua Cole and Hervé Le Bras have shown, the collection of French population statistics was never merely factual.[26] Bibendum's statistics mask the extent to which the German birthrate also fell precipitously in the early twentieth century. Moreover, although the Alliance claimed that France would soon lose its empire if there were not enough French to defend it, it is interesting that births in the colonies, that is, births among nonwhite subjects, were not included. Even France's relationship with Britain and the United States could be placed in jeopardy; instead of those countries' populations being seen as guarantors of French sovereignty, as they were in the First World War, the Allies themselves would supposedly desert a France that did not reproduce in sufficient numbers, that *s'abandonnent,* a verb that can also mean to give in or to give oneself to in a sexually passive sense. Although the language as a whole was not gendered, there is nevertheless the suggestion, far more pronounced in Alliance Nationale propaganda, that French men could regain a supposedly jeopardized masculinity by fathering more children, and thus by extension asserting their authority in the world.

Bibendum's rhetoric further referred both to Clemenceau's own doubts about the Treaty of Versailles if France did not have more babies and to the wartime propaganda that equated the German invaders of 1914 with earlier Germanic tribes: "No barrier, even that of the Rhine, has ever prevented the Germans from arriving in Gaul. And no scrap of paper will ever stop the invasion of a prolific people in the deserted field of its neighbor. There is no humanitarian theory that stands up to the imprescriptible laws of Nature. . . . *Depopulation is the suicide of France* [emphasis in the original]."[27] This social Darwinistic interpretation of world affairs, however seemingly correct it might have seemed from a French national perspective, overlooked the role of diplomacy among allies in winning World War I, not to mention maintaining the peace.

The solution to France's problems, according to Bibendum, was as simple as the statistics that he provided. The Alliance Nationale, "founded and directed with the faith of an apostle by the Doctor Bertillon, leads a good fight for this cause. It works to increase natality and fights for an active propaganda against harmful, destructive [neo-Malthusian] elements by energetically forcing before public opinion and the Parliament projects that will improve the situation of large families." The article concluded with an exhortation for the reader to join the group, listing the address and various membership levels, beginning at a simple 5 franc membership and climbing to that of 5,000 francs.[28]

Advertising for the battlefield guides also quickly transformed the pictorial representation of Bibendum. With the exception of a few pictures advertising the *jumelés,* or dual tires, that featured small twin Bibendums accompanied by their father, prewar advertising always portrayed Bibendum as a single man, often in the company of one or more beautiful women.[29] After the war, the women literally disappeared from sight, just as pronatalist discourse (at least before the 1930s when the "modern woman" [*femme moderne*] became a subject of derision in the pages of the *Revue de l'Alliance Nationale*) tended to

SAVING THE FRENCH NATION

135

ignore women in its many appeals to men to father more children. Bibendum found himself suddenly surrounded by hordes of young French children, up to two dozen in some advertisements, becoming a big, smiling father figure. Yet Michelin left Bibendum's paternity somewhat ambiguous; actual children were not specifically referred to as his own.

In one short printed advertising sheet entitled "Souvenir du salon de l'aviation" of Christmas 1919, as at the Aviation salon itself, Michelin described Bibendum as a father, perhaps even a mother, of sorts. With six actual children gathered around Bibendum, the text reminded the reader that the profits from Michelin's *guides des champs de bataille* would be paid to the Alliance Nationale. But here the guidebooks themselves, rather than actual children, were Bibendum's progeny. "For the sake of French repopulation Bibendum has in seven months brought twenty-four children (fourteen French and ten English) into the world, who will show you around the battlefields. And he continues! Next year, he will give birth to [*donner le jour à*] thirty new children."[30] Bibendum could thus be compared to the men listed in the pages of the *Bulletin de l'Alliance Nationale*, where the number of children that they "have had [*a eu*]" was regularly noted, with accolades implicitly given to large numbers. Bibendum fulfilled his role as a French man fathering his own *famille nombreuse*, even though the expression *donner le jour à*, at least as much so as the verb *a eu* (rather than simply *a 'x' enfants*) is normally used for women giving birth.[31] The actual children featured in the illustration are of uncertain origins. They could be Bibendum's. Or they could be the result of Michelin's many efforts to encourage men, including its employees, to father more children.

In the end, the Michelin brothers resorted to direct donations to support the organization's propaganda. In 1919, the operating budget of the Alliance Nationale consisted of 128,717.14 francs. Profits from sales of the guides totaled 467.60 francs, whereas membership dues brought in 25,684.10 francs and the state subsidy consisted of 13,000 francs. Although the organization ran a surplus of 59,113.69 francs, the Alliance's costly propaganda was only made possible by direct grants from Michelin of 54,420 francs and from the Fondation Cognacq (established by the head of the Samaritaine department store) of 18,000 francs.[32] Michelin's own contribution totaled more than twice the amount from dues, and it represented over a third of the operating budget of the Alliance.

Throughout the 1920s, the Michelin brothers continued to fund the associa-

16. Advertisements for guides to the battlefields frequently featured many children along with multiple volumes of the guides, suggesting that Michelin's prolific production of guidebooks would result in the proliferation of French babies. Here Bibendum is passing out copies of the guide to Verdun, the children's "traveling companion during summer vacation." From "Le Samedi de Michelin," *L'Illustration*, 27 December 1919.

tion. In 1920, the Michelins gave the Alliance 50,000 to match the subsidy from the French state. In the electoral campaign of 1924, by April they had donated 100,000 of the 125,000 francs that the Alliance had at its disposal. Their contribution helped to make possible the production and distribution of pamphlets and a poster, drawn by Jean Droit, entitled "La dépopulation, c'est la ruine du pays" [Depopulation is the ruination of the country]. The poster laid out the group's demands and featured a small French baby trying to protect a kougelhopf (the typical Alsatian cake that thus symbolized the territory of Alsace-Lorraine) from a much larger German baby. The Michelin brothers offered another 200,000 francs later in 1924 as well as 200,000 francs yearly in 1925 and 1926, mostly for the Alliance's attempts to convince the French teaching corps of the importance of covering pronatalist themes in class. Michelin brothers donated yet another 100,000 francs during the 1928 elections.[33]

17. Bibendum, the prewar *mondain* ladies' man, took on the air of papa in advertisements for the guides to the battlefields, adored here by French children of various ages. There is an equation of the guides themselves with children, a reminder that the profits went to the Alliance Nationale. At the bottom of the advertisement (not shown), babies sit and crawl out of cabbages, a symbol of newborn boys (girls often growing out of roses) frequent in pronatalist iconography. From "Le Samedi de Michelin," *L'Illustration*, 21 February 1920.

In 1922, just one year after the end of publication of the guides to the battle-fields, Michelin further supported the Alliance Nationale by founding the Prix Michelin de la Natalité (the Michelin natality prize). In announcing the prize in June, the Alliance noted that it had long wanted to "get in a good hit likely to awaken both the Nation's conscience and the apathy of the government." The Prix Michelin, awarded to the author of the best brochure describing the "problem" of depopulation, was supposed to make that strategy possible, first by encouraging people, particularly members of the association, to study the issues in detail, and second by paying for the mass distribution of the winning essay. The Michelin prize would be the best publicized natality prize awarded by the Alliance in the 1920s, to be rivaled only in the 1930s by the numerous prizes paid directly to large families by the Fondations Etienne Lamy and Cognacq-Jay.[34]

Michelin offered the Alliance 270,000 francs, 120,000 of which was for headline-grabbing prizes. The first-place prize consisted of 50,000 francs, the second-place 10,000, the third-place 8,000 with the top forty-six finalists earning at least 1,000 francs each. The additional 150,000 covered the costs of the competition, the largest of which was the publication and distribution of 500,000 copies of the winning entry.[35] Although it is impossible to know exactly how many people would read the winning pamphlet or any other publication, the average press run for a novel by a popular writer was only 15,000 copies (the exceptionally successful *La Garçonne,* which first appeared in July 1922, had sold 300,000 copies by the end of the year).[36] The prize money itself drew the attention of the national press, which in several cases had promised to print large sections of the winning essay before the Alliance even announced the competition. To ensure that attention, the Alliance named several well-known politicians, military leaders, and newspaper editors to its honorary prize committee. Léon Bourgeois, the president of the Senate, Marshals Foch and Lyautey, former ministers, senators, and deputies shared the stage with the editors of mass-circulation dailies *Le Matin, Le Journal, L'Echo de Paris,* and *L'Intransigeant.*[37]

The actual selection committee included Jean-Louis Breton, the former minister of health who had created the Conseil Supérieur de la Natalité, other former ministers, and several members of the Conseil Supérieur, including Fernand Boverat and André Michelin. The selection committee specifically asked contestants to submit brochures with no more than thirty-two easy-to-read pages with "an energetic style" that would summarize in "short and vigorous sentences the arguments capable of striking most vividly the imaginations of the peasant, the worker, and the employee, thus refuting neo-Malthusian assertions." The essays were required to focus on the fact that the French state should equalize the financial burden on large families and "those without children."[38]

By January 1923, after receiving 1,055 entries, the committee chose "La vie ou la mort de la France" [The Life or the Death of France] by Paul Haury, former student at the Ecole Normale Supérieure and a secondary school teacher from Lyon.[39] The winning essay featured on the title page the notation "Prix Michelin de la Natalité" and appeared in the requisite five hundred thousand copies. Haury, who along with Boverat would become a leading propagandist for the Alliance in the late 1920s and 1930s, used the essay to summarize the agenda of the Alliance, suggesting several themes that would increasingly dominate the group's propaganda. In the meantime, Michelin's place as donor was yet another reminder of the Michelin brothers' role as generous benefactors of French nationalist issues.

Using medical terminology frequently employed by pronatalists who not only were sometimes doctors themselves, as in the case of Jacques Bertillon, but were also accustomed to addressing the many doctors sitting in French parliaments, Haury described "the illness that is killing France: . . . the suicide of a great people," then offered the carefully chosen statistics that were such a trademark of the Alliance. Whereas France had been the largest and most powerful country in Europe at the time of Louis XIV, by 1924 it was the least populated of all of the major powers. The threat was real: "Before the war, seven years saw more deaths than births of French people; and during the hostilities, whereas Germany replaced through births its civilian population, **through a shortage of births, France doubled the losses caused by enemy fire: she thus lost not just 1.5 million dead but three million human lives** [bold in the original]. . . . If the fertility of our young households does not increase right way, it will soon be the end."[40] And Germany would soon be back.

Haury then noted the consequences of the declining birthrate in apocalyptic terms. "Like the Greeks and the Romans of the empire, [France] is undermined by sterility: like them she will know civil war and foreign conquest."

Like the Romans, France would fall to the Germanic hordes. In addition, there would be no laborers to farm, no workers for factories, and no consumers to "permit factories to undertake cheap mass production in order to compete against foreigners." For those who believed that foreigners might fill the empty positions, Haury noted that "foreign conquest in peacetime becomes dangerous in wartime." Rather than proclaim the necessity of assimilating foreigners and offering naturalization, as sometimes argued in the pages of the *Revue de l'Alliance Nationale* in the early 1920s, Haury assumed that foreigners would always keep their primary allegiance to their place of origin. As a result, France would soon have "colonies but no colonists" and "an army without soldiers."[41] Here, too, Haury rejected out of hand the notion that colonial subjects could assimilate or that immigrants might fill the void, although the *Revue* itself was more ambivalent.[42] Without more French babies, France would face another onslaught from Germany for the same reason that it faced one in 1914. Repeating the Alliance Nationale's interpretation of the causes of the First World War and that suggested by Michelin, Haury maintained that "**we don't need to look any farther for the essential cause of the war of 1914: because we had fewer and fewer children, the Germans saw us as a 'dying nation.' They believed that the moment had come to finish us off** [bold in the original]."[43]

In the final section of the pamphlet, Haury explained the "remedies" for France's "illness." France needed to have more babies, a message that "religion, schools, and the press" were charged with delivering; Haury, like the interwar Alliance, moved closer to the Catholic Church than Bertillon had been, comforted by its focus on reproduction. Above all, Haury believed that the traditional family would save France. Paraphrasing the *abbé* Sieyès' famous revolutionary tract, Haury asked, "**What is the family? Everything. What has it been for too long in the eyes of the state? Nothing. What does it want to be? Something** [bold in the original]." In contrast, households with singles and childless couples were merely, again using a medical analogy, "cells that don't reproduce themselves," that leave behind only tombs. Repeating the Alliance's frequent interwar characterization of all childless people as "sterile," Haury argued that "without doubt the childless can be individually respectable, especially if their sterility is involuntary or has been imposed by their religious vocation or by health [problems]. The fact is however there: they received life and do not give it. In that regard, they are in general parasites ... so too are the households with one ... or two children parasites because it requires three births on average to replace the father and the mother [given French infant mortality levels at the time]." According to Haury, only those people with three or more children have the right to shout "Vive la France."[44]

Getting rather close to the antirepublican, proto-fascist arguments that Haury and Boverat used in the 1930s, when both Hitler's and Mussolini's pronatalist policies would be held up as models, Haury claimed that the French state neglected the family in favor of the individual, construing "liberty merely as the right to have or not to have children, equality as the equality between single men and fathers, so that the latter have to face a heavier financial burden." Only briefly mentioning the necessity of lowering infant mortality, Haury emphasized that "the essential is not to prevent death, but to cause birth." The state needed to make payments of at least 300 francs at the birth of the third child in all of France, and not just in the departments that had instituted such plans. Moreover, those without three children needed to make payments to be distributed to those with large families. The private entrepreneurs who had not already granted allowances to employees with large families needed to do so. The state needed to provide such coverage for all of its functionaries. Above all, France needed electoral reform in the form of the familial vote, awarding a vote for a wife and for each child to every father. The basis of the French republic itself thus needed to be changed. "*The only electoral regime that conforms to equality and justice as well as to the national interest is universal suffrage accompanied by the family vote* [his emphasis]." France, according to Haury, who was writing during the craze surrounding *La Garçonne*, was really just a "Republic of Garçonnisme," where "everyone is getting used to counting on the children of other people." For Haury, the Third Republic was therefore no different from the much-dreaded, supposedly sexless *garçonne* of the interwar years.[45]

Making the connection between depopulation and the war that characterized pronatalist works in general, including Michelin's own ads for guides to the battlefields, Haury used his trump cards in addressing French veterans and French women. Veterans needed to realize that "it was to assure the future that our brothers died, that we suffered; our sacrifice demands children." In more psychologically manipulative terms, he asked young women to "listen to the poignant regret of so many sad mothers. They lost in the war their only son and think back to the period when his young voice made vain entreaties for 'a little brother'; they sob before an unchangeable fact 'if I had known!'"[46] Haury's appeal to form traditional families, utterly out of sync with the levels of women's work outside the home in late-nineteenth and early-twentieth-century France, reinforced a fundamentally gendered worldview. Men needed to make babies, especially sons, just as they had made war. Women needed passively to accept a role as wife and mother or risk losing their only son. Women, whose second and third sons might also have died in the slaughter of the war, were implicitly to blame for the war for limiting the size of their families.

Haury's essay went further rhetorically than Michelin's own public pronouncements or advertising; like Alliance pamphlets generally, it was perceived as a wake-up call, an apocalyptic warning of what could happen to France. But Michelin must have approved the pamphlet. André Michelin's place on the selection committee and Michelin's willingness to have its name proudly displayed on the title page indicate that Michelin was more than content to have the name of the company associated with Haury's ideas. In its own pamphlets Michelin made fewer direct criticisms of the French Republic, and its language was somewhat less gendered, in part because its own pamphlets were more factual and less rhetorical. In instituting family allowances and energetically publicizing them, the company did, however, work to reinforce paternal authority in the homes of its employees just as it offered a series of measures to increase their rates of reproduction.

MICHELIN'S ALLOCATIONS FAMILIALES

In June 1916, in the midst of its wartime expansion, Michelin established an early system of family allowances. From the beginning, Michelin and the Alliance attempted to ensure the public's knowledge of Michelin's beneficence and patriotism. As early as October 1916, Michelin's allowances received an entire two and a half pages of description in the *Bulletin de l'Alliance Nationale* (two other allowance programs announced in the same issue received a quarter-page each). As would be the case throughout the 1920s, the Alliance portrayed Michelin as a real innovator, whose "magnificent example should be followed by French industry [generally]." Michelin itself devoted a section of its wartime pamphlet "Michelin pendant la guerre" to the allowances, offering them as proof of the company's dedication both to its employees and to France.[47]

Contrary to the impression given by Michelin's interwar publications, the Michelin company was not the first French employer to offer payments to its employees with children. A handful of prewar employers had offered family allowances since the 1890s. Reacting to Pope Leo XIII's *Rerum Novarum*, which called on employers to pay a just wage, Léon Harmel, the social Catholic cotton manufacturer, set up a fund to pay allowances to families with children. In response to a decree of 1889 forcing companies involved in state public works to institute allowances, French railway companies adopted such benefits in the 1890s. The companies used allowances to limit across-the-board wage hikes in favor of raising the wages of employees who most needed higher salaries. In 1890, the Compagnie du Chemin de Fer du Nord (the Northern Railway Company) began to offer allowances for employees with at least three children; just before the war it used the program to dole out cost-of-living increases to men with families, those most hit by inflationary pressures without formally increasing wages and thus all workers' expectations of what a just wage should be.[48]

The French state itself offered family allowances before the Great War. The French navy had given additional benefits to those with children as early as 1860. By 1908, primary school teachers received an annual allowance for children under sixteen, and employees of the Ministry of the Treasury, such as port and customs' officials, received similar benefits in 1909. The French army also granted special benefits to men with children. Moreover, in 1913, prompted by the Alliance Nationale's pronatalist campaign, the French government began to provide family allowances for any needy family with at least four children, whoever the parents' employers. Although parents had to prove need, thus placing the allowances in the category of special aid, rather than a right for all, this law was a crucial precedent for the later expansion of family allowances. While amounts provided under the 1913 law, like those given to state employees, were quite low, it is nevertheless true that the notion of family allowances—informed by social Catholicism, the French state's policies for civil servants, and employers' attitudes to wage hikes—certainly existed before 1916.[49]

Other private employers also undertook the development of allowances during the war. Without any apparent coordination of plans with Michelin, the engineer Emile Romanet convinced Régis Joya, a hydroelectric builder from Grenoble, to begin offering substantial family allowances in 1916 as well.

Romanet had been heavily influenced by social Catholicism, including the as-sistance that Harmel had given his workers. In November 1916, the Grenoble metal syndicate adopted similar allowances for all of its member firms. This gave an incentive for some firms to avoid hiring or retaining heads of families, a problem resolved when the metal syndicate created in May 1918 the first *caisse de compensation*. The *caisse*, or account, pooled the money of the various employers (at the rate of 5 percent of the wage bill) for the sake of allowances, thus equalizing the cost. In the 1920s, most French businesses offering allow-ances did so by means of a *caisse*, usually organized by region and sector, such as metallurgy.[50]

Joya and Michelin were hardly alone among public or private employers during the war. The French state expanded the number of recipients of allow-ances to include "all government employees below a particular wage level in 1917 and to all government employees in 1919." The state also encouraged the payment of allowances among mining firms and railways. When the Ministries of Labor and Munitions began to settle labor disputes, they heavily favored the introduction of family allowances; some fifty-two of such wartime settlements included a provision for the creation of allowances.[51] Like cost-of-living bo-nuses, which were doled out individually and not added to the base wage rates, family allowances eased the burden caused by wartime inflation without forc-ing employers to enact across-the-board wage hikes. Allowances were not only cheaper because they did not include all employees, but they also allowed employers to assist workers without creating higher expectations or admitting that wages were inadequate for workers to make ends meet.

Michelin's family allowance program thus fit squarely into a larger context in which allowances melded together employers' wage strategies and prona-talism. Although Michelin was by no means alone, its allowance program was in several respects unique: allowances were higher than those of either the French state or other private employers; they provided greater inducements to have large families; and they were better publicized. In 1916, whereas Joya offered a payment of 20 centimes daily for each child under 13, Michelin only began offering a payment upon the birth of the family's second child, when it consisted of 20 francs monthly (the equivalent of about 20 hours of work) for the first year and 10 francs monthly for the second through the sixteenth years. As the numbers increased, so too did Michelin's allowances. For a family with 3 children, Joya offered 18 francs monthly whereas Michelin provided 45

Table 1. Comparing Family Allowances, 1929–1930

	Michelin (1929)	National Average (1930)	Civil Servants (1930)	*Caisses* (1929)
1 child	100 F	28 F	55 F	60 F
2 children	200 F	70 F	135 F	138 F
3 children	405 F	123 F	265 F	276 F
5 children	675 F	270 F	586 F	558 F

Source: André Gueslin, "Le système social Michelin (1889–1940)," *Michelin, les hommes du pneu: Les ouvriers Michelin, à Clermont-Ferrand, de 1889 à 1940,* ed. André Gueslin (Paris: Editions de l'Atelier/Editions Ouvrières, 1993), p. 111.

francs. Michelin paid even more if the head of the family died, and yet more if the father died defending France.[52] By the late 1920s, when both the French state and various *caisses* had increased amounts to cover inflation, Michelin had altered its own program but still offered considerably more than other French employers, despite the relatively low cost of living in Clermont-Ferrand.

Besides the obvious largesse, however, what most distinguished Michelin's benefits from those of other institutions was that they were better publicized. Beginning with the extended notice in the *Bulletin de l'Alliance Nationale* in 1916 and the brochure "Michelin pendant la guerre" of 1917, Michelin held itself up as a model for other French firms. In 1923, Michelin produced a free brochure entitled "Allocations et rentes pour les familles nombreuses," designed to convince other employers of the necessity of adopting similar programs. Although the information in the brochure was technical, laying out precisely what Michelin paid, its justification evoked national duty, while also hinting at social responsibility for any employers inclined to the arguments of social Catholicism: "It is just to aid families who have a lot of children, and also widows. It is a duty. This duty is even more imperative since France is experiencing depopulation."[53]

By the mid-1920s, the company became more assertive, tying its family allowance program even more tightly to the pronatalist movement. In yet another free brochure, "Une expérience de natalité," Michelin began with a rhetoric quite similar to that of the Alliance Nationale and to Paul Haury's prize-winning brochure. According to Michelin, France would soon be "a desert or a colony." Examining the birthrates of their own employees "a few years ago," Michelin decided that "if this continues, our factory will slowly become empty or we will have to fill it with foreigners. And since we are no exception, either France will become a desert or it will become a colony of other countries richer in children." After listing the amounts that Michelin offered its employees with varying numbers of children, the company exhorted employers to join a *caisse de compensation* like that created by Joya and Romanet in Grenoble. Just as the *voie sacrée* (sacred road) at Verdun had led to victory, employers needed to join Michelin on the *voie féconde* (fertile road) in 1926.[54]

Placing pronatalism at the center of its argument, Michelin attempted to convince other employers that family allowances were effective. The company asked statistician Georges Risler to study birthrates in Clermont-Ferrand and in nearby villages where Michelin employees also lived. His results, summarized in "Une expérience de natalité" and in an article by André Michelin in the *Revue de l'Alliance Nationale* (of which thirty thousand offprints were mailed to French physicians by the Alliance),[55] revealed that Michelin employees did indeed have a much higher birthrate than "non-Michelin" couples, up to two and a half times more on average.[56] In an effort to address the contemporary criticism of its figures, particularly the fact that its employees were younger, and thus only seemed to have higher rates of fertility since their children were not yet raised,[57] the company continued to argue that its programs had made a difference and that its statistics had made allowance for differences in age.[58]

The point here is not to conclude definitively that Michelin policies did or did not raise the birthrate, because other factors, such as Michelin's higher rates of pay and the fact that working-class couples were generally associated with higher fertility rates, may also have had an influence. What is significant is that Michelin firmly *believed* that the allowances made a crucial difference, finding that to be the most persuasive argument to convince fellow employers to adopt such policies. Michelin assumed that other French employers shared a commitment to the pronatalist cause. In all its publications, the only argument the company used to justify allowances was their pronatalist impact. Employers' responsibility to provide a just wage was not evoked, nor was the possibility that allowances might allow a more flexible wage strategy for employers. Although Michelin did note the relatively low cost of allowances in

UNE
EXPÉRIENCE DE NATALITÉ

Peut-on faire naître plus d'enfants en France ?
Les résultats obtenus chez Michelin
permettent de répondre :

oui !

Cette brochure
est envoyée gratis et franco
sur demande adressée à

MICHELIN & Cie
Clermont-Ferrand

—

1926

18. "An experience with natalism: Can one encourage more babies to be born in France? The results achieved by Michelin make it possible to respond *yes!*" Greatly simplifying rather complex demographic comparisons of the fertility of its workers in Clermont-Ferrand with other inhabitants, Michelin could prove the efficacy of family allowances to other French employers. The result of widespread adoption of allowances would presumably be yet more of the babies like those held by Bibendum on the cover. "Une expérience de natalité," 1926.

its brochures (1.56 francs per worker per day since Michelin paid allowances for only one or two children), the company's sole justification for allowances was that they would effectively increase the French birthrate.[59] According to André Michelin, the "sterility of [French] households was not an inevitable law." But without measures like Michelin's, "France will be half its current size in 30 years. Reduced by half that means half the fields lying fallow, half the factories closed, and our splendid colonies, capable of feeding several populations like France, abandoned to foreigners."[60]

Michelin's approach to foreigners closely resembled that of the Alliance Nationale. On the one hand, given the labor shortage during the 1920s, Michelin relied heavily on immigrant labor; approximately 10 percent of its workers were foreign in 1926. On the other, the company took a nationalist approach to pronatalism, so foreign employees did not receive any family allowances. Moreover, Michelin's definition of "foreign" meant everyone but white Frenchmen: included as foreign were immigrants from the colonies such as Moroccans and Kabyles (from Algeria) as well as Italians, Spanish, Belgians, and Swiss.[61] Both the Alliance Nationale and the Conseil Supérieur de la Natalité were adamant that allowances should be used only for French couples. When the French state negotiated an agreement with Belgium, Luxembourg, Italy, and Poland that allowed nationals from those countries working in France to receive only the additional payments for birth and nursing offered by the prefecture of the Seine, the Alliance Nationale denounced the measure, noting that "they can't find enough money to raise the pathetic national aid for large families, but they can find some to encourage foreign births on our soil."[62] There was of course never a question of paying actual allowances for the children of foreigners.

The ways in which Michelin paid allowances implicitly asserted a primary role for men as heads of families. In 1920, the Conseil Supérieur de la Natalité's proposed law called for allowances to be paid to the mothers of children, even when they resulted from men's work. In actual practice, many *caisses* used payment of allowances to mothers both to limit the potential for men's wasting of the money, particularly on drink, and to provide an at-home inducement against men's labor activities and absenteeism.[63] In contrast, although Michelin's allowances could be suspended for alcoholism, absenteeism, or negligence on the job, Michelin made the payments to the father of the children. The language and illustrations of Michelin brochures reinforced men's place as heads of families. In one case, a large burly man holding his baby comforted his fawning wife with the words, "Even if I die, you will be able to raise him" as a way of making the point that women continued to receive allowances after their husbands' deaths. Michelin not only assumed men's primary role as breadwinners in large families but also women's position either at home or as secondary wage earners; while the company specifically noted that if the father died before the children reached age sixteen the mother would receive the allowances, no such provision was noted in the case of a mother's death.[64] Moreover, although Michelin, like other interwar pronatalists, helped widows, there was no such systematic benefit for divorced or single mothers. That is, the solidarist notion shared by many pre–World War I reformers that any and all children needed to be helped took a back seat to the importance of the traditional family structure.

Nevertheless, when compared with *caisses* that attempted to use the allowances to keep women and children from seeking employment outside the industries involved in the *caisses,* thus retaining cheaper women's and children's labor as well as more expensive men's labor,[65] Michelin's program was considerably less restrictive. The only couples barred from receiving allowances were those in which the spouse of the Michelin employee worked for one of Michelin's competitors in Clermont.[66] In the end, although the company provided minimal short-term aid to women who decided to stay home with their babies in order to nurse them, women could earn far more by joining their husbands at Michelin. In the company's desperation to recruit enough workers in the mid-1920s, Edouard Michelin specifically told his staff to examine what tasks in the factory might be done by women or teens with an eye to increasing their employment more systematically.[67] Michelin needed women in its workforce, a fact that tended to undermine assumptions of separate roles for men and women; as a whole, Michelin brochures suggest that men should be heads of households but that women might well be working as well, preferably at Michelin. An internal *crèche* and Michelin schools could look after the children.

Of course, Michelin's public pronouncements and its underlying strategy were not necessarily always the same. Susan Pedersen has found a wide divergence between other employers' claims that allowances resulted solely from employers' generosity and their internal arguments that the allowances were essentially strategies designed to hold down overall wages by only paying additional amounts to those most in need. Similarly, André Gueslin has suggested that Michelin used the allowances to attract a stable workforce and avoid high rates of turnover. The company was indeed desperate to hire and retain employees. But unlike the Consortium de Roubaix-Tourcoing, Michelin did not use high allowances to pay substandard wages; average wage rates at Michelin were higher than those of any local industrial employer and were just below wages in Paris itself, which had a much higher cost of living.[68] In addition, in its efforts to recruit more employees, Michelin used allowances in addition to subsidized housing (with priority and reduced rent offered to those with large families), subsidized recreation programs, and medical insurance, as well as profit-sharing. The company, which had adopted "American" production methods of Frederick Winslow Taylor and Henry Ford by the 1920s, opted not for a cheap workforce but for a more productive, healthy, and loyal one. Family allowances fit easily into that agenda.

Like most employers, Michelin used the term *allowances* (*allocations*) rather than the expression *salary supplement* (*sursalaire*) to describe its payments. In the early 1920s, employers dedicated to social Catholicism argued that payments to those with children constituted a duty, so that supplemental salary and not allowance was the more accurate term. Employers establishing *caisses* overwhelmingly preferred the term *allowances* (*allocations*) because it implied that the payments were freely granted out of employers' generosity, much as a generous father would grant allowances to his children.[69] The employer, or a group of them as in the case of the *caisses,* remained firmly in control, capable of determining how much the allowances would be and who should receive them, and reserving the right to withdraw them.

Yet it would be anachronistic to assume that Michelin or other interwar employers did not combine paternalism with a genuine concern, of course within carefully defined parameters, for workers. As Henri Hatzfeld has asserted, industrial self-interest and genuine concern about workers were not mutually exclusive motivations for the introduction and development of family allowances.[70] Nor can Michelin's own nationalist pronatalism be easily separated from other motives. In Michelin's case, the company's very public, attention-getting advocacy of pronatalism should still be taken seriously as statements of the brothers' own worldviews. Michelin did have something to gain from its family allowance policies, yet more than better retention rates or reasonably grateful workers; by offering very high allowances and by trumpeting their effectiveness, Michelin also received public attention. In addition to more general coverage in the national press and its own brochures, Michelin received accolades in the *Revue de l'Alliance Nationale*. The Alliance itself sent out at least sixty-five thousand copies of Michelin's brochures in 1926–27 alone.[71] Even once various *caisses* developed across France, Michelin maintained its own system and continued to praise its efficacy. Although a cornucopia of academic works appeared in the interwar period trumpeting the *caisses* of various industries, what set Michelin apart was its own widespread popular advocacy of the allowances as well as the Alliance's seal of approval.

In the end, Michelin gained more publicity from allowances than was possible for French employers joining the *caisses*. There were few issues of more importance to nationalists in the interwar period than pronatalism, as the influence of the Alliance Nationale corroborates, so Michelin could associate itself, no doubt partly inadvertently, with what became widely perceived as a French national interest. Because Michelin was the separate brand name of a product whereas the *caisses* put companies in the same sector together, Michelin could conduct a civic campaign for pronatalism. Automobile and bicycle owners could buy a *French* product, the Michelin tire produced by those pronatalists from Clermont-Ferrand, rather than a *foreign* tire produced by Dunlop or Goodrich (even though both companies produced tires in their French factories). Given that the potential buyers for tires in interwar France remained mostly bourgeois and petit bourgeois, Michelin's pronatalism appealed to the company's likely clientele. Although the Alliance Nationale itself proclaimed political neutrality, before the mid-1930s the only political parties even lukewarm about pronatalism were the Socialists and the Communists, neither of whom had a constituency that could afford to buy many tires (and by 1936 even the Communists adopted a pronatalist language not far from that long spoken by the Alliance). In short, while Michelin's family allowances were far more than marketing, resulting no doubt from the brothers' heartfelt patriotism, the very public nature of the company's advocacy did remind bourgeois tire buyers that the Michelin brothers were French patriots and that Michelin tires were quintessentially French tires.

PRONATALISM TRIUMPHANT

Although French employers' participation in *caisses de compensation* grew phenomenally in the 1920s to the extent of including nearly 2 million employees by 1930, it would be foolhardy to credit that growth to Michelin's constant entreaties.[72] Even employers' own claims of their generosity masked a wage strategy that served their economic interests. Yet the fact that French employers resorted to family allowances rather than to cost-of-living bonuses or other means reveals the power of the pronatalist discourse in interwar France. While Michelin was neither alone nor uniquely persuasive in making pronatalist arguments in favor of allowances, its support of the cause contributed on some level to the reinforcement of a certain way of seeing France in the interwar years. That ideological contribution, however difficult to measure, had some influence on the fact that employers chose allowances over other alternatives to wage hikes.

The Alliance Nationale and the Conseil Supérieur de la Natalité had both argued throughout the 1920s that all Frenchmen, and not just those working for large firms that participated in *caisses*, deserved family allowances. Along with structural political reform of the Third Republic in the form of the many unsuccessful proposals for the family vote, pronatalists perceived of allowances as one of the single most important means of increasing the French birthrate. But because the nationalist constituency most supportive of allowances covered the political spectrum from the centrist Radical party to the far right, pronatalists had a fundamental problem. The political parties and constituencies most desirous of a higher French birthrate, those on the Right, were the same ones who resisted any major increase in public expenditure for fear of higher taxes. A fundamental irony graced the pages of the *Revue de l'Alliance Nationale* throughout the 1920s: the group frequently denounced postwar tax hikes as the cause of death of the French family, but it advocated expensive solutions. The political Right in France, however desirous of increasing the birthrate, was loath to raise taxes to help pay the bills for working-class children. Even the conservative Bloc National government of 1920–24, capable in 1920 of stringent legislation against abortion and contraception and in favor of medals to be awarded prolific mothers, did little to increase family allowance coverage. In 1923, all French parents with at least four children received the right to family allowances paid by the French state, provided that they paid no income tax, did not already receive an allowance as civil servants, and did not already qualify for assistance under the 1913 law. Even once people met these requirements, the amount received was 90 francs, at a time when Michelin offered 400 francs for families with four children, an amount that did not include Michelin's other subsidies.[73] As the Alliance

Nationale repeatedly proclaimed, the French state's amount was too small to induce hesitant would-be parents to take the plunge into parenthood. In this context, it is not surprising that family allowances administered and paid for by employers and their *caisses* offered a politically acceptable alternative to higher taxes. The issue became how to spread the allowances to cover all French parents.

Before and after the legislative elections of 1928, the Alliance lobbied for a national system of family allowances. Socialists and Communists preferred a system administered by the French government, and deputies on the Right, preoccupied by pronatalist visions, pushed for a solution that did not raise taxes. In the political give-and-take with large business, the compromise that became law in 1932 simply mandated that all French industrial and commercial employers belong to a *caisse* and set minimum rates of allowances. Those companies that already had their own systems, such as Michelin, could request exemption. The compromise was made more palatable to big business by keeping the minimum allowances low, generally lower than what *caisses*, and far lower than what Michelin, were already paying in 1932. Michelin allowances remained 3 to 5 times higher than state minimums, depending on the number of children couples had.[74]

Slowly, the system came to include small businesses and peasants, despite widespread resistance. In 1936 half of all employers had still not joined a *caisse*, and fewer than half of all parents eligible received any allowances.[75] The Popular Front (1936–38) and Daladier's subsequent right-wing government resolved the problem by including state subsidies to offset the cost to farmers and small business, then Daladier's government increased the size of allowances. Although not fully implemented before the German invasion of France in 1940, the system remained largely in place under the Vichy regime and postwar governments. Quite substantial family allowances, one of the defining characteristics of the post–World War II French welfare state, thus became law.

In her comparative study of the development of family policy in Britain and France, Susan Pedersen has persuasively argued that family allowances so favored by pronatalists in France thus came to be seen as "common sense" in the political discourse of the late Third Republic. Michelin's own role declined somewhat after the death of André Michelin in 1931. Although his second wife and son Jean accepted positions on the board of the Alliance Nationale, neither they nor Edouard maintained the pronatalist profile that André had. More-over, Michelin brochures in the 1930s abandoned the question of family allow-ances, focusing instead on promotion of the automobile. Edouard Michelin did join luminaries, including among others Alexandre Millerand, Raymond Poincaré, Edouard Herriot, Marshal Lyautey as well as Robert Peugeot, in calling on all French families to have at least three children, but he did little more publicly.[76] In part, the company did not need to argue on behalf of the pronatalist movement in the 1930s. The ideas espoused by the Michelin broth-ers, so close to those of the Alliance Nationale, had achieved a political con-sensus on the eve of the Second World War.

Yet despite Michelin's relative silence in the 1930s, its influence in promot-ing pronatalism in interwar France was undeniable. Although the pronatalist movement had certainly existed before 1913, Michelin's financial participation occurred at the very point in time when the Alliance Nationale was gaining national prominence. The company's very public association with the prona-talist cause offered the movement, and the Alliance Nationale in particular, greater credibility. Advertisements for the guides to the battlefields informed readers that the Great War could have been averted if French couples had only listened to pronatalists before the war. Michelin also joined the Alliance in asserting that increased French births would avoid any future war with Ger-many. And while Michelin may not have been the first to establish family al-lowances, its publicity encouraged a growing French consensus for a form of redistribution of wealth from the childless to parents. Here Michelin found a place at the center of yet another transformation in early-twentieth-century France. While the introduction of family allowances and other pronatalist measures were economic, social, and political developments, the assumption that all French people should receive them was a cultural one that slowly emerged from the middle of the nineteenth century to the 1930s. As in so many other domains, Michelin's actions definitely reflected this new cultural development and, to a certain extent, helped to shape it.

5

ADVOCATING AERONAUTICS

Modernity and French Elan

ON THE EVE of the First World War, Jean Arren noted that the Michelin tire company had already spent more than 800,000 francs (160,000 dollars) sponsoring aviation, calling it a "campaign for the national interest," a "sacrifice made by Michelin for the general [public] interest."[1] Given that Michelin did not stand to gain directly from flight because it did not produce any of the important airplane components, the company's sponsorship was clearly more than an effort to sell more tires; the Michelin brothers, particularly André, had much enthusiasm for flight already in the first decade of the twentieth century. In the 1910s and the 1920s, as the military potential of airplanes became more apparent, André's interest grew and merged with the brothers' advocacy of modernization and their patriotism. In the 1920s, André devoted many of his final days prophetically warning the French public that without a strong air force, France would be doomed.

Although clearly originating in the Michelin brothers' patriotism and André's enthusiasm for flight, Michelin's very public promotion of French aeronautics simultaneously associated the Michelin name with aviation, a symbol of progress in the early twentieth century. Interestingly, huge volumes of press clippings about aviation in the Michelin archives themselves carry the label "Publicité" and include thousands of newspaper accounts of Michelin's aviation prizes and André Michelin's pronouncements. As in the case of early automobile races, newspapers used reports of early flights to sell newspapers and covered early pilots' feats in minute detail. Each time that an aviator attempted to win a Michelin prize, the major and even smaller Parisian, provincial, and specialized newspapers gave at least some coverage to the event, thus keeping the Michelin name before the public; Michelin employees dutifully cut out each article and pasted it into the thick "Publicité" volumes as evidence of the firm's successful publicity campaign. Michelin's support of aviation was not primarily advertising, but in keeping the Michelin name before the French public, it was an indirect, perhaps partly inadvertent, form of marketing. Michelin's advocacy of French aviation shows how thoroughly a company, even when it was not advertising in a direct way, could help to define the interests of the French nation, reflecting a much larger discourse that paired success in flight with national strength.

THE PRIX MICHELIN

In 1896, André Michelin took a ride in a hot-air balloon dubbed "Le Touring Club," one of the many such airborne escapades sponsored by the seemingly omnipresent Touring Club de France. Fascinated by the experience as by so many innovations, André Michelin became a founding member and part of the steering committee (*comité de direction*) of the Aéro-Club de France, a group of elite aviation enthusiasts, in 1898.[2] Although André Michelin immediately took an avid interest in aviation, the firm itself did not undertake the production of the huge balloons, sticking instead to production for the fast-growing market for bicycle and then automobile tires.

In contrast, the German firm Continental took an early lead in production of fabrics necessary for flight. Because of Michelin's dominance of the French tire market, not to mention Michelin's early ability to tie its name with a certain idea of France, Continental's biggest struggle had been "to affirm with force its place in the French industrial landscape." It needed to "constantly impose a [French] national image" for itself, a difficult task a mere generation after the Franco-Prussian War. Its *Guide routier*, first appearing in 1904, had been

Continental's attempt to serve the French motorist just as Michelin did. Continental's best opportunity, however, for associating its French affiliate with the French nation came with its production of the rubberized fabric that covered dirigibles before the First World War. Between 1905 and 1909, government-owned French dirigibles with such names as *La Patrie* (the Nation), *La Ville de Paris* (the City of Paris), and *La République* (the Republic), had been built with Continental's rubberized fabric (*toile caoutchoutée*), a product manufactured in its French factory at Clichy.[3] The product allowed Continental both to affirm its Frenchness and to proclaim its technological superiority through its contribution to flight.

In 1907 and 1908, the airplane increasingly captured the interest of the French public. In December 1907, Alberto Santos-Dumont, a Brazilian living in France, managed to keep his airplane aloft for all of 50 meters; in January 1908, Henri Farman traveled 1.5 kilometers in an officially monitored closed circuit. Aviation enthusiasts and newspaper publishers offered prizes to advance aviation and to sell newspapers. Various organizations offered some 372,000 francs in prize money during 1908 alone, although much of it was simply not attainable given early technology. Among the many early prizes, the well-known aviation patron Ernest Archdeacon had offered a prize to the first aviator to stay off the ground 25 meters in an airplane, a prize won by Santos-Dumont in October 1906. Archdeacon and petroleum magnate Henry Deutsch de la Meurthe had provided the 50,000 franc prize won by Farman in his January 1908 flight.[4] Prompted by Archdeacon to fund a prize,[5] on 6 March 1908 the Michelin brothers announced in a letter to the Aéro-Club de France, which would oversee the competitions, the creation of two separate Michelin aviation prizes.[6] Archdeacon, who was trying to find more patrons of aviation, proudly told a reporter for the *New York Herald* that "so far I have succeeded with M. [André] Michelin. I demonstrated to him that he could secure by the offer of a substantial prize a world-wide advertisement."[7]

The first award consisted of a yearly competition for the Michelin Cup (Coupe Michelin) to be held from 1908 to 1915. The winner each year received 20,000 francs in cash and a replica of the cup, the original of which, worth 10,000 francs, would be given to the final winner in 1915. Each year, the Aéro-Club de France would lay out the circuit with the necessary turns and minimum altitudes. The winner would be the aviator who had flown the farthest by December 31 (thus keeping up the flow of publicity throughout the year),

MARKETING MICHELIN

158

and who had at least doubled the previous year's distance. If no one won the prize in a given year, the following year's prize money doubled. The Aéro-Club de France defined the special requirements for 1908: each entrant needed to pay a fee of 50 francs to the local aéro-club. In addition, takeoffs could take place only between 10:00 and noon or after 2:00 P. M., thus preserving the sacrosanct midday meal for spectators and judges. In 1908, the winner of the cup would be the aviator who had gone the farthest and who had flown at least 20 kilometers, some 10 times the distance that Farman had flown by March 1908. The second award announced in 1908, the Grand Prix Michelin, consisted of 100,000 francs in cash to be given to the first pilot to leave the Paris area, circle the Arc de Triomphe and fly approximately 400 kilometers to Michelin's hometown of Clermont-Ferrand, circle the cathedral, and land on top of the 1,465 meter-high local peak, the Puy-de-Dôme. For both prizes, Michelin invited aviators of any nationality to participate.[8]

Michelin was able to use the Michelin Cup to generate publicity in a variety of ways. As an annual international competition with contestants from various countries, the company could play on the notion of international rivalry built into the recently reestablished Olympic Games and into Gordon-Bennett's automobile races. Michelin also sponsored a competition of sculptors for the design of the cup, to be done in bronze and to be worth approximately 10,000 francs. Michelin widely advertised this competition, open only to French sculptors, noting the additional 14,800 francs in prize money offered to the finalists. In the weekly "Lundis," Michelin reminded sculptors that their work not only had a chance at capturing the 8,000 francs as first prize but that they would also receive widespread exposure. All contestants would have their work displayed at the automobile salon beginning on 27 November, an event that, according to Bibendum, regularly had thirty thousand in attendance whereas salons devoted to painting and sculpture only had around ten thousand. A jury chosen by the company and the contestants (ultimately 113) then made a selection and the winner and other finalists had their works again displayed, this time at the second automotive salon, featuring trucks, at the end of December 1908.

The sculpture contest offered Michelin yet more opportunities to point out its generous support of aviation and also provided abundant visual images, in the form of photographic reproduction of the sculptures, that illustrated the dream of flight. The winner, Paul Roussel, submitted a representation of

Icarus (the mythological figure most associated with flight among the classically educated early-twentieth-century French bourgeoisie) lying on the ground while a woman, representing victory or liberty, ushered a biplane into the air. When Wilbur Wright ultimately won in 1908, Roussel adapted the cup to eliminate what had obviously been Farman's biplane rather than Wright's *Flyer*. Michelin further promoted the event by producing a series of a dozen postcards featuring photographs of the top-ten sculptures and that were available free of charge to anyone willing to cover the 65 centimes for postage.[9]

Michelin's sense of timing was superb. In 1908, competition for the Michelin Cup, as for other prizes, was fierce. At the time of Michelin's announcement in March, European aviation experts and the press remained quite skeptical of Wilbur and Orville Wright's claim that they had flown as early as 1903 at Kitty Hawk. In May 1908, as part of their ongoing efforts to secure a European military buyer for their airplane (with a price tag of $200,000), Wilbur Wright traveled with the plane to France to prove that it did indeed fly. Headlines in the late summer and fall featured Wright's flying machine and reports about this "bird" from America. Wright consistently bested any distance traveled by French pilots Henri Farman and Léon Delagrange. Wright reported to his father in early January 1909 that he even remained in France over Christmas 1908 as a deterrent to any aviator who tried to beat his record. To be sure of his victory and to make the superiority of his machine clear to potential military buyers, Wright made his final flight on 31 December, the last possible day of competition; he flew more than 124 kilometers in a little over two hours, establishing a new record and taking home the Michelin Cup, after it had been recast so that it no longer featured Farman's plane.[10]

The year 1909 was another year of close competition, providing Michelin with considerable press coverage. Pilots Latham, Rougier, Paulhan, Château, Legagneux, and Farman all competed throughout the year for the Michelin Cup, which Farman ultimately won. André Michelin's collaborator Fernand Gillet described the competition: "The company gets a very respectable amount of publicity out of it, which extends to America" [an accomplishment aided, no doubt, by Wright's victory the previous year]; several newspapers reproduced a long interview with Jules Hauvette 'manager of the Michelin Tire Company' [Hauvette's title was in English in the original]."[11] Clearly, Michelin hoped to gain publicity from the competitions and was pleased to have gotten notice even in the United States. Having just established a tire

19. "The Triumph of Aviation": just after Henri Farman managed to fly one kilometer, Michelin proposed the Coupe Michelin. Paul Roussel designed the trophy, featuring a plane resembling Wilbur Wright's Flyer. Photograph of the sculpture, 1908.

plant at Milltown, New Jersey, in order to serve the rapidly expanding American tire market, it was quite fortuitous to have one of Michelin's managers quoted in national newspapers.

Each year after 1908, the cup seems to have garnered somewhat less attention as long, successfully controlled flights were more common, even though Michelin continued to promote the cup as evidence of the company's dedication to both progress and France.[12] The competitions for other prizes frequently overshadowed that of the Michelin Cup. The conditions of Michelin Cup competition themselves became increasingly complicated because of the ground rule that each year's winner needed to double the distance covered by the winner of the previous year. In 1912, when the distance became more than 2,400 kilometers, bad weather made it impossible for anyone to fly the necessary three to four weeks in order to win. In 1913, Emmanuel Helen spent thirty-nine consecutive days flying 2,893 kilometers, finally winning the cup. As a result, the Aéro-Club de France adapted the rules in 1914 to feature speed over distance. The winner in June 1914, Eugène Gilbert, took two days to cover a distance of 3,000 kilometers (there were no more attempts once the war began in early August). It was a high-speed Tour de France, covering a circuit from Versailles to Péronne, Reims, St. Dizier, Gray, Joigny, Beaune, Vienne, Nîmes, Pau, St. André-de-Culzac, Romorantin, Angers, Evreux, Calais, and then a triumphant return to Versailles.[13] In the face of somewhat less press coverage, Michelin proudly proclaimed in 1913 that the competition for the cup "summarizes the entire history of aviation and the marvelous march of progress. Bibendum, King of Space, after conquering the earth has favored the conquest of the air."[14]

In 1909, attempts to cross the English Channel frequently upstaged the Michelin Cup competitions in the media spotlight. Both Hubert Latham and Louis Blériot attempted to fly across the English Channel and win the 1,000 pounds (5,000 francs) offered by the publisher of the *Daily Mail*, Lord Northcliffe, as well as 25,000 francs offered by the champagne producer Ruinart and the Parisian daily, *Le Matin*.[15] Latham, who had quickly learned to fly his Antoinette airplane in February and March 1909, was the object of much positive publicity in 1909, particularly once he announced his decision to cross the channel. In July 1909, he went 7 miles across the channel before his engine gave out, requiring his rescue. Less than a week later, Louis Blériot successfully crossed the channel in one of the many airplanes he had built, thus winning the prize. The reaction of the French press, particularly that of *Le Matin*, bordered on delirium: a Frenchman in a French plane had again proven his mettle. While Lord Northcliffe used the event to sell more newspapers and to force the British government to take aviation seriously, himself playing the national card, *Le Matin* used the event to sell newspapers by pressing home the point that Blériot had shown French potential, comparing him to the Gallic hero Vercingétorix, whose reputation as a national symbol had recently been established. As Robert Wohl puts it, "a nation that was capable of producing a Blériot was not a nation in decline, as so many pessimists were claiming." [16]

Blériot's success and Latham's failure had implications for both Michelin and Continental. Although the Michelin company later claimed that it never produced fabric for airplanes or dirigibles, in early 1909 Michelin briefly advertised its production of *tissus caoutchoutés* (rubberized fabric) for dirigibles and airplanes, noting that it provided the fabric used on Latham's plane.[17] The company quickly abandoned production as well as any advertising that associated the company with Latham. Continental, in contrast, had long been in the market, and it had supplied Blériot. Blériot's successful flight allowed Continental to tie itself with a French pilot flying his French plane, further evidence of the company's service to France. The company could boast that it provided the *fortes toiles* (strong fabrics) used on Blériot's plane just as it provided the fabrics for planes built by Delagrange, Breguet, De Caters, Esnault-Pelterie, and even Farman. So just as the French public learned that one of its own, Louis Blériot, was the first pilot to cross the channel in July 1909, so too did it learn that Blériot's plane had been outfitted with Continental fabric. Given the nationalist spin that Blériot's flight received in both France and England, where it spurred fears that England would be vulnerable to air attack, presumably by the Germans, there could not have been within France a better symbol of Continental's French credentials. Continental went so far as to display Blériot's plane at the aviation salon held in Paris in the fall of 1909.[18]

Of course, as a producer of fabric for airplanes and dirigibles, Continental could be blamed for failure as well as praised for success. In September 1909, a French military dirigible, *La République,* crashed. At the time, the French military bought motorized dirigibles for reconnaissance just as it had used balloons in the nineteenth century. Because the French dirigible consisted of a single large balloon, when a blade broke off the propeller and punctured the fabric of *La République,* the gas exploded out of the craft. Although the propeller was an obvious culprit, Michelin claimed that Continental's fabric had been to blame because it consisted of juxtaposed strips of fabric. Throughout the fall and winter of 1909–10, Michelin led a campaign against its competitor. In October 1909 the company published a public letter to the French War Ministry arguing that Michelin had developed a better fabric, more resistant to puncture. Somewhat disingenuously, Michelin wrote that

> it is not our intention to continue industrial production of a product that we undertook only as an experiment and with a uniquely patriotic goal. However, not wanting the results of our efforts, done without any desire for profit . . . , we are offering the French state a free license to produce or have produced by any truly French company (and not by the subsidiary of a foreign company hiding behind a French facade [cachée sous un faux-nez français]) the fabric developed by us that is covered by the secret patent no. 5,923, filed on 4 June 1909. We want to add that we offer our experience and advice to help either officers or the company chosen by you. We hope, Sir, thus to permit the French military establishment to free itself from the tribute paid to foreigners and to perfect its own operations.[19]

Without doubting Michelin's patriotism, the company's public statement was a somewhat misleading attempt to show its service to France in contrast to Continental, even though Continental produced such fabrics in France; had the purpose of the letter been solely patriotic, one wonders why it needed to be public and why the point would continue to be made in later "Lundis" and other advertisements. Moreover, it also seems doubtful that the French military would award the rubber contract to a firm other than the developer, Michelin. At the aviation salon in 1909, while Continental displayed the Blériot airplane, the Michelin booth compared Continental's supposedly inferior dirigible fabric with its own, explaining the cause of the *République* catastrophe. When the military ignored Michelin's offer, the company used a "Lundi" to point out that as disinterested, *bons français,* Michelin offered the patent to the French army whereas Continental sold an inferior product to the French military and a better-designed one to the German army. Raymond Poidevin has attributed Michelin's pieces, printed in several newspapers, as having initiated the increasingly frequent attacks by the Parisian press on German products in the prewar years.[20]

At the same time that Michelin continued to offer the cup and to criticize

Continental's dirigible fabric, the Grand Prix Michelin of 100,000 francs received somewhat less attention. When Michelin first announced the 400 kilometer trajectory in 1908, Farman had just flown all of 2 kilometers. Several periodicals, including *L'Illustration* and Georges Clemenceau's *L'Aurore*, accused the company of using the feat solely for advertising since no flyer could possibly win it. To quell such criticism, Michelin went so far as to wager 5,000 francs against any and all persons who were willing to bet that no aviator would within six hours go from the Arc de Triomphe in Paris to the cathedral at Clermont-Ferrand before landing on the Puy-de-Dôme before 1918.[21] Open to pilots of any nationality, the voyage had a certain symbolic value for Michelin. It began at the Arc de Triomphe, a powerful Napoleonic symbol of the French nation, included a symbol of the Catholic faith, and ended on top of the highest peak in the mountain range of their native Auvergne, a reminder that Michelin was both the provincial Auvergnat tire firm, practical and tightfisted to the core, and a Parisian-based, seemingly disinterested and generous promoter of French national interests.

Although there were no attempts, and thus little publicity in 1908 or 1909, in 1910 Charles Weymann made it to Volvic, just 10 kilometers away from Clermont before having a mechanical breakdown. In October 1910, the Morane brothers nearly killed themselves in pursuit of the Grand Prix Michelin. In fact, fear that pilots were planning a crash landing on Puy-de-Dôme caused the Michelin brothers and the Aéro-Club de France to specify that to win the prize the airplane had to be capable of continuing to fly. In March 1911, Eugène Renaux, accompanied by the explorer and meteorologist Albert Senoucque, covered the necessary itinerary, landed the Farman biplane successfully on top of the mountain, and won the Grand Prix. Marcel Michelin, the son of André Michelin, met the aviator on the mountain and brought him down to Clermont-Ferrand for a series of celebrations. At the various banquets and receptions, duly reported in the press, André Michelin cast Renaux's success in strictly nationalist terms. He called on the minister of war, Maurice Bertaux, to devote the amount spent on building a battleship, 100 million francs, on building five thousand airplanes, a demand that he would make repeatedly in the years before the First World War. He noted that "military aviation has entered into the practical domain and will permit us to overcome, if it is sufficiently developed [by the do-little government and the press], the danger created by our weak birthrate."[22]

Clearly, André Michelin was a French patriot, but he and Edouard were also businessmen who understood the multitude of ways that one could sell a product, including the usefulness of nationalism. They established yet another prize in 1909, this one in Britain, which further reveals the ways that the company used the nationalistic orientation of many early advocates of aviation to promote its name. Because of the prewar dominance of the French in both the production of airplanes and in breaking records, Michelin's international cup and Grand Prix provided excellent opportunities to remind potential tire buyers that Michelin represented both progress and France by encouraging development of the airplane (rather than the dirigible, seen as that most German of machines). In Britain, however, where the aircraft industry was considerably less advanced and where Michelin was attempting to increase its market share at the expense of Dunlop, Michelin founded a separate "British Empire Michelin Cup." This British cup included 500 pounds (5,000 francs or 2,500 dollars) in yearly prize money in 1909 and 1910, and beginning in 1911 two prizes worth 2,500 pounds each. These competitions were open only to pilots of British nationality, flying British planes, of which all of the principal components needed to be British.[23] The competitions gave the company a chance to do in Britain what Continental did in France, that is, to associate its name with the national interest of the country of implantation. Clearly, the aviation prizes were at least in part a marketing strategy.

Michelin used its advocacy of aviation to promote its name, associating itself with progress, with technical novelty, and, as in so many other cases, with the French nation. It tapped into widespread cultural assumptions about flight and contributed to their redefinition. Early flight, particularly the fast movement of early airplanes, seemed to many contemporaries to embody modernity. But by 1908, in the increasingly nationalistic tenor of public discourse, international competition in the realm of technology became an important accompaniment of advances in aviation. Although the Wrights' very attempts to sell their machine to any government, American, British, French, or German, willing to pay the price indicate a motive of pure profit rather than patriotism, the European press's response to the Wrights was heavily laced with nationalist overtones. Skepticism about the Wrights, grudging admiration, and then the competition to outdo them were all manifestations of the nationalistic context in which aviation emerged in prewar Europe. Similarly, the development of Count Zeppelin's German dirigibles came to symbolize the grandeur of that nation.[24]

This is not to say that there were not many idealists in the early ranks of aviators, people who learned Esperanto and believed that aviation could guarantee peace. International competition in aviation could be compatible with peaceful agendas, provided that the planes were not developed by European militaries as yet another weapon in the intense arms race that characterized the prewar years. Already, however, events of the year 1908 hinted at the future: the Wrights recognized that the European militaries were likely buyers, and Lord Northcliffe, Ernest Archdeacon, André Michelin, and others sponsoring aviation prodded their governments to develop airplanes and dirigibles as military weapons. Just as a mania for Zeppelins swept Germany, British promoters of military aviation, including the Harmsworth brothers and Lord Northcliffe, founded the "Aerial League" and their French counterparts founded the Ligue Nationale Aérienne (the National Aerial League), a group that included André Michelin and Henry Deutsch de la Meurthe and had the backing of *Le Temps, Le Matin,* and *L'Auto.*[25] In 1908 and 1909 immensely popular writers H. G. Wells in Britain and Rudolf Martin in Germany, among others, imagined bombs that might rain down from dirigibles, primarily Zeppelins, in the time of war.[26] In the ever-more frenzied years before World War I, particularly after the second Moroccan crisis of 1911 and the Balkan Wars, Europeans came to see airplanes as well as dirigibles as weapons. The efforts of the Michelin brothers reflected that transformation and, to some extent, helped to make it happen.

BOMBING BEFORE THE WAR

Beginning in 1911, Michelin spearheaded the effort to persuade the French military to foster the development of airplanes for offensive as well as reconnaissance missions. In a letter to the Aéro-Club de France in August 1911, Michelin expressed a desire to found two new prizes for air targets, to be called the Prix de l'Aérocible Michelin and to be awarded in the years 1912 and 1913. 75,000 francs would be awarded each year. The Michelins' letter was quite specific about the purpose of the prizes.

> There is much argument over the question of whether the military airplane is a simple instrument of reconnaissance or whether it might soon become a terrible machine of war. Could it incapacitate bridges and railway links, cutting in two the mobilization of a nation, annihilate a fortress, or sink a

battleship? At Waterloo could it have crushed the Holy Alliance or wiped out [*exterminer*] the English guards who waited, crouching in the fields? Maybe it could do even more: destroy arsenals, supply centers, and powder magazines of the enemy, thus making its cannons and rifles useless?[27]

According to the Michelin brothers, the prizes would prove whether such widespread bombing, associated in popular literature with dirigibles and increasingly airplanes, might be possible. They obviously foresaw both battle-field and longer-range strategic targets. The actual description of the rules further revealed likely targets. One prize, for 50,000 francs, was for the aviator and passenger who dropped the most 7.1 kilo projectiles, out of fifteen total, within a 10 meter circle from an altitude of 200 meters. The second, for 25,000 francs, would go to the crew who dropped the most projectiles into a space measuring 100 by 10 meters from an altitude of 1,000 meters. Clearly, the competition to bomb a specific circle had a variety of uses; the rectangular competition from a higher altitude seems strange until one realizes that those were roughly the dimensions of the dirigible hangar that housed the much-feared Zeppelins.[28] Unlike the earlier Michelin prizes, both the pilots (although not the passenger, as it turned out) and the airplanes needed to be French.

Michelin generated considerable positive coverage in a press long accustomed to feeding the nationalist frenzy before the war. Hundreds of articles reprinted the Michelins' letter in its entirety, many adding words of approbation, although the Michelin announcement did not always receive first-page coverage because of the recent theft of the *Mona Lisa,* a fact much-bemoaned by the company. In *L'Auto,* Henri Desgrange noted that Michelin realized the "solution of the conflict [with Germany]. Doesn't it consist, almost always, in flexing one's biceps?" *L'Aurore's* Henri Lavergue noted that such bombing techniques might be necessary but reminded its readers of what the bombardment of Paris in January 1871 had been like for innocent civilians. Michelin's prizes came to embody patriotic lobbying for bombing. In *Le Journal,* Georges Prade asked readers to imagine war with Germany in which the Germans won the early battles, but that the French would ultimately win, particularly after cutting the communications and transport of the enemy. The airplane would even allow penetration into Alsace-Lorraine. "The tactical superiority of the airplane is such that it can pass over even victorious forces, that it can bring the war to the doorstep of the enemy, who thought he was at peace, defended by a buffer of fortresses and millions of bayonets. What could one do with an airplane at the foot of the Vosges? Cut communications, railroads, bridges, and fly over several cities not that important for the Empire and of which the majority are in Lorraine or Alsace."[29] Michelin further fed the flame by issuing a series of postcards with designs featuring airplanes bombing military targets. Different postcards illustrated the destruction of a railroad, a sky full of French aircraft, the annihilation of an enemy headquarters, and the sinking of a battleship. One showed a French plane flying over a peasant's field and village as Bibendum assured the peasant in clogs that the airplane was preserving the peace. The company even sponsored a competition for "Lundi" readers who wished to send in their own drawings for postcard production.[30]

The Michelin brothers' prizes and André Michelin's public pronouncements kept the Michelin name in the press. To critics of the Michelin target prizes, Michelin explained that its motives were patriotic and universal. "We have

20. A postcard promoting Michelin's prize for bombing is a reminder of the classical education bourgeois automobile buyers had before the First World War. Bibendum proclaims that "before it was the sword of Damocles, now they are the droppings of Bibendum." That is in lieu of the sword held over Damocles' head by a single hair, Bibendum, a virtual bird, can now, due to the progress of aviation, leave his droppings as terrorized people scatter. The classical analogy seems forced except to the extent that people living under threat of bombardment were much like those with swords suspended above their heads. Postcard by Hautot, 1912.

traveled a lot [in selling tires]. In examining foreigners closely, we have better sensed the qualities of our own race. The French obviously have major faults, but their good qualities are even grander. It is a good race. It would be deplorable for the history of the entirety of humanity to see it disappear or diminish."[31]

In an open letter to the French government calling for a national subscription to sponsor military aviation and printed in *Le Matin*, André Michelin argued in December 1911 that the National Assembly was utterly negligent in its duty of providing for the national defense. Michelin deplored that the Assembly had voted fewer credits for aviation than had the German *Reichstag*. Michelin and other advocates of aviation, more knowledgeable and more responsible, thus needed to mobilize the French population and fill the void. "Since the government thus forgets its duty [and] since not one of our deputies could find a word to remind [the government of its responsibility], don't you think that public opinion alone, with its robust good sense, is capable of understanding the gravity of the current situation and of demanding that we act?"

21. Using the title of André Michelin's brochure, "Our Future is in the Air," this postcard features Bibendum hanging from a steeple topped by a Gallic cock. The image of an air filled with planes capable of attacking from afar was frequent in prewar books on aviation. In the foreground, the bombardier waves to Bibendum, that great promoter of French aviation. Postcard by Hautot, 1912.

He called for the creation of a national subscription to buy airplanes and provide airfields, in order to prompt the French government to do more.[32] By February 1912, the Comité National de l'Aviation Militaire (National Committee of Military Aviation) was born and André Michelin served as treasurer. A great many manufacturers of airplanes joined. The mass-circulation dailies *Le Matin, Le Journal, Le Petit Journal,* and *Le Petit Parisien,* and the municipal councils each contributed 50,000 francs to the campaign. Michelin contributed 100,000 francs. The participation of the widely circulated Parisian dailies ensured that the French public would notice Michelin's generosity and its patriotic fervor.[33]

In February 1912, Michelin stepped up the struggle by publishing a million copies of a 40-page brochure entitled "Notre avenir est dans l'air" [Our future is in the air]. The brochure featured a series of military experts and commentators who vouched for the necessity of more airplanes, more pilots, and, above all, better funding. They praised Michelin's patriotic, disinterested sponsorship of aviation. Michelin repeated its own demands as an easy formula: France needed "5,000 airplanes, 5,000 pilots, and at least 50 million francs yearly."[34] The brochure further claimed that airplanes needed to be used for bombardment, and not mere reconnaissance. The cover of the brochure sported the blue, white, and red of the French flag and the title in large letters. This is the brochure that appears on the right side of Picasso's cubist painting *Nature morte: Notre avenir est dans l'air,* a reminder of the interest that modern artists have had in aviation.[35]

Michelin's public campaign continued right up to the beginning of the war. In January 1914, André Michelin went so far as to criticize publicly the president of the Republic himself, Armand Fallières, for only making a contribution to the national subscription of the Comité National d'Aviation Militaire once he had seen the map of sixty airstrips purchased and hangars built with funds from the committee. Claiming that advocates of aviation represented the country better than the government, Michelin announced the "country did not understand why [the government] was hesitating." What France needed was quite simply an aerial fleet (*flotte aérienne*) in order to "shelter [it] from all attacks." The fleet needed to be like "the naval fleet of the English, which must be *stronger* than the two next largest fleets combined [emphasis in the original]."[36] According to Gillet, Michelin's work on the Comité National d'Aviation ended in late January 1914, when the committee offered the French army 120 planes and seventy air fields.[37]

During and after the First World War, the Michelin company forcefully asserted that its advocacy of aviation and its wartime production of Breguet airplanes resulted solely from the Michelin brothers' patriotism. Without doubt, they were very patriotic, but the interests of the company were quite compatible with what the Michelin brothers perceived French national interests to be. Even during the war, Michelin stood to gain a fair amount in profits from an array of products for the army, in maintaining a skilled workforce ready to be reconverted to civilian production at the end of hostilities, and in reinforcing its image as a servant of the French national cause.

After the war, Michelin noted with pride and outrage that its warnings since 1911 had not been heeded by the French government. Whereas Michelin had called for 5,000 airplanes and 5,000 pilots to do surveillance, to fight, and particularly to bomb, the company claimed that France only had 120 airplanes when war broke out in 1914. Moreover, the company claimed that these planes had been built by 14 different manufacturers, making them difficult both to maintain and to fly, because of the peculiarities of each kind of aircraft.[38] In fact, the French military had since 1911 been actively sponsoring competitions among airplane manufacturers, attempting to standardize the fleet, and purchasing planes. France had more than 120 planes; estimates range up to 450, not including those in reserve or those used for training purposes. This was of course still a far cry from the number 5,000. Michelin's wartime and postwar complaints about the military use of planes at the beginning of the war were generally accurate. Airplanes completely replaced dirigibles as France's airborne reconnaissance craft in 1914, but they had little role in combat, only being outfitted with machine guns in 1915, and the army general staff initially foresaw little real role for aerial bombardment.[39]

Michelin announced to the press on 20 August 1914, a few short weeks after the outbreak of the war, that it would give 100 airplanes to the French army. In negotiations with the army, Michelin defined carefully the parameters of its offer. It was not an offer of cash, but a donation of 100 airplanes to be built by Michelin with motors, accessories, and rolling equipment supplied by the French government. The 100 were to be deployed together as a Michelin bombing squadron. One critical senator, the Count d'Aubigny, reported that only about one-fifth of the cost of the squadron would actually be covered by Michelin.[40] After the initial 100, any additional airplane bodies would be provided by Michelin at cost. Michelin specified that the airplanes needed to be those designed by Breguet; Michelin contracted with Breguet to build the planes with 4 percent of the cost of the first 100 planes and 10 percent thereafter being paid to Breguet. Michelin further promised to halt such production at the end of the war; given the growing market in tires and the unstable market for airplanes, not to mention that Michelin established no research and development of aircraft, it was a good deal for Michelin as well as for Breguet. Michelin specified the big, lumbering Breguet because it was big enough to carry 40 8 kilogram bombs, thus fulfilling the wartime use for aircraft that Michelin had advocated since 1911. Negotiations with the government and production delays led to the deployment of only a handful of Breguet aircraft by August 1915. The entire 100 were placed in service by June 1916. Michelin produced additional Breguets in 1916 and 1917.[41]

In the early years of the war, there was a divergence between André Michelin's vision of the future of aviation, largely realized in the Second World War, and the army's appraisal of the technical capability of aircraft in the years 1914–16. Michelin firmly believed that the bombing of strategic targets, such as railway lines and industrial plants, could seriously impede the Germans' ability to move munitions and men to the western front. At the same time, pilots and the general staff operated in the realm of quotidian reality. They found the early Breguet-Michelin, like other aircraft, utterly incapable of successful bombing raids. French attempts to bomb German industrial plants in 1915 failed abysmally, causing the army to limit widespread bombing until the last two years of the war. By the time of Breguet-Michelin's deployment in 1915–16, the Germans already had Fokker and other fighters that were much faster and capable of easily outmaneuvering and downing the Breguet-Michelin. Moreover, pilots complained that the Breguet-Michelin climbed too slowly and was difficult to maneuver, making it a sitting duck. By the time of their deployment, the early Breguet-Michelins, too slow for bombardment during the day, were instead relegated to night raids. Although Michelin led a concerted defense throughout the war against army complaints, claiming that they originated at general headquarters where incapable bureaucrats held up their initiatives, it is clear that the general staff did not make up pilots' concerns.[42] Moreover, when the Breguet XIV produced by Michelin made its debut in 1917, the army used it for successful daytime bombing missions throughout 1918 and the airplane received widespread praise, indicating that earlier resistance to the Breguet-Michelin planes was motivated by more than unfair political manipulation or inefficiency within the army general head-quarters.

Throughout the war, the French army used airplanes for reconnaissance and observation more than for any other purpose, but this was particularly true in 1914 and 1915. While Michelin argued for a strategy of bombardment, the army gave airplanes the more limited role of reconnaissance, claiming that no aircraft could bomb effectively. Airplane reconnaissance had, after all, given the French crucial information for positioning its artillery and halting the German advance in the battle of the Marne. In 1915 and especially 1916, one-seater fighters increasingly came into vogue, a reality underlined by the importance of flying aces in those years, circling the skies over the battlefields of Verdun and the Somme and participating in dogfights. By 1917, however, as bigger planes also became fast planes due to ever-larger engines, bombard-ment increasingly became a priority for the French army. When Minister of Armaments Louis Loucheur consolidated airplane manufacture within France in 1917, Breguet was one of the few models chosen to remain in production. A later model, the Breguet XIV, produced by Breguet, Michelin, and five other firms, came into widespread use in 1918, realizing for the first time the prewar dreams of André Michelin that intensive bombardment could contrib-ute to the war effort. Even Michelin's preferred strategy of using several large bombers to bomb from on-high, rather than individual dive-bombers, became a reality by March 1918. In 1917 and 1918, Michelin produced 1,584 BM XIV airplanes. By October 1918, Michelin alone produced 6.8 percent of new French airplanes.[43]

Of course, airplanes did not win World War I; infantries and artilleries did, with the assistance of reconnaissance airplanes that dominated the French air arm. In February 1916, the French army had 1,149 planes, of which 826 were for observation; even at the end of the war, the army used one-half of all French planes solely for reconnaissance. The fighters could hurt morale (as when widely used by the Allies against the Germans at the Somme), and bombers were useful on the field of battle. But the French did not have the capacity for successful strategic bombing of German cities and industrial plants by the end of the war.[44] André Michelin's prewar and postwar visions that 5,000 airplanes could win the war without the use of the infantry remained a dream, a prescient one perhaps, but a dream nonetheless.

Michelin gained a great deal from its wartime industrial production. Annie Moulin-Bourret has convincingly argued that Michelin's aircraft production, particularly after 1917, allowed it to receive shipments of scarce resources. For example, in December 1917 Michelin received a promise of 2,000 tons of coal to fire its power plant, a plant that powered Michelin's civilian tire production, its for-profit production of other war material, and the production of airplanes.[45] Moreover, the company developed and produced bombs, bomb launchers (*lance-bombes*), and the scopes to see targets. It did earn profits on these other materials necessary for bombardment. Michelin boasted after the war that it had produced not only 1,884 Breguet aircraft, but also 8,600 bomb-launchers and 342,000 bombs of various sizes.[46] Above all, airplane and bomb production during the war associated the Michelin name both with the modern phenomenon of flight and with French dominance of it. However necessary for the war effort other goods produced by Michelin, such as tents and sleeping bags, may have been, they did not spark the imagination; they possessed considerably less marketing potential than the airplane.

"OUR SECURITY IS IN THE AIR"

In December 1919, Michelin continued to combine its projection of the Michelin name with its advocacy of aviation in the publication of "Notre sécurité est dans l'air," a forty-page brochure that referred specifically to the company's prewar "Notre avenir est dans l'air." Sporting a tricolored cover like its predecessor, the brochure used simple language and avoided the literary and historical references that had characterized Michelin's prewar advertising. Michelin printed more than one million copies, and the company ingeniously used the distribution network of the guides to the battlefields in order to ensure widespread diffusion; Michelin wrote to all of the booksellers who sold the guidebooks, enclosed a free copy of the brochure, and offered them a reasonably good deal. Booksellers would be allowed to sell the otherwise free brochure (for anyone covering postage) for 25 centimes but would pay Michelin only 1 franc for 10 copies (10 centimes each), just enough to cover Michelin's cost for postage.[47]

After a brief introduction in which the company announced that its calls for five thousand French airplanes had been ignored by the French government before the war, Michelin noted that brochure would show the verity of "the American precept: experience is the unique source of truth."[48] The brochure had five parts. In the first, Michelin traced its own contribution to aviation, including the establishment of the various Michelin prizes, including the prewar target prize. The public letters, newspaper accounts, and the publication of "Notre avenir est dans l'air" were all recounted, with repeated notations that the company's agenda was ignored by the French government.

The second section, which constituted about half of the total length of the brochure, discussed Michelin's contributions during the war years. After claiming that France had only 120 airplanes at the start of the war, the brochure then reprinted the company's multitude of letters to the army general staff and the president of the Republic himself (who by 1916, was Raymond Poincaré, the lawyer turned politician who had earlier represented Michelin against Dunlop) during the war. Michelin praised Minister of War General Joseph Galliéni for permitting Michelin to establish an airstrip and testing ground at Aulnat, where Michelin built the first cement runway and a school of bombardment. The fickleness of early aircraft meant that pilots needed to learn the individual eccentricities of each aircraft, in this case the Breguet-Michelin, and Michelin also wanted to instruct navigators in how to use their scopes and bomb-launchers. In addition, the brochure described the various bombing demonstrations that Michelin held near Aulnat for an array of army officers and deputies from the National Assembly. Not surprisingly, "Notre sécurité" argued that the Michelin bombing materials and the bombing method it advocated were effective on the testing ground. The brochure even included photographs taken of the target from the air. The photographs themselves reveal just how few bombs actually fell inside the target, which had an outer diameter of 100 meters. Nevertheless, the company also included a graph of where its bombs and those of its competitors landed; in this case, the Michelin bombs, launched with Michelin equipment, appear to have been much more effective. More significant for Michelin's overall argument, however, was its overlay of a map of the Parisian Saint-Lazare train station over the testing ground at Aulnat; the illustration made the simple point that the bombs (launched by some fifteen different aircraft, each carrying about forty 8 kilogram bombs) could have hit much of the surface of the train station. To conclude the section, Michelin argued that after it became clear, at least according to Michelin, that bombing could be effective in 1915, it took three years before the army's general staff was fully convinced. Thus in 1918, precise, effective bombing was at last employed by the French army with successful results. Michelin noted that it was "better late than never."[49]

The third section included a tirade against the three nameless officers who had opposed Michelin's plans and an attack on "Monsieur Lebureau" (Mr. Office), the inefficient "retrograde" bureaucrat who slowed adaptation of Michelin's superior notions of how to employ aircraft during the war. Only the common sense of a few good men, such as General Galliéni, a willingness to overlook regulations, and the toughness to fire the inefficient, as General Joffre had done early in the war, could have solved the problem.[50]

Further bolstering the argument that Michelin dutifully served France, the brochure then reprinted an open letter the Michelin brothers penned in March 1919, in the midst of the negotiations of the Treaty of Versailles. In the letter, they claimed that the "suppression [of German military aviation] offers the allies only a platonic satisfaction and the danger of an aerial attack remains more formidable than it has ever been, if the Germans are authorized to possess a civil aviation." (The implicit reference to sexual "satisfaction" reveals much about early-twentieth-century assumptions of conquest, both by victors and by men over women.) They noted on the basis of their own experience that within an hour, any civilian aircraft could become a military aircraft. The letter ended with a warning in bold and italics: "*Germany must be absolutely forbidden to construct or to use a single airplane or a single dirigible. If our generation does not abolish the German civilian aerial fleet, we will carry into History a crushing weight of responsibility.*"[51] The brochure reported that, yet again, their warning had not been heeded.

"Notre sécurité" closed with a section entitled "What France Must Do." What France needed, according to Michelin, was an aerial fleet that was stronger than Germany's and that would "inspire a salutary fear."[52] Doubting that the government would ever vote a sufficient budget, Michelin wrote that what France needed was a "commercial aerial fleet that will give us power, prestige, and wealth, and that the State will be able to use for the eventual mobilization and armament." Michelin asserted, however, that there were still far too many accidents, particularly during landings, for commercial craft to be considered truly viable. To spur the development of such commercial aviation, Michelin announced at the end of the brochure its latest aviation prize, this one for 500,000 francs. The Prix Michelin de la Sécurité would be awarded to the first French pilot who could before 1930 fly an entirely French plane at an average speed of 200 kilometers an hour from Versailles, to Reims, and back to Versailles, landing in the space of 5 meters from touchdown to complete stop. As a postscript to the brochure, Michelin reminded the reader that "it is as Frenchmen that we have spoken out, not as manufacturers of airplanes."[53]

In December 1919, hundreds of French newspapers announced Michelin's latest prize in articles duly clipped by Michelin employees. Coverage was overwhelmingly positive. Most articles simply quoted large sections of "Notre sécurité," thus allowing the Michelin brothers to speak for themselves. The *Journal des Débats* also featured an interview with André Michelin, who again repeated that Michelin no longer constructed airplanes so it was "as a Frenchman and as a patriot that I have been alarmed by the state of our aerial fleet." Too many Germans were dreaming of a future war that would be characterized by "germs, sickness, flame-throwers, tanks, and airplanes." It was to be a veritable "scientific war between races." He claimed that the Germans were still building planes and would soon have twenty thousand to thirty thousand commercial aircraft. France's only option was industrial development, including that of aviation.[54]

In the same article, Michelin then abruptly changed the subject and elaborated for the first time a theme that would become the leitmotif of Michelin's advertising after the Great War. André Michelin claimed that

I will do everything in my power to invite French industry to augment its production and to move ahead always, without hesitation. I am dazzled by the results obtained by American industrialists. We have been outpaced, us and other Europeans to an unimaginable extent. A businessman from the United States told me the other day that there are now 6,250,000 autos over there! In 1920, 2 million new vehicles, cars for individuals and trucks, will come off the line. But, in the rest of the world, there are only 2 million cars total: 180,000 in Britain, 100,000 in France. Another example: in 1913 the Americans built 19,500,000 tires (we produced 1 million). Today, they build 40 million. . . . We can only with difficulty imagine the progress made by American business during and since the war. Only the Japanese have made a similar effort.[55]

At first, Michelin's segue from advocating French aviation to discussing American industry seems strange. But Michelin was implicitly arguing that only American-style industrial progress could allow France to compete commercially, and thus militarily, with the Germans and even with its former allies.

Michelin hinted at a commercial strategy that the company used with aplomb by the mid-1920s. At a time when "America" represented progress, the Michelin company cast itself as French to the core, in fact so French that it had not only adopted "American" methods of production but did so very publicly. Even in "Notre sécurité," Michelin praised Galliéni for using a stopwatch to improve efficiency in the bureaucracy of the War Ministry, just as Michelin trumpeted its own use of Taylorized production methods in the production of bombs and Breguet aircraft. Michelin even enhanced its own credibility as a technologically advanced company by noting in "Notre sécurité" how many of its bombs and planes had been manufactured for the American army in France. If the presumably more technologically advanced Americans wanted Michelin planes and bombs, that fact alone must prove their superiority. The company was especially proud that American officers had been trained in Michelin's bombing school at Aulnat during the war.[56] At a time when America seemed to many Europeans to represent the future, Michelin thus managed to associate the company with progress.

After 1919, Michelin's Prix de la Sécurité received little attention. The requirement that the heavier-than-air apparatus land within a 5 meter stretch remained technologically impossible until the development of helicopters. The resurrected Coupe Michelin for international pilots with a prize of 20,000 francs did have a winner most years during its existence between the years 1921 and 1936, although competitions no longer dominated the domestic and foreign news coverage as they had in 1908. Michelin itself did little to advertise the prizes. International competitions had lost some of their luster by the interwar period, and given inflation, 20,000 francs was not the handsome amount it had been before the war. The average speed of winners—80 kilometers an hour in the 1920s, but 320 kilometers per hour in 1936—reflects the changes in airplanes during the period, but unlike the prewar years, it would be difficult to attribute the advances primarily to the influence of aviation prizes.[57]

ANDRÉ MICHELIN'S POSTWAR CRUSADE

After the announcement of the prizes, sponsorship of aviation became less of a marketing strategy for the company, remaining above all a personal campaign of André Michelin. Throughout the 1920s, André Michelin took an exceedingly active role in raising money and propagandizing on behalf of the airline industry. His name certainly reminded readers and listeners of the firm, but neither he nor the company's publications regularly noted the connection. Even though the Michelin company published dozens of mass-diffusion brochures in the 1920s and 1930s on everything from automobiles to breastfeeding, not one brochure after "Notre sécurité" took up the subject of aviation or mentioned André's efforts. The time and considerable money André Michelin spent in the 1920s reveal that even though advocacy of aviation and Michelin's marketing might have been inseparable before and just after the war, André in particular ultimately believed in aviation and its importance for France beyond any potential profits.

In early 1920, the board of the Aéro-Club de France invited André Michelin to become its president, replacing the illustrious Henry Deutsch de la Meurthe.[58] He immediately began to use the group to further the agenda laid out in "Notre sécurité." His style was direct and even brash for the president of the elite organization. In his concern that Germany could easily carry out a chemical attack on Paris, Michelin publicly told the minister of war, Louis Barthou, how to do his job: "Sir, it is necessary that you use your admirable eloquence to get parliament to vote a sum that, however enormous it must appear to you, will be considered paltry by our sons for the equipping of our factories producing nitrogen and poisonous gases and those building airplanes, more airplanes, ever more airplanes."[59]

Beginning in September 1920, André Michelin spearheaded a subscription to raise funds to support French aviation. He personally offered 100,000 francs and promised to add 400,000 more if the group managed to raise 2 million. Petroleum and armaments magnate Basil Zaharoff did the same. Others who contributed considerably less included the brother of the late Henry Deutsch de la Meurthe, Emile (also known as a benefactor to the Université de Paris for the construction of the Cité Universitaire, where the first *maison* was named for him), the Peugeot brothers, Gustave Eiffel, the Prince de Broglie, and Fernand Boverat as well as leaders of the Rothschild, Crédit Lyonnais, and Société Générale banks.[60] In 1921, when Michelin was convinced that the Aéro-Club could not have the impact that he had hoped to make, he abruptly resigned as its president. Although he cited his doctor's orders to reduce his activities, Michelin then literally threw himself into the foundation of the Comité Français de Propagande Aéronautique (The French Committee of Aerial Propaganda), a group funded by the earlier subscription. The French state recognized the group as an association "d'utilité publique" in November 1921.

By November, the organization had only forty-six members but a war chest of almost 1.5 million francs.[61]

Although Michelin convinced the former aviation commander, General Duval, who had adopted bombing in line à la Michelin in early 1918, to serve as titular head while he took the title of vice president, Michelin remained the driving force in the group. Its board (*conseil d'administration*) boasted major donors and other luminaries, including Léon Auscher, vice president of the Touring Club de France. For the next ten years, the Committee made public presentations and published newsletters and an array of brochures warning of German air potential. It insisted on a strong French response. The group even packaged presentations on a variety of subjects that could be read by presenters too busy to do their own homework. The Committee further established a range of prizes for technological development of aircraft, meteorology, piloting, and, eventually, bomb shelters. Although the group proclaimed constantly that it was mobilizing public opinion, it generally construed public opinion to include primarily the educated elite. For example, it did not undertake widespread use of cinema, and its presentations were directed at elite audiences in places such as the Sorbonne amphitheater. When it addressed schoolchildren, the group focused on *lycées* for boys, the secondary schools attended almost exclusively by the sons of the bourgeoisie, and not primary schools.[62]

André Michelin himself made innumerable public presentations warning French audiences about the threat from Germany, elaborating the themes first addressed in "Notre sécurité" and raising new issues. In particular, Michelin became preoccupied with German potential to mount a successful chemical attack against Paris and other major cities. In "Le danger allemand: Aviation et guerre chimique" (December 1922), Michelin argued that the Treaty of Versailles itself, by monitoring only traditional armament production as at Krupp's, would encourage Germany to abandon "old methods of warfare and to throw itself into the study of modern ones."[63] He laid in detail how large the German chemical industry had become. He foresaw that the restrictions on German aviation, even the Allied ban on new civilian aircraft (in effect until July 1922), would have no long-term impact. Michelin specifically worried that gliding, which was becoming immensely popular in Germany, in part because it represented German willingness to fly in spite of Allied restrictions, would enable pilots and manufacturers to perfect skills and airplanes. He noted as well that Germany used its cooperation with the Russians to construct aircraft and to give its pilots more flying experience, especially in covering long distances. According to Michelin, Germany would later need only to combine its expertise and material developed in the civilian realm with its chemical capacity; it would then be easily capable of mass gas attacks on Paris and other major cities. France had only one option: "the dominance of the air, or at least a redoubtable aerial fleet."[64]

Skeptical that the French parliament would ever vote the necessary credits, he made an appeal for contributions to the Comité National. These themes were repeated in numerous other speeches by André Michelin between 1921 and 1926. In a replay of wartime propaganda, Michelin warned frequently that the *Boches* were, despite their technical and organizational excellence, barbarians. In October 1923, after the Allies had lifted the ban on German civilian aircraft production (all restrictions would be lifted in 1926), Michelin closed a speech at the Sorbonne by comparing planes to vultures and arguing for the complete elimination of any civilian German aviation. "Nothing can protect [us] from a fleet of a thousand vultures that quickly sweep down from the sky in order to kidnap [our] small children, gouge out their eyes, and smash in their skulls. Here, Ladies and Gentlemen, is what we must say, repeat, and shout out to our own leaders and to those of our friends the English, 'Don't let the Kraut have the vulture.'"[65]

Despite the Committee's quite vocal demands and its various competitions, the association did little to organize existing French civilian aviation before 1926, when Michelin convinced Marshal Louis Lyautey, former minister of war and longtime resident general of Morocco, to lead it. Lyautey obviously provided the sterling nationalist and colonial credentials honored by the nationalistic Committee. Lyautey called on General Boucabeille, who had served with both Lyautey and Galliéni at Tonkin and Madagascar, to assist him. After 1926, Lyautey and Boucabeille steered the Committee away from Michelin's focus on likely chemical warfare and toward the development and organization of civilian aviation. They wanted French airlines to compete with the rapidly growing, state-subsidized Luft Hansa in Germany, making advances that might one day be used in time of war. Between 1926 and 1931, without abandoning earlier initiatives, the Committee worked closely with constructors, local chambers of commerce, and the PTT (the French postal, telegraph, and telephone service) to subsidize a veritable network of local airports and to establish flight patterns between French cities.[66]

After 1926, the Comité National de Propagande Aéronautique argued that the driving force for the advancement of aviation would have to be travel for business, communications (such as the PTT), and tourism. Greater use would lower costs, making flights as common as train trips. The rapidly developing network of air strips sponsored by the Committee was a necessary step in the development of aerial tourism.

Although the Touring Club had sponsored flights of hot air balloons and featured regular reports of such tours since the 1890s, tourism by airplane had been virtually nonexistent before the war. In 1911, Léon Barthou, president of the TCF's committee on aerial tourism, noted that "aerial tourism [by airplane] has been so little practiced that it does not really exist" even for the rich.[67] The problem, he noted, was the utter lack of landing fields and hangars for planes, a responsibility that municipal governments were reluctant to accept. For all, even the wealthy, cost was also a factor. The Aéro-Club de France had counted only six airplanes kept for tourism in 1913; one of them belonged to Henry Deutsch de la Meurthe himself, who rarely flew. Emmanuel Chadeau estimates that it cost at least 50,000 francs yearly to buy, maintain, and fly an airplane before the war, 2 times the price of maintaining two luxury automobiles with maintenance, chauffeurs, and other necessary expenses.[68]

In those heady early days of heavier-than-air aviation, Continental had attempted to build its image as a manufacturer of dirigible and airplane equipment by calling its *Guide routier* (road guide), which first appeared in 1904, a *Guide routier et aérien* (air and road guide), beginning in 1910. Despite the title, however, Continental did not offer any substantive information about the location, altitude, or conditions of air fields, which did not yet exist in most of France. In fact, there was absolutely nothing that distinguished the Continental guide as an "air guide" except the implicit claim in its title. For pilots, the Continental guidebook had no practical use whatsoever; its purpose seems to have been solely to remind users that Continental represented progress in the air as well as on the roads of France.[69]

In the 1920s, the Touring Club de France began to sponsor tourist flights in airplanes and to publish accounts of them. In July 1928, one member reported how a tourist might use existing flight patterns to see France. With wonderful photographs of the panoramas that one might see, a Lieutenant Thoret described his flight over and around Mont-Blanc.[70] In October 1929, the Touring Club organized the first group tour by airplane and included a long account in the *Revue*. The group flew to that symbol of interwar French patriotism and military might, the city of Strasbourg, for a day trip. The report proudly noted that the group left in the morning around 7:00, climbed into six different airplanes to go to Strasbourg, and landed at the Strasbourg-Neudorf airstrip, in time for their midday meal. Signaling the precedence of the trip, the minister of the air, M. Laurent-Eynac, met them to dine. Employing the regional and culinary language that came to dominate interwar tourist publications, the report pleasingly noted that the group had "an exclusively Alsatian dinner [*déjeuner*]: pâté de Strasbourg, choucroute [sauerkraut], jambon et saucisses d'Alsace [Alsatian ham and sausage], tarte aux quetsches d'Alsace [plum tarte], all accompanied by beer from Strasbourg and delicious wines from the east side of the Vosges, Zwicker de Ribeauvillé and Traminer-deux-clefs. The servers were young Alsatian women in regional dress." After a brief ceremony, the group got back in the airplanes at 2:30 in the afternoon and headed back to Paris. Four of the planes arrived back in Paris between 6:00 and 7:00 (they were flying from east to west, that is, against the wind current). One of the airplanes could not pass through the high winds at the pass through the Vosges at Saverne, and another had mechanical problems; their passengers were forced to take the train and arrived home much later.[71] Clearly, no one saw much of Alsace, but they did participate in the longstanding TCF ritual of dining together, in this case eating typical Alsatian cuisine. The details of the trip, such as the fact that one-third of the passengers returned home by train, are a reminder of the limits of aerial tourism in the interwar period.

In 1930, after the mass building of French airstrips assisted by the Comité Français de Propagande Aéronautique, Michelin issued its first aerial guide to France and northern Africa "under the auspices and with the collaboration of the Ministry of the Air." The guide itself contained no mention of the Committee, further evidence that Michelin was not using André Michelin's energetic efforts after 1919 in order to market its products. The introduction of the guide optimistically stated that lower prices of aircraft and other improvements had made possible and necessary this first guide destined for tourists and business travelers. The advances were bound to continue: "Aeronautical progress is accelerating at an increasingly rapid pace; soon the density of airplanes cutting across the sky will be greater than that of automobiles on the roads."[72]

The guide was quite literally the aerial counterpart of the Michelin red guide for motorists. There were short notations of sights to see, ratings of hotels and restaurants, and lists of practical information, such as the location of post offices, in the cities. Guides included the views or panoramas that the aviator should not miss, listed aviation experts in each locality, and noted meteorological stations. It laid out itineraries much as the Michelin regional guides did: in 1935–36, the guide described Paris–Marseille, Bordeaux–Marseille, Paris–London, Paris–Brussels, and Paris–Strasbourg routes. It listed lighthouses and two-way radio bases. Moreover, Michelin enumerated alphabetically the airports, airstrips, and landing fields near each French city or town. The company included notations of the latitude, longitude, and altitude of the air fields. The length and width of the runway, obstacles such as trees, and possibilities for fueling and repair of one's plane all received mention. Guides even listed the best means of getting into local towns and cities, be it by rail or taxi. The guide noted the landing fields where one would need to touch down near the border so that passengers and the plane could pass through customs. Finally, the guide delineated those zones of France, obviously above army and navy bases, where flights and photography were forbidden.[73]

There was little marketing potential for the guides given the paucity of individual tourists traveling by air before the Second World War. Like André Michelin's other interwar efforts after 1919, the guides were an attempt to advance the cause of aviation for the sake of Michelin's notion of French national destiny. Any people wealthy enough to travel by plane certainly had heard of Michelin and would have already associated the company both with France and with flight. Interestingly, unlike Michelin's other guides and maps, the *Guide aérien* did not advertise Michelin tires. In fact, it did not even contain the multitude of advertisements for Michelin's other tourist guides and maps. For those bound to travel on the ground, maps featured ads for regional guides and red guides; pages of the red guides advertised maps and regional guides; and the regional guides plugged red guides and maps. The *Guide aérien* practiced little such cross-referencing. A further indication of the limits of the tourist market was the cessation of new editions after the mid-1930s.

FLIGHT FOR FRANCE

In 1931, the seventy-eight-year-old André Michelin died. Within a year, the Comité Français de Propagande Aéronautique folded, finding itself reabsorbed by the Aéro-Club de France, although the Committee maintained a financial independence within the association. The virtual collapse of the Committee is a reminder of the centrality of André Michelin's own contributions. Only a year after André's death, Etienne Michelin, Edouard's eldest son and heir apparent, died when his plane crashed on the top of the Puy-de-Dôme. Although Michelin continued to produce aerial tourist guides and to sponsor the Michelin Cup until 1936, Edouard Michelin avoided any new initiatives or a public role.[74] The Michelin company's very public encouragement of aviation, so strong until the 1920s, waned.

Much like its support of the pronatalist cause, the Michelin company's support for flight resulted from a patriotism widespread in early-twentieth-century France, in this case a patriotic devotion that envisioned a technologically advanced and militarily superior France reliant on the latest weaponry. Whereas the company's pronatalist arguments assumed that a large population alone would save France, Michelin's arguments in favor of aviation claimed that only a superior air force would do so. The two positions, at first glance contradictory, complement each other; both were potential means for saving France from foreign, particularly German, incursions. Both also had ramifications for Michelin's marketing. In advocating aeronautics as in advocating pronatalism, Michelin's support was not anonymous; the company proudly proclaimed its actions in publications and advertisements. This is not to say that Michelin used aviation, any more than pronatalism, as a mere cover to sell tires. Rather, this family firm loyal to its vision of France took for granted that its patriotic actions should be appreciated by the French public.

By sponsoring aviation, Michelin placed itself at the center of an important development in early-twentieth-century France. Before 1914, in a country that dominated early flight, that produced more airplanes than any other country, and that exported more than any other country, to encourage aviation was to trumpet the strength of France. In the interwar years, as the threat from Germany seemed to loom, and as French industry clearly fell behind the productive capacity of the Americans, to call for the resurrection of French aviation was to call for the reemergence of France as a major power. Michelin accepted and promoted a patriotic vision of a modernized, competitive France, economically and militarily secure from foreign domination.

6

ADVOCATING AMERICANIZATION?

Taylorism and Mass Consumption in the Interwar Years

IN 1929, Georges Duhamel, physician by training and writer by trade, published *Scènes de la vie future,* the account of his recent trip to the United States. Duhamel's book became an instant best-seller and has remained among historians the best-known interwar French commentary of life in the United States. The appeal of the book then and now is simple: it is an unambiguous, humorously oversimplified portrayal of "America," that summarized the major interwar European critiques about an "American materialism" menacing the very existence of European culture, including French *civilisation.* Duhamel described with disgust a society without a soul, without time for contemplation, where people sat mesmerized by cinema, awash in automobiles, refrigerators, and vacuum cleaners, deprived of wine and spirits, living in indistinguishable houses, and forced to eat the mass-produced beef of the Chicago slaughterhouses. This disaster of a life was created above all by companies that sold on credit and advertised massively, creating hitherto unrealized "needs" among people, drowning them in material goods, especially cars.[1] Duhamel devoted much of his attention to automobiles and the changes they had wrought, just as industrialists visiting the United States went as on pilgrimage to Detroit to see Henry Ford's factories.

Like other Europeans, including industrialists as well as intellectuals, Duhamel had traveled to the United States to get a glimpse of what Europeans held to be the modern world, the future of Europe. Although the title of Duhamel's book would be translated into English as *America the Menace*, thus playing on the "anti-Americanism" such criticisms of the United States have been assumed to be, his original title and the subtitle of the English translation, *Scenes of Life of the Future*, better indicates the interwar context in which the book appeared. For Duhamel and other critics of "the American way of life," as well as for Western European businessmen and labor leaders, American industry and society represented "modernity," the very future that Europe might eventually experience, for good or ill. Whether Europeans liked that future, at least certain aspects of it, or sounded the alarm to avoid a similar fate, many agreed: the capitalist future could bring to Europe a society and an economy like those of the United States. Interwar Europeans thus paired "America" with "modernity." Similarly, "pro-American" or "anti-American" positions were in large part commentaries on "modern" mass production and consumption as they emerged in the United States in the early twentieth century.

The American "threats" to traditional French culture and the French economy were quite palpable after World War I. European architects traveled to cities like Chicago to study and argued about the applicability of American models and building techniques, particularly as regarded the reconstruction of northeastern France.[2] American "talkies" competed fiercely with more theatrically oriented French films, dominating an industry in which France had been a leader before the war.[3] Most obviously, the mass production of automobiles best exemplified the relative strength of American production techniques and marketing strategies. Until 1904, France was the largest producer of automobiles in the world, and it remained second only to the United States in 1914, exporting much of its production. By the interwar period, however, the divergence between the American and French industries was huge both absolutely and in proportion to their respective populations. In the years 1919–23, American auto makers produced over 4 million cars, while French firms produced just under 300,000. Although French production grew dramatically in the 1920s while American production steadied after 1924, by 1930, there was one car or truck for every 27 French people but one for each 4.6 Americans.[4] Clearly, America represented a model to which French industrialists might aspire, in large part because it also presented a threat; for French auto makers or a tire maker hoping to recapture prewar positions as leading world exporters, it became clear that the French automotive industry would have to compete with the Americans.

It was in this context that Michelin became the foremost French industrial proponent of Frederick Winslow Taylor's scientific management and Henry Ford's ideas about the mass production and consumption of automobiles. Although Europeans would also adopt the term *rationalization* (*rationalisation* in French, *Rationalisierung* in German) to describe Taylor's ideas and Ford's production processes, Michelin itself referred constantly to "Taylorism," and repeatedly argued that French automobile manufacturers needed, in essence if not in name, to follow in Ford's footsteps; Michelin thus tied its own name to that of these American industrial gurus. Although French auto makers never rationalized production sufficiently in the interwar years to match the Americans or to suit Michelin, they too publicly displayed their willingness to adopt American production methods, in part by publicizing their own trips to see Ford's production facilities. André Citroën not only made several well-publicized trips to the United States and posed for photos with Henry Ford, but he also appears to have been proud of being dubbed "the French Ford" by the press.[5] At a time when the United States represented technical superiority and economic expansion more generally, to be associated with the United States was a superb, yet ironic, marketing ploy. Citroën, that grand master of advertising in the 1920s, used it with flourish, as would Michelin. Michelin stood to profit in two ways from its very public advocacy of Taylorism and Fordism in the interwar years: first, Michelin could sell more tires in the expanding market for automobiles; and second, Michelin might beat its European and even its American competitors by tying its name so tightly and ostensibly to American methods of production and consumption. While continuing simultaneously to set itself apart from Dunlop and the American tire companies as *the* French tire firm, using nationalism as it had before the First World War, Michelin's public adoption of *Taylorisme* could also associate the Michelin name with the perceived technological superiority of things American in the interwar years. To buy a Michelin tire was thus to buy a "modern" as well as a "French" tire.

Clearly, Michelin constructed an American model for France, but Michelin's vision was a far cry from the "Americanization" of France so feared by French commentators, including Duhamel. For Michelin, the automobile could offer convenience, mobility, and economic growth. But unlike Duhamel, Michelin never publicly entertained the widespread contemporary assumption that its version of "Americanization" or "modernity" had the potential of upsetting existing social hierarchies in France; the social conservatism of Michelin's potential clientele, increasingly petty as well as upper bourgeois, allowed for an easy selection of those perceived features of "America" that were acceptable for France. Whereas Duhamel firmly believed that materialism erased differences between the sexes, making women more masculine while emasculating men, Michelin portrayed a modernized France firmly in the hands of men, where differences between the sexes remained undisturbed by the automobile. Moreover, in creating a paternalistic social system for its employees, the company diverged from both Taylor's ideas and the practices of American tire giants who offered such stiff competition in the interwar years, further indicating that Michelin did not want to "Americanize" France so much as it wanted to modernize French industry, to promote techniques of production and mass consumerism for the sake of France, and, indirectly, for the sake of Michelin.

THE COMPETITIVE CONTEXT: TIRE MAKING AFTER WORLD WAR I

From the perspective of the interwar years, the period before the First World War represented a veritable Belle Epoque for Michelin. Technologically, the company was at the cutting edge, at least until 1908, and it remained quite competitive up to 1914. Michelin had been the first producer of pneumatic tires for automobiles in the 1890s and remained an innovator, as witnessed by its success on the racing circuit in the first decade of the twentieth century. Michelin had contracts with the major French automobile producers to be their exclusive supplier of tires. Before the war, France's position as primary exporter of automobiles in the world made it possible for Michelin to profit handsomely from the success of French automobile exports. Michelin supplied approximately one-third of all of the world's tires in 1914. Within France, consumers' penchant for replacement tires of the same brand as their original equipment and the quality of Michelin's products allowed its list to dominate other firms' pricing policies. In 1914, about 60 percent of all French cars were equipped with Michelin tires, with most of the remainder being produced by French subsidiaries of Continental and Dunlop. The American firms, particularly Goodyear and Firestone, were quickly gaining market share internationally, in large part as a result of their own contracts with American auto makers. Nevertheless, a sign of Michelin's global competitiveness was the successful establishment of a tire-producing facility in Milltown, New Jersey, in 1908–9, before the first American firm, Goodrich, established a plant in France at Colombes in 1910–12.[6]

The war exacerbated the divergence between the productive capacity of American firms and that of Michelin. Although the French market was closed to imports during the war, Michelin struggled to retain its export market in the face of wartime controls on both exports and raw materials. In the meantime, American firms had expanded rapidly along with both the market for Ford's Model T and military demand. American companies, particularly Goodyear, also captured a substantial part of Michelin's overseas market. In 1920, employment in 3 of the 4 largest American tire firms located in Akron, Ohio, peaked at 73,500 (before precipitously falling to 19,600, but recovering to 58,188 by June 1929), with Goodyear and Goodrich both employing more than 20,000 and Firestone about 15,000. In contrast, Michelin employed only 12,000 employees in Clermont-Ferrand; these numbers further understate the difference because of the higher productivity of American rubber workers just after the war.[7]

Technologically, the gap was even larger. European armies had demanded tires during the war, but they needed truck tires far more than automobile tires. Within France, that fact meant that Bergougnan's solid rubber tires received important military contracts. More significant, there was little need or opportunity for even major European tire companies to focus on the development of new passenger tires or the production process. As a result, all of the major national "first-movers" in Europe—Continental, Dunlop, Michelin, and Pirelli—fell significantly behind the American firms technologically. The Americans marketed a stronger "Cord" tire during and after the war, which Goodrich introduced through its French subsidiary in France as the "Souple Corde" in 1920. Firestone pioneered the development of a low-pressure "Balloon" tire in the early 1920s. Michelin then offered the "Corde" and "Confort" tires in an effort to keep pace. The American firms also made a series of advances by altering the chemical composition of tires. They began the use of "carbon black," a substance that not only made the tires jet black but also made them much more resistant to destruction. American companies also began to add various chemical accelerators that not only improved tires but also low-ered the cost of production, thus increasing productivity. By the early 1920s, the Americans produced better tires more cheaply than their European coun-terparts, including Michelin. In 1920, the situation was particularly critical, when American companies facing excess capacity reduced their stocks by dumping tires onto the world market, further driving down tire prices within Europe. Michelin, because of its long-term concentration on tires rather than other rubber products, was in a particularly difficult position. Smaller French firms, notably Bergougnan and Hutchinson, survived the crisis as a result of their diverse product lines. Michelin, by contrast, remained independent, forced temporarily to copy the Americans but redoubling efforts to offer tech-nological innovations of its own.[8]

In the end, Michelin's strategy worked. Stiff French tariffs on tires produced in the United States essentially closed the French market to the American manufacturers who focused on tire production, notably Goodyear and Fire-stone. In the meantime, whereas the automobile market within the United States stabilized after 1924, that of France grew quickly throughout the 1920s; in 1920 there were 156,872 automobiles owned by individuals in France, but there were 1,109,006 such vehicles in 1930.[9] Until 1929, Michelin maintained contracts as the primary supplier of tires to French automobile producers who themselves retained dominance of the French market. Michelin's primary com-petitors within France quickly became the subsidiaries of Dunlop and to a lesser extent Goodrich, who both avoided tariffs by producing in France. Al-though Michelin's products may not have been the best or the cheapest in the 1920s, the rapidly expanding market for automobiles allowed the firm to grow along with the market itself.

The late 1920s brought new challenges. As early as 1929, pricing pressure led major French automobile manufacturers Citroën, Peugeot, and Renault to limit the exclusive contracts with Michelin for new automobiles. While Mi-chelin attempted to play the national card by reminding Louis Renault, who negotiated on behalf of the car companies, that he had an obligation to support a French company over foreign ones, Renault predictably responded that he was attempting to save the French automobile industry by cutting the cost of production. The car companies came to rely increasingly on Dunlop as well as Michelin as a primary supplier, although tariffs and the two firms' size made it possible to shut out other competitors, save Goodrich, which was less willing to increase production to supply tires near cost. Although Michelin became the exclusive supplier to Citroën in 1934, Michelin's takeover of the firm further strained relations with Peugeot and Renault, prompting the latter to begin, like Ford, to produce its own tires. By 1938, Michelin and Dunlop supplied 80 percent of the tires on new French cars. Michelin remained the larger firm within France, but it had lost its overwhelming dominance and its ability to set prices in France. Significantly, the Michelin workforce in Clermont, some 18,000 strong in 1926, had shrunk to 7,000 by the end of 1936. Dunlop, which had 3,500 employees at its Montluçon plant in 1926, had nearly 4,000 in 1939.[10]

Michelin's other problem, French loss of the export market for automobiles and automotive components, was one faced by French auto makers. Although France's automobile exports remained second only to those of the United States throughout the 1920s, rising by some 200 percent between 1921 and 1925, they fell some 51 percent between 1925 and 1930. The decline coincided with the stabilization of the franc (the advent of the *franc Poincaré*) in 1926, with new British tariffs in 1926, and then with the worldwide Depression. In the 1930s, the French political elite's preoccupation with maintaining the *franc Poincaré* by not devaluing the currency left French automobile manufacturers uncompetitive outside France. Only in the protected market of the colonies did French automobile exports rise. French tires faced high tariff hurdles across Europe, made worse by the excess capacity and dumping on the part of American tire companies. Michelin countered by establishing factories in a host of European countries and remained competitive within other markets also protected from American imports. Michelin could not, however, compete with American firms' low prices on their home turf; in 1930, the Milltown plant closed.[11]

Despite short-term problems, however, Michelin maintained a long-term strategy focused on technological innovation in the conception and production of tires. In the 1920s and particularly the 1930s, the firm invested heavily in research and development, and the results began to show. In 1931, Michelin produced the "Micheline," a railway car that ran on rubber tires; although not in widespread use when the war broke out, the Micheline was in many respects the prototype of later Parisian *métro* cars. A series of new products for individual market segments appeared in the 1930s, including a "Pneu Confort Poids Lourd," a low-pressure balloon tire for trucks. The "Stop Superconfort"

Table 2. National Shares of World Tire Exports, 1912–1929 (%)

	1912	1919	1923	1929
France	33.8	23.7	32.1	15.5
Germany	12.7	0.9	5.4	3.2
Italy	20.3	7.7	10.4	9.1
United Kingdom	15.8	17.5	8.4	13.5
United States	7.3	36.2	29.0	31.3
Canada	1.3	6.7	10.2	18.3
Japan	—	5.6	2.9	2.4
Belgium	—	1.6	1.5	6.9

Source: Michael J. French, "The Emergence of a US Multinational Enterprise: The Goodyear Tire and Rubber Company," *Economic History Review* 40, no. 1 (1987): 69.

Note: Numbers have been rounded, and smaller producers have been excluded; and numbers do not include tires built by foreign manufacturers in the target market.

offered a new nonskid technology perfected in the development of the Michelin tires on rails. In 1938, Michelin introduced the "Metalic," a new tire for trucks in which steel belts replaced the usual cotton bands.[12] By maintaining and upgrading its independent research facilities, the company laid the foundations for the wartime and postwar development of what became known as the "pneu x," the steel-belted radial tire that would prove to be the firm's salvation after World War II.

Michelin's advocacy of both Taylorism and the mass consumption of automobiles in France fits squarely into the context of the global tire and automobile markets in the interwar years. In the early 1920s, Michelin worked at breakneck speed to reorganize its own production facilities in accordance with Taylor's ideas; given the company's concentration in tire manufacturing, distinguishing Michelin from Bergougnan in France and Goodrich and U.S. Rubber in the United States, it was imperative that Michelin remain competitive in the tire market.[13] The company's fate also depended heavily on that of the French automobile industry generally. Not coincidentally, Michelin argued for the adoption of Taylorism by French companies, and French automobile producers in particular. By the late 1920s, Michelin pushed harder than ever to expand the French automobile market by holding up the United States as an example. As before the First World War, the firm attempted to prosper not by producing a sufficiently good product more cheaply than its competitors but by simultaneously leading in innovation and attempting to expand use of the automobile. Its warm embrace of Taylorism and Fordism were the latest signs of Michelin's long-term commitment to put more cars on French roads. Michelin optimistically assumed that those cars would need Michelin tires.

TAYLORISM

Frederick Winslow Taylor (1856–1915) was not the first proponent of the scientific study of either workers or the organization of work in Europe or the United States.[14] In the interwar years, however, Europeans identified his name not only with various attempts to increase industrial and household productivity but also with his country of origin, the United States, at a time when even nonindustrial images of America had a certain cachet. That is, Taylor's ideas did much to further the image of an efficient, productive America, but contemporary European interest in Taylorism also resulted from a broader interest in "the American way of life," represented by mass consumption of new consumer goods.[15] The ideas of Henri Fayol, in many ways so close to those of Taylor, had in contrast some appeal in Fayol's promotion of them as natively French. Interestingly, Michelin associated itself with Taylor and thus with the United States, essentially ignoring Fayol.[16]

Taylor's ideas themselves were relatively specific; along with an array of associates considered "disciples," Taylor developed a system of "scientific management" that he began to popularize in earnest in the first decade of the twentieth century. The bedrock of the system was time-motion study done with a stopwatch to determine how quickly a given worker could complete a given task. After careful measurement of movements, the industrial engineer could reorganize the placement of materials and then choreograph the order of a worker's precise movements. The implementation of Taylor's ideas took time and training. Time-motion study, like the planning of the pace of production, was to be carried out by experienced engineers housed in a central planning department within the factory. When the whole of the production process was mapped, the planning department would then issue instruction cards to be filled out by workers as they completed individual tasks. The engineers could thus track production, eliminating the reliance on foremen's scheduling of work. Once an individual worker's productivity could be tracked, the worker's salary was to be determined in part by his or her own productivity, thus providing an incentive for workers. Taylor called for a piece-rate system that would award substantially higher salaries to the most productive workers and that provided an inadequate wage for less capable workers. Taylor, like the many "disciples" who followed his ideas and attempted to implement them in American industry, earnestly believed that his scientific management—as opposed to the presumably helter-skelter organization of work by foremen—held the promise of resolving the conflict between management and labor coming to a head in turn-of-the-century America. Increases in productivity, according to Taylor, could increase wages, while at the same time lowering the cost of production and thereby increasing profits. Management and labor would thus no longer compete for finite resources but would instead prosper cooperatively.[17]

Before the Great War, French engineers had been the most receptive continental European audience for Taylor's ideas. The innovation of automotive and automotive-related industries, in which France remained the largest European producer, labor strife, and a long-standing technocratic interest in the organization of work were all factors in preparing the way for Taylor's scientific management. As early as the first decade of the twentieth century, French engineer Henri Le Châtelier translated Taylor's works and propagandized his ideas. At least a handful of French entrepreneurs and engineers corresponded with Taylor, paid visits, and received him in Europe. Although the French military's own armaments factories had earlier reorganized work to increase productivity and General Abaut adapted several of Taylor's ideas in the artillery workshops he ran, the best-known prewar example of the self-conscious implementation of Taylor's ideas in French industry was Louis Renault's adoption of time-motion study in order to increase the pace of production in 1912–13; at best a pale imitation of Taylor's method and condemned by Taylor himself, Renault's early use of time-motion study offered no additional compensation for workers and met stiff resistance among foremen, resulting initially in strikes.[18] Before the First World War, there was in France great interest in the ideas of Taylor and his fellow travelers among engineers and a few entrepreneurs, but French firms did not implement scientific management to any noticeable extent.

Armament production during World War I led to the first widespread adoption of scientific management in France. Socialist Minister of Armaments

Albert Thomas pressured both industry involved in war-related goods to use Taylor's methods and skilled workers to accept the inevitable increase in unskilled labor. In late 1917, his successor, the industrialist Louis Loucheur, backed away from Thomas's accompanying emphasis on higher wages and collective bargaining, but he nevertheless continued to promote the reorganization of work in order to increase productivity. Etienne Clémentel, minister of commerce and industry, and, by the end of the war Premier Georges Clemenceau himself also recognized that Taylor's methods were crucial for making the best use of scarce wartime resources, particularly labor. Unlike the American firms that produced munitions for the Allies, French manufacturers had to increase production without a commensurate increase in skilled laborers; better organized and mechanized production allowed firms to employ women and other workers earlier closed out of skilled industrial work.[19] After the war, the efficiencies demonstrated in munitions production would be seen by many entrepreneurs as a model of how civilian products might be manufactured, a means of addressing a labor shortage in many parts of the country in the early 1920s.

The Michelin brothers' interest in Taylor dated from the Belle Epoque. André Michelin heard Taylor speak in Paris, and the Michelins corresponded with Taylor. By 1912, a Taylor disciple, Horace K. Hathaway, had toured Michelin's Milltown, New Jersey plant (built in 1908) and advised Edouard Michelin's nephew Marcel about how it might be "Taylorized." However, when advised that his engineers would need to complete a substantial training period before scientific management could be implemented, Edouard Michelin, like most French and American entrepreneurs interested in Taylor's ideas but preoccupied with the bottom line, scoffed. Except for the sporadic use of stopwatches, little about the Milltown and Clermont operations reflected Taylor's ideas in 1912, and the same would be true in 1914.[20]

During World War I, Michelin began to use an assembly line, that brainchild of Henry Ford, in the production of gas masks and then Breguet airplanes, much as Citroën and other munitions and arms manufacturers installed assembly lines in their plants.[21] After the war, Michelin installed a line for the production of inner tubes and then began an intensive series of time-motion studies. In adopting Taylor's ideas, Michelin resembled several other French firms in rapidly evolving manufacturing sectors, namely mining, munitions, and the new chemical and electrical industries.[22]

What came to distinguish Michelin was not the adoption of rationalized production and Taylor's scientific management itself but the company's quite vocal support for Taylorism. While Michelin was of course not alone in calling for the adoption of Taylor's methods, the company took a very public role, one more public than that of any other French firm.[23] In the period 1921–25, Michelin propagandized Taylor's ideas of organizing work, in essence following on the heels of Henri Le Châtelier's publications from before the war. Le Châtelier convinced Michelin to bankroll a committee whose charge would be to spread Taylor's ideas in France. The committee worked to ensure that the engineers trained in the elite Grandes Ecoles for French engineers were specifically trained in Taylor's methods. In addition to sponsoring presentations on Taylor's ideas at the Ecole des Ponts et Chaussées, the Ecole des Mines, and the Ecole des Arts et Métiers de Paris, Michelin funded students' training in how to do time-motion studies, the centerpiece of Taylor's scientific management. In addition, the committee sponsored 2-week internships in companies that employed Taylor's methods. Michelin even organized a study trip to the United States, spending more than 80,000 francs in 1922–23 and 100,000 francs the following year. In 1931 the company claimed that under the auspices of the Comité Michelin, 860 students of Grandes Ecoles had done internships in "Taylorized factories" and that 300 prizes, representing a total of 135,000 francs, were distributed to the authors of the best individual studies done during these internships.[24]

Despite later claims, the Comité Michelin does not appear to have had the desired success. Students at the Grandes Ecoles remained generally uninterested in the specifics of time-motion study and the detailed analysis of industrial work. Among students doing internships, one actually hauled in an easy chair (*fauteuil*) to sit in while using a stopwatch to time a worker's movements. Another was spied reading a newspaper in the workshop. In the end, only three or four of the students went on to use the skills learned during the internship to reorganize the production of other French firms. In the mid-1920s, Michelin ended its support of such attempts to reach elite engineering students and settled on subsidizing an association of engineers interested in Taylor's methods.[25]

After 1925, Michelin attempted a new strategy, shifting its focus to publicizing Taylor's methods and their use in the form of free (for one copy) and low-cost (for multiple copies, with the price falling as quantities increased) pamphlets, targeted largely at the management and workers of other French businesses. Many of these pamphlets appeared as part of the well-known company series called *Prospérité* [Prosperity] that appeared as a periodical in the late 1920s and 1930s. By 1931, the company had already produced ten such brochures with a combined circulation of nine hundred thousand.[26] Each pamphlet considered a separate angle or argument for adopting Taylor's methods to increase production. Designed for as large an audience as possible and including illustrations, the pamphlets explained why French firms needed to adopt Taylor's methods. The company frequently noted the importance of either competing with the Americans or producing the consumer goods available to Americans. While Taylor received the credit for transforming American society, Michelin tried to spark interest in his rather technical methods by associating Taylor with mass consumerism. Even the assurance that Taylorism would avoid labor conflict included an implicit notion of mass consumerism by promising workers higher wages, lower prices, and thus greater access to consumer goods.

In "Comment nous avons taylorisé notre atelier de mécanique d'entretien," the Taylorization of Michelin's internal mechanical maintenance workshop was at issue. Whereas production itself could be most improved by use of assembly lines, the manual labor of maintenance could be made more efficient only through the better organization of materials and the orchestration of workers' movements. In addition to laying out the organization itself, the brochure justified why readers should take an interest in such technical matters. "We are writing this brochure because we believe that it is in the application of Taylor's methods that we will find the salvation [*salut*] of French industry. We know to what extent these methods have penetrated American industry. We know that it is thanks to them that the Americans produce automobiles, automatic machines, sewing machines, and typewriters, etc. in large quantity and cheaply. And it is thanks to the low cost of production that they can sell without difficulty everything that they produce." Thus, while associating a specific method of production not just with Taylor but with American mass production, Michelin traced the future deliverance of French industry in appropriating Taylor's ideas. Michelin did not, however, forsake its own loyalty to France in the process but instead reasserted it. The brochure continues, "we are persuaded, and it was Taylor's opinion: no people is better gifted than the French to apply these methods. If Frenchmen—bosses, engineers, and workers—work together, they will be superior to the Americans themselves [since Taylor died in 1915, he could not have argued otherwise]."[27] That is, France needed to adopt Taylor's "American" ideas in order to compete with and surpass the Americans.

In "Pourquoi et comment chronométrer," Michelin laid out why French firms needed to undertake time and motion studies. To oversimplify the message, Bibendum appears on the inside of the front cover holding a stopwatch with money flowing out. The caption reads, "Le temps, c'est de l'argent [Time is money]." The pamphlet proceeds to note who can do the time-motion studies. According to Michelin, after being trained himself, the boss (*patron*) or director could then train the operator of the stopwatch, whose most important quality was to "love progress," that is, who was convinced that "method is superior to routine."[28] In "Sur le tas, ou Conseils pour débuter dans la méthode Taylor" [To the task, or counsels for beginning with Taylor's method] Bibendum describes for an engineer all of the advantages of Taylor's method, providing four examples of gains in productivity as a result of Taylorization. Bibendum concludes with a pronouncement of full faith in Taylor: "If you can Taylorize yourself, you can Taylorize anything."[29] For Michelin, there was no question about the necessity of adopting Taylor's ideas as quickly as possible, lest it lose ground to the Americans. As Edouard Michelin told his engineers, "We have to go fast. We have to gallop. Our direct competitors are the Americans . . . [who] do not waste time. . . . We have to do the same thing here."[30] While associating the notion that time is money with the Americans and calling for its adoption in France, Edouard Michelin nevertheless maintained that France was quite capable of "galloping" as well as the United States.

Even Michelin's account of its construction of subsidized housing for workers credited Taylor with substantial savings of labor and thus costs. As the first French company to adopt American practices of standardizing building designs and materials, organizing production into successive teams of unskilled workers directed by an experienced one, Michelin appears to have set a precedent later adopted in the construction of low-cost, subsidized public housing after the passage of the Loucheur law in 1928.[31] Inspired by the American builder Frank Gilbreth, follower and sometime nemesis of Taylor, Michelin used time-motion study to facilitate bricklaying and faucet installation, among other tasks.[32] In "Deux exemples d'application de la méthode Taylor chez Michelin," the company claimed that "we were able to organize work in such a way that our bricklayers get less tired, earn more, produce more, and our houses can be rented more cheaply." Michelin then scrupulously documented time, wages, and savings in order to prove that "thanks to the use of Taylor's method in the construction of workers' houses, we have been able to save 50% of labor costs." [33]

By increasing productivity in the factory as at building sites, Michelin argued that wages could be raised while profits increased, thus avoiding labor conflict. This was a central tenet of Taylor's philosophy of management. Unlike the vast majority of French industrialists, including Citroën and Renault, Michelin did not assume that the primary advantage of rationalizing French industry would be to lower prices. [34] Rather, Michelin seems to have accepted the notion that increased salaries were a fundamental part of the adoption of Taylor's "mental revolution." Using the example of a presumably typical French housewife (*ménagère*), enthroned under the words "Honni soit qui mal dépense [Evil to her/him who spends badly]," an alteration of the royal "Honni soit qui mal y pense," Michelin claimed that "it is the client who is King" whom employer boss and worker both must "obey." If both worker and employer work together to reduce the costs of production, "her majesty [the consumer] pays less." The result is that "the worker earns more"; "the employer sells more"; and "everyone is happy." [35] Although careful assessment of Michelin's accounting even within the *Prospérités* shows a greater proportionate gain for the employer than for the employee, even while reducing the sale price, the firm did consistently advocate higher salaries for workers as a result of gains in productivity.

In general, Michelin's claims in the brochures corresponded with the firm's own practices. Within the Michelin factories, the company used comparatively high salaries by the 1920s in order to recruit and retain workers, in a sense practicing the high wage rates that it preached to other firms, making up for the increased cost by increasing productivity yet further. The average hourly wages at Michelin throughout the 1920s and 1930s were considerably superior to those of other rubber firms in Clermont-Ferrand, up to one franc more per hour, and nearly matched daily wages within metallurgical (including auto) industries in the Paris region, where the cost of living was higher. [36] Michelin's need to recruit a workforce that grew from 10,385 in 1919 to 18,000 in 1926 and to minimize turnover were certainly factors that led to Michelin's higher wages.[37] Nevertheless, Michelin's abrupt change from the first decade of the twentieth century, when the company's salaries lagged behind those of its Clermont competitor Bergougnan, makes it clear as well that Michelin deliberately paid higher wages after the First World War in order to attract better employees more capable of increasing the firm's own productivity.[38] Edouard Michelin admitted as much during the shrinkage of Michelin's export markets in the late 1920s, when he rejected the suggestion that all salaries be cut, noting that it was better to pay more for an elite workforce capable of constantly increasing the productivity of the firm.[39]

Michelin did not institute piece rates according to Taylor's model largely because the firm had already established a system of bonuses having a somewhat similar effect. As early as 1898, the company instituted a profit-sharing plan that annually rewarded employees a portion of profits, amounting to as much as six weeks of salary. The money earned an interest rate up to 2 percent higher than that of a savings account. Initially, only a small portion of the workforce qualified, but by the late 1920s as much as 40 percent of the workforce received such bonuses. Employees could not, however, have more than one-quarter of the amount until retirement or separation from the firm, provided that the employee did not undertake work at one of Michelin's competitors (in which case, the remaining three-quarters was forfeited to the company). Trade secrets could thus be kept, and employees encouraged to stay at Michelin, a system that does not seem to have incurred the wrath of workers, at least not before the layoffs of the 1930s. As a result of the Depression, workers laid off from Michelin could not then find work at other rubber firms without losing their profits earned at Michelin. In the strikes of 1936, a change in the profit-sharing system was one of workers' primary demands of Michelin; Michelin abolished the requirement against working for a competitor, paying the profits to anyone over fifty who left the firm.[40] Michelin thus managed as early as the turn of the century to encourage employees with monetary benefits in addition to salary; surely, those workers most open to Michelin's notion of productivity were the most likely to gain the right to share in the company's profits. The company provided an inducement to increase productivity for the long-term health of a firm in which workers had, quite literally, a vested interest. Whereas Taylor's bonuses offered an immediate incentive for workers, Michelin offered a comparable long-term one that further bound the employee to the company.

Michelin took its focus on productivity to the point of setting up a Service of Suggestions in the Clermont plant in 1927, making it one of the first French firms, along with Renault and the Chemins de Fer de l'Etat, to use workers' suggestions to further increase productivity.[41] The practice coincided closely with the decline of Michelin's and French auto makers' export markets after 1926. In "Suggestions: Comment nous avons amené notre personnel à collaborer avec nous à la recherche des progrès et des économies," Michelin defined the practice as a model for other French firms. Michelin encouraged any employee with an idea for eliminating waste or generally improving the production process to write the idea down. Engineers charged solely with handling suggestions were to help workers otherwise hesitant for fear that another employee might be punished for an earlier oversight, that colleagues might make fun of the worker, or that a foreman might feel threatened. As for the fear that someone might lose a job if the company became more efficient, Michelin responded that if the factory could not remain prosperous, it would be beaten by the competition and would need to lay off workers.

Suggestions ultimately implemented brought a payment from the company to the worker. The stick accompanied the carrot for those who preferred not to make suggestions; the brochure on suggestions noted that an employee seeing waste and not saying anything "is a bad egg [*un mauvais esprit*] and we are not interested in even a punctual and reliable employee who puts up with obvious waste, and he won't last long in our shop." According to the company a total of 82,029 suggestions were made between 1927 and 1933. In the first year, 30 percent were considered good, while in 1933, 55 percent were good, and thus received the bonus.[42] The amount to be received was determined by engineers who assisted the worker in defining how a change in procedure might produce a savings. The foreman then gave the payment to the worker, a practice that, Aimée Moutet maintains, reminded the worker of the hierarchy of the firm and generally kept the foremen involved in the process.[43] Within the company Edouard Michelin frequently repeated to his employees the importance of suggestions, reworking the rules in order to make it easier for workers to offer them.[44]

At the end of the brochure "Suggestions," Michelin laid out the full value of suggestions for the firm; above all, suggestions helped to reinforce a certain mental unity, an *esprit* within the company that, Michelin hoped, maintained the close relationship between ordinary workers and upper management in an enterprise too large for Edouard Michelin to know personally the employees. "The savings and measurable progress that have resulted from the suggestions that we have received more than justify in our eyes the efforts that we have made. But in a workshop or an office where one makes regular suggestions, we believe that the state of mind created has worked to develop significantly the three following ideas: 1) Progress must be continuous; 2) There should not be impenetrable compartments in the factory. Everyone works for the same boss. Or more precisely for the same client, that is the buyer of Michelin tires; 3) We do not have the right to make the buyer of a Michelin tire pay for unnecessary paperwork, an unnecessary man, or an unnecessary supervisor."[45] Clearly, the Michelin company was supposed to be a united block, taking on waste in materials or labor during the production process. A paternalistic firm in several respects, Michelin nevertheless stamped that paternalism with a notion that Taylorism was what Taylor himself had called a "mental revolution."[46] Suggestions encouraged, even prodded, employees into thinking about the larger production process, reducing waste and increasing productivity.

In "Cela vaut-il la peine de s'occuper de la méthode Taylor," and its later reworking as "Aux dépens du gaspillage," the company laid out why the elimination of waste was crucial for increasing productivity. It began with the two sides of a potential labor dispute. On the one side, "workers want a fat paycheck. Put yourself in their place. They are not wrong, for the slum is a terrible thing and the furniture even worse; what could be worse for a married man than to be in a situation where it is impossible to care for his sick wife or raise his children?" On the other side, "the owners want a low price of labor." According to Michelin, if you put yourself in their place, you would realize they must keep their prices low to maintain demand for their products. The resolution of the differences is not obvious: "Thus, some want to earn more and others pay less. How do we get out of this contradiction?" Michelin considers Taylor's attempts to cut time from the production process as the elimination of waste. The results resolved the supposed conflict between worker and boss, "Thus we repeat it because it can't be said too many times, the better the pay for the worker, the better the profits for the owner. Owner and worker have become united [solidaires]; they march to the beat of the same drummer; their interests are no longer opposed."[47] Conflict between labor and management could thus be resolved by cutting wasted labor. The reduction of waste within industry had been popular among American industrialists (not to mention the American secretary of commerce in the 1920s, Herbert Hoover) and several French entrepreneurs including Louis Renault, but Michelin's spin was decidedly both Taylorist and solidarist: less waste would lead to greater productivity, and greater productivity would allow for both higher salaries and higher profits, not to mention lower prices.[48]

Michelin's constant, self-proclaimed "struggle against waste" (*lutte contre le gaspillage*) moved beyond the focus on manpower to include the elimination of seemingly insignificant material waste in production. In "Sam et François," Michelin compiled minute statistics to illustrate how a firm might further shave savings off the cost of production. "François," pictured as a cobbler, can save much money by cutting as many soles and heels out of a piece of leather as possible. "Françoise," a secretary, can save by not using a whole sheet of paper when a half-sheet will do. In each case, Michelin calculates the total lost. "Françoise thus wastes 10 half-sheets each day, 1,500 each year. In a factory that employs 100 secretaries, this waste costs 1,500 francs each year." The lesson continues with the example of soap, nicely illustrated on the same page. "François believes that he can wash his hands better by using a lot of soap while half would suffice to get the same result. Even if he only wastes 5 grams of soap each day, that is 1 kilo, 500 grams each year. If there are in François' factory 2,000 workers who do the same thing, that is 3,000 kilos of soap wasted in the year, **an unnecessary expense of 6,000 francs** [bold in the original]." In another example, "François never turns out the light when leaving the workshop. It burns unnecessarily for an hour. It is a form of waste easy to avoid [in turning out the light]." The calculation follows, "the light costs, in one hour, about 4 centimes, or 12 francs a year. If in an automobile factory 5,000 lights are left on for one hour each day unnecessarily, that wastes 60,000 francs a year. **That is the price of 3 new automobiles** [bold in the original]." Through these intricate calculations, the reader was supposed to realize just how crucial the elimination of waste was for gains in productivity to be achieved.[49]

As inducements for other French businesses to accept Taylorism, it is altogether unclear whether Michelin's brochures were any more successful than the Comité Michelin had been. At least one railway company, the Compagnie de l'Est, adopted Taylorist methods as described in *Prospérité*. Michelin's influence was often more direct: the company pressured both suppliers and clients to accept some rationalization of their production processes. Ironically, it was during the 1930s, when laid-off Michelin engineers sought work in other industries, that Michelin's own scientific management seems to have spread most easily to other companies. Other firms accepted at least some aspects of Taylor's methods, although not necessarily as a result of Michelin's efforts. Barring certain innovative sectors, however, most French companies remained relatively small and did not embrace Taylorism, which could have been partially adapted to small-scale enterprise. Even the major automobile producers did not undertake the kind of detailed, systematic study of the organization of work that Michelin promoted.[50] In the end, one of the most important effects of Michelin's brochures on Taylor may well have been as advertising among the managers and workers of France. A company that had long fashioned a certain national image for itself by promoting tourism, aviation, and pronatalism now associated itself with a modern vision of France informed by, yet distinct from, the modern America of Taylor and of the omnipresent automobile.

ADVOCATING THE AUTOMOBILE

In the Belle Epoque, before learning much about Taylor or Ford, Michelin had been a forceful champion of the future of the automobile. Lobbying for the numbering of roads and the promotion of automobile tourism were the focus of Michelin's prewar efforts to foster the development of a market for cars and thus tires. But before the war, the French market for automobiles had been limited to the deep pockets of the upper bourgeoisie. After the war, potential buyers grew quickly to include better-off rural artisans, farmers, shopkeepers, and assorted tradesmen. Those unable to buy a car increasingly could choose to take a bus, which usually had pneumatic tires by the late 1920s (Parisian buses were equipped with pneumatic tires beginning in 1924).[51] As a result, the tenor of Michelin's promotion of the automobile changed. While continuing earlier tourist initiatives designed for the elite few, Michelin also claimed to represent the entire spectrum of motorists, most of whom did not do much touring but instead used their autos and trucks for business. Michelin's publications lost their literary references as well as their more overt references to social class and to differences between urban and rural populations. Combining its interest in increased production à la Taylor with its long-term desire to encourage as large a market in automobiles as possible, Michelin attempted in the interwar years to convince manufacturers, the government, and individuals that a mass market for automobiles could and should emerge in France. In the guise of providing a public, even national service, Michelin was a vocal promoter of the automobile. In the process, the company continued to associate itself simultaneously with the progress that America supposedly represented and with the French nation, melding the two into an image of a modernized France that could presumably emerge once France had as many cars per capita as the United States.

Although Michelin's interwar publications rarely mention Henry Ford himself, references to Ford and his autobiography abound in the typewritten transcripts of Edouard Michelin's pronouncements within the factory.[52] Given that Ford motorcars were considered imports by Michelin in its publications (even when assembled in France) and that they did not regularly come equipped with Michelin tires or wheels, Michelin clearly had cause to accept the message but to avoid repetition of the messenger's name.[53] Yet without using the name, Michelin's brochures nevertheless related a vision of how to create a market for automobiles by lowering the cost of production, raising wages, and selling ever more cars. "Sam et François" (1927), the first brochure to include the word *prosperity* in the title, combines visions of mass consumption as well as Taylorized production. In "Sam et François," Michelin began by comparing a typical French worker, "François," with a typical American worker, "Sam." First, the company established its credibility. "A lot of people talk about America without knowing anything about it. That is not the case with us. We build tires in Clermont-Ferrand. Our subsidiary, the Michelin tire company, builds them in Milltown, near New York." Using its subsidiary in Milltown as evidence that the company knows of what it writes while reinforcing the connection between Michelin and America, the pamphlet notes the gap in the standard of living between French and American workers, embodied by "François," earning 45 francs daily and "Sam," earning 7 dollars.

By saving 2 days of salary, François can buy a pair of shoes for 90 francs, whereas Sam can buy a pair of shoes for 6 dollars, a shirt for 2 dollars, and 2 hats for 3 dollars each. By saving 43 days of salary, François can buy a small motorcycle for 2,000 francs, while Sam can afford a car that seats 4 for 300 dollars. If each saves 500 days of salary, François can buy a small house with 2 rooms for 22,500 francs whereas Sam could buy a cottage with 4 rooms, central heat, and a bathroom for 3,500 dollars. Thus, "Sam, compared to François, is rich."[54] The remainder of the pamphlet points out why Sam is richer than François, focusing on less waste in the American production process and higher overall productivity.

The company's arguments in favor of the automobile itself abounded; everyone could live a better life if equipped with a car. "Cheval et auto" [Horse and auto] contrasts the lower price of maintaining a car with the price and upkeep of a horse, attempting to convince any farmers still reluctant to purchase a car or truck. An intricate calculation of all of the food, shoeing, veterinary costs, and taxes for a horse resulted in the figure of 10 to 12 francs daily to maintain a horse. An automobile's gas and oil, "Confort" tires (produced by Michelin), repairs, tune-ups, insurance, and taxes gave the total cost of 6 to 9 francs daily. Most of the brochure, however, consists of one-page testimonials from farmers who have made the switch, with their names, towns, and departments listed in order to reinforce the credibility of the sources. Not surprisingly, the advantages of the automobile or truck were supposedly legion: they included greater access to markets to sell products, the possibility of saving the time normally spent on chores to maintain horses, easier access to the doctor in case of one farmer's son being injured by a horse in the fields, quicker delivery of fruit and milk, easier attainment of replacement parts for implements during harvest time, quicker access to the veterinarian (still necessary for draft animals in the fields), the possibility of selling directly to consumers in towns without working through a middle man, and—lest leisure be forgotten—the ease with which one might spend Sunday in town rather than at home, letting the horses rest after a hard week.[55]

Michelin continued to expound on the value of automotive transport for agriculture in "L'automobile, source de la richesse: L'exemple de Aveyron." Aveyron, a department situated between Auvergne and Languedoc, contained a poor, agricultural region known as the Ségala. The brochure recounts the words of a local peasant, named François Lacombe. Lacombe remembered that in 1900

you would not have found me sitting at the table after a good dinner, with coffee, cake, and a glass of rum. It was tough. We lived in an old, one-story house with a kitchen and two rooms. There were three pigs in the basement. Adjacent to the house was a stable with three cattle and fodder, with a big pile of manure in front of the door. I did not have more or less

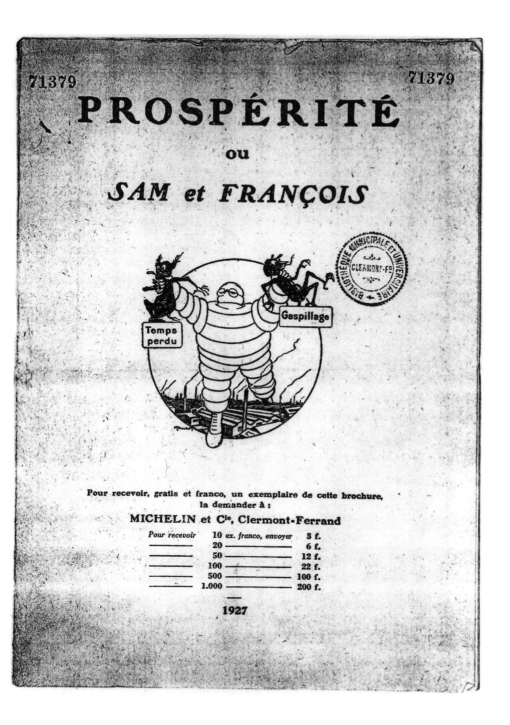

22. "Prosperity, or Sam and François" relies heavily on the comparison between mythical American and French workers. On the cover, much like Henry Ford and a host of American industrialists, Bibendum has rooted out "lost time" and "waste," veritable cockroaches preying on the Clermont factory at his feet. "Prospérité ou Sam et François," 1927.

La rentrée des ouvriers

à l'usine de Clermont-Ferrand. *à l'usine d'Amérique.*

23. By picturing the entrance of Michelin workers both into the Clermont-Ferrand factory and into the company's Milltown, New Jersey plant, Michelin gained credibility in its descriptions of the reasons why American workers had greater spending power in the 1920s. The illustration shows the difference: whereas only two automobiles were entering the Clermont works, most Americans had cars. As Michelin proclaimed, "the American worker is rich if one compares him to a French worker." "Prospérité, ou Sam et François," pamphlet, 1927, p. 2.

property than today: 11 hectares [26.5 acres] in 26 parcels. The Lacombes have worked it from father to son forever and forever with the same methods. Half of the land was wasteland [*étaient en landes*]. The rest gave us some potatoes, a little rye, and some oats. What did we eat? Rye bread that we cooked in the house every two weeks, potatoes, cabbage soup for every meal, and a little bacon at the noon meal. Some wine and meat, once yearly, for the local festival. The land in those days could not support large families.

Since only rye could be raised for market, there were real limits to the agricultural diversity in the Ségala, a term that literally meant land of rye. All of that changed because trucks could haul large amounts of lime to local fields, something impossible earlier due to the distance of the railway station and the practical impossibility of hauling it by cart or wagon. Lime, soon accompanied by volcanic ash, nitrates, and sylvanite, enriched the soil. Improved soil made it possible for Monsieur Lacombe to grow 19 quintals (1 quintal equals 100 kilos) of wheat, 180 quintals of potatoes, and 70 quintals of hay. He could maintain 16 cows, 7 calves, and 16 pigs. Moreover, the expansion of bus routes eliminated the isolation of the area, making it possible to haul goods to market and to purchase consumables more easily. The brochure then describes the transformation of the area at large, which could grow more potatoes than ever and could now effectively grow wheat; the average peasant of the Ségala who had produced 7 hectoliters of rye could now grow 24 hectoliters of wheat. The brochure ends by reminding the reader that Aveyron, transformed from a *pays pauvre* (poor region) into a *pays riche* (rich region), is the model for other areas, provided that the automobile is not encumbered by unnecessary regulations or taxes, "because if peasants pay too much for their lime, the Ségala will return to the Ségala of the old days and progress will be killed."[56] In short, according to Michelin, the automobile had the potential to transform the rural world, if only the French state would let it. Here the battle is clearly on two fronts: one objective is to convince potential users, in this case farmers, and the second is to put pressure on the French state to minimize regulations and, in particular, taxes on the automobile and its fuel.

Michelin also made the argument that anyone needing to travel for business would profit from use of a car. In "Ce que l'auto coûte réellement: Les bénéfices qu'elle procure," the company reported on the basis of thirty-one of its own traveling employees the cost of maintaining either a 5 cylinder or a 10 cylinder automobile, depending on the number of kilometers driven. Costs included the interest that the tied-up capital might have been earning, taxes, insurance, depreciation, gas and oil, repairs, and tires. The brochure then describes the amount of business that might be gained by being able to reach towns not served by the railroad, about 50 percent of towns in rural areas, and the time saved in comparison with the use of public transport. The brochure is careful to note the additional costs of meals away from home and hotel rooms due to the slowness of the train and bus when compared to the car, claiming—after the usual intricate calculations of all possible expenses—that the personal automobile permitted the augmentation of profits by 39 percent. For the timid or hesitant, the brochure then makes the argument that one could learn to drive a car in a week and that cars were, contrary to common belief, quite safe, with the risk of having an accident only once in every twenty-seven years. The final argument is that of convenience, with two possible scenarios nicely illustrated for the reader. The first has a man, alone in a hotel room with an alarm clock waking him at 5:00 A.M. as he jumps out of bed. The second has a man in bed with his wife, apparently sleeping, while the baby lies asleep in a nearby cradle. The caption reads, "With the car, you can more often enjoy the pleasures of family life."[57] It is a theme that runs through Michelin's pro-automotive literature: the automobile will not only make you more money but it will also improve the quality of your life. A similar brochure, "Voyageurs de Commerce," also showed how the automobile would make businessmen's working lives both more profitable and easier, tracing on a map two trajectories: the limited one represented those towns accessible by train and the complex one the towns that one could reach by car. Clearly, the automobile was supposed to generate more business.[58]

After the Depression hit France in 1931, Michelin still had arguments for why people should buy cars. In "L'auto contre la crise," a reader could learn that the best way for small businessmen to combat the crisis was not to sell or mothball their vehicles in a period of retraction but instead to use them for home deliveries, since customers were venturing out less frequently. Those small shopkeepers without trucks or cars needed to buy them if they wished to have any customer base whatsoever, given that in the economic crisis rural folk were reported to have stopped coming as frequently to nearby towns in order to shop. As in other brochures, Michelin included the inevitable testimonials of individual shopkeepers with their addresses for any skeptical reader who thought Michelin might merely be pursuing its own self-interest. Across the board, Michelin had an argument for why virtually everyone was better off with a truck, a car, or a bus. Cobblers, grocers, jewelers, lumberyard owners, accountants, leather sellers, fruit sellers, bakers, cinema owners, and hairdressers all attested to the usefulness of loading their products or equipment into trucks in order to provide services in or near customers' homes.[59] There was, according to Michelin, no reason not to continue to extend automobile usage despite, or even as a result of, the Depression.

All told, reluctant consumers were not Michelin's greatest *bête noire* in the struggle to increase usage of automobiles in interwar France. The earliest Michelin brochure considers the increased taxes proposed in the French parliament in 1923, declaring in the title that "On veut tuer l'automobile" [They want to kill the automobile]. *On* in this case is the French state. Michelin claims that the problem was not that insufficient taxes had been collected on automobiles, gas, and oil, but that they were diverted to purposes other than the maintenance of roads, hence a proposed law to increase further the tax on automobiles. The pamphlet, which was free, was an effort to mobilize as many as possible on behalf of a product that did not at the time belong to many French people, as Michelin's list of affected parties who should be concerned makes clear. The list stretches those with an interest to include "all current car owners; all of tomorrow's car owners—those who would buy a car if tax relief permitted the realization of the popular automobile; all French who realize the social and economic role of the automobile; [and finally, the trump card in the wake of the war] all the patriots who know what importance the automobile has for national defense, what services it fulfilled in the last war, and what hopes army leaders have for it in the future."[60] With its usual penchant for statistics, Michelin attempted to show with careful calculations that adequate taxes were already being paid and ends the brochure with a *formule de protestation*, a form letter that local groups could copy, sign, and mail. Michelin distributed some six hundred thousand copies of the brochure to anyone who would take one. The proposed law was defeated.[61]

In the late 1920s and particularly the 1930s, new taxes on automobiles and gasoline increasingly made the French state Michelin's primary whipping boy. Brochures, including those that were part of the *Prospérité* series, show the evolution over time. Initially, as in the pieces on Taylor, automobile producers

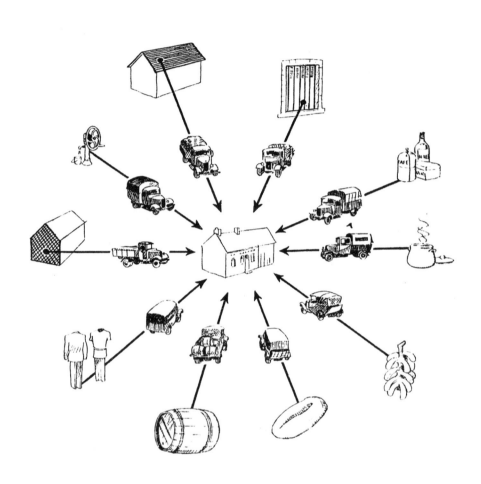

24. Claiming that small businesspeople needed to buy more, not fewer, automobiles during the Depression, Michelin illustrated all of the products that French households could buy from suppliers capable of delivering. They range from barrels of wine, loaves of bread, bananas, and coffee, to building materials, clothing, and water pumps. As the legend pointed out, "everything that they buy is brought by guys with automobiles [a term that could also mean truck]." "L'auto contre la crise," 1935, p. 16.

and potential buyers were guilty of having clay feet. In 1927, "L'avenir de l'industrie automobile française" laid out Michelin's dream: to have as many automobiles per capita as the United States. "If France followed the example of the United States . . . annual production in France would in five years be 800,000 autos. . . . This annual production would represent: an invested capital of 9 billion; an annual sales of 13 billion; and for 600,000 workers and employees, a salary of 8 billion." According to Michelin, if producers lowered prices and implemented scientific management, France could achieve this goal. Michelin quoted "one of the major builders of automobiles in America: 'Don't say we have a lot of autos because we are rich but that we are rich because we have a lot of autos.'"[62] The vision and words were of course those of Henry Ford, even if his name does not appear.

By the early 1930s, however, the French state was the guilty party. After 1928, Michelin offered a special issue of *Prospérité* each year called "Des faits et des chiffres," which gave facts about automobiles in France and elaborate statistics about production, ownership, and taxation levels with some comparative information about other countries, particularly the United States. In 1933, the company notes that the "facts and figures presented in this brochure do not really justify the title '*Prospérité.*' . . . "*If the auto is to keep its brilliant promises for the future, the taxes that it supports must cease to increase* [emphasis in the original]." The company warned that taxes had risen in both 1932 and early 1933, asking whether "they want to kill this magnificent branch of our national economy? . . . The automobile, an essential cause of prosperity must live! [underlining and emphasis in the original]."[63] Michelin made the claim that automobiles were virtually indispensable for the French nation:

There are in France this year 200,000 more automobiles than last year, and that in spite of the [economic] crisis. Why? Because there is not one single Frenchman out of 40 million who does not need a car. Twenty-five years ago, only the rich drove around in cars; today, everyone uses the automobile. The auto is thus not a luxury; it is now an essentially democratic tool for work. When they put their hands on cars, they hurt everyone and everyone consists more of workers than of rich people. Everything that can get in the way of the automobile—be it the grip of government, excessive regulation, or taxes that are too high—will be detrimental not to a small group of Frenchmen but to the entire population.[64]

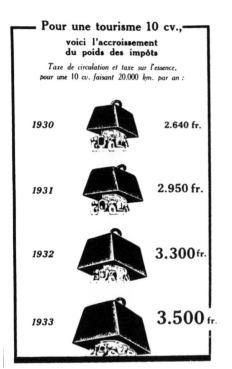

Pour une tourisme 10 cv.,

voici l'accroissement
du poids des impôts

*Taxe de circulation et taxe sur l'essence,
pour une 10 cv. faisant 20.000 km. par an :*

1930   2.640 fr.

1931   2.950 fr.

1932   3.300 fr.

1933   3.500 fr.

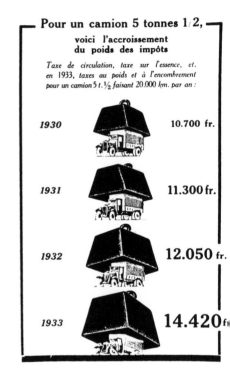

Pour un camion 5 tonnes 1/2,

voici l'accroissement
du poids des impôts

*Taxe de circulation, taxe sur l'essence, et,
en 1933, taxes au poids et à l'encombrement
pour un camion 5 t.½ faisant 20.000 km. par an :*

1930   10.700 fr.

1931   11.300 fr.

1932   12.050 fr.

1933   14.420 fr.

25. In several of the interwar pamphlets, Michelin criticized the French state for high taxes. By the early 1930s, Michelin frequently illustrated governmental attempts to generate revenue by increasing taxes on motorists, particularly in the form of the gas tax. Here both the automobile and the truck were quite literally crushed by tax hikes. "Des faits et des chiffres," *Prospérité* 5 (July–September 1932): 8–9.

Grossly overstating the centrality of cars, trucks, and buses for all sectors of society, Michelin thus attempted to construct a reality that corresponded to its business interests. In the brochure as a whole, the company repeatedly asserts that automobiles were not only in widespread use but also, as the quotation above illustrates, were downright democratic. The position is an interesting one, because the company was clearly lobbying the French state with a populist message that a nonessential material good—one that all people had lived without a generation earlier and only the well-off actually owned in the 1920s—was now presumably crucial for the nation as a whole. In the brochures, Michelin repeatedly claimed that the automobile was "democratic" because in one sense or another it affected the lives of everyone, even though it was of course only automobile owners who actually paid the taxes.[65] The company's constant references to the United States, that huge democratic nation to the west, served two purposes: they linked democracy with the automobile, thus bolstering Michelin's argument for low taxes; and they stoked French nationalism by constantly pointing out that France was in most respects second only to the United States.

In its attempts to prod the French government, auto makers, and potential buyers into following its lead, Michelin's greatest strength may well have been the facts and statistics that it so artfully presented to the public. Before the First World War the company constantly exhorted users of its tourist office and readers of the "Lundis" and red guides to write the company with information, questions, or comments. Immediately after the war, envisioning a bigger market for automobiles, Michelin changed its tack. In 1923, the company did what was in essence an early market survey in an effort to gauge the potential market for automobiles. At root, Michelin wanted to know how powerful, how big, and yet how cheap a car would need to be in order to entice those without cars to take the leap. In posters announcing the survey, the model was of course the United States, where "there is one car for 10 inhabitants" whereas in France "there is one car for 150." Using a national logic not unlike that employed by tourist groups after the war, Michelin noted that "if, instead of 260,000, France possessed a million cars, proportionately that would still be four times less than the United States, it would be easy to imagine the *élan* that all of the country would experience. The economic crisis, born of war, would be quickly overcome. *That is in the national interest* [in huge letters in the original]."[66] Michelin mailed out some 4 million copies of the "enquête nationale de l'automobile populaire," including some 220,000 to known automobile owners, 200,000 peasants, 160,000 hotel owners, 90,000 small business people (*commerçants*), 80,000 entrepreneurs, 75,000 salesmen, 65,000 teachers and clergymen, 60,000 professionals, and 40,000 automobile and bicycle dealers. The company also sent out some 30,000 posters advertising the survey.[67]

Louis Baudry de Saunier, that longtime promoter of the automobile and friend of the Michelins, summarized the results in *L'Illustration*. The consensus seemed to be that there would be a sizable market for an automobile for four people, even one that could not surpass 100 kilometers per hour, provided that it did not cost more than 7,000 francs. That was a price far below anything produced by French auto makers. Even Citroën's early adoption of the installment plan did not alleviate the problem because people did not earn enough over time. The value of the survey was clear; the French public did not yet have the resources to buy automobiles en masse.[68] Similarly, the yearly "Faits et des chiffres" (1928–35) provided a gold mine of information—still used by historians—about automobile production, exports and imports, social and regional breakdowns of ownership, as well as the costs (including of course taxes) of owning a car in France.[69]

Clearly, Michelin propagated a Fordist vision of the future of France, although before the mid-1930s the vision appears to have been premature. The market for automobiles did grow dramatically in interwar France, but auto makers did not believe there to be a market for a smaller, less expensive French variant of the Model T. Wealthy buyers expected innovation and styling, while the less well-heeled could always buy bigger used cars. Michelin itself had published "L'auto d'occasion" [Used car], to convince people unable to afford a new car that there was much value and reliability remaining in used ones.[70] Only in 1934 did Fiat, through its Simca subsidiary, announce the sales of the first small vehicle priced under 10,000 francs and targeted at a mass market. Not coincidentally, as Michelin took control of Citroën in 1934, it conducted a series of market surveys. The new president and vice president of Citroën, Pierre Michelin (son of Edouard) and Pierre Boulanger, became convinced that there was a market for a small, utterly functional, underpowered four-seater, provided that vast changes in design and production lowered costs. Like the small popular cars under study at other French auto makers, the resulting Citroën model, the Deux Chevaux, would not make its debut before the French public until after the war.[71]

MICHELIN

L'auto d'occasion

Cette brochure
est envoyée gratuitement
sur demande adressée à
MICHELIN & Cie
Clermont - Ferrand

—

1927

26. Unlike auto makers themselves, for whom the used car market was problematic, Michelin encouraged the buying of used cars, featuring testimonials of satisfied buyers. As Bibendum states, "the used car is within reach of *everyone*." As if to illustrate that point, he leads these little people, who appear to include petty bourgeois men, a peasant, and a priest, driving little toy cars. Interestingly, the problem for auto makers desirous of building a small car in order to increase the market for cars was the existing stock of bigger, more solid used ones, a particular problem in France because of its early lead in automobile production. "L'auto d'occasion," 1927, front cover.

Although the mass market in automobiles sought by Michelin did not develop in France until after World War II, throughout the interwar years, by calling constantly for more automobile purchases, Michelin aggressively attempted to create a larger market for automobiles and thus its tires. In the brochures, Michelin focused on the potential usefulness of cars, trucks, and buses for increasing profits and daily convenience. Wielding statistics of car ownership and testimonials of car owners, Michelin attempted to prove how important the automobile industry was to the very future of France by holding up the United States as the ideal. If France could just be more like the United States, the brochures seemed to say, France would be better off.

MODERNIZING FRANCE

Despite the images of America offered by Michelin, the company did not suggest that the automobile necessarily brought with it any change in the existing social order within France. France could have a mass consumer market without threatening preexisting social roles, particularly between the sexes and between employers and workers. While offering favorable portrayals of American life, defined implicitly as high rates of productivity and automobile consumption, Michelin avoided any of the contemporary assumptions that France would come to resemble the America described by Duhamel and others—except to the extent that men would drive automobiles instead of horses.[72]

In *Scènes de la vie future*, Duhamel devoted a chapter to "The Automobile, or The Law of the Jungle." The chapter describes him riding in the front seat of a car driven by a beautiful "Mrs. Lytton." Writing as though there were no better way to warn French readers of the danger of an American society strewn with cars, Duhamel notes that in the backseat sat "Mr. Lytton, an athletic and profoundly silent gentleman, who was smoking a cigarette—like the criminal just before his execution." When Duhamel asks Mrs. Lytton whether her husband drives, she laughs: "Oh, Mr. Lytton hasn't the time. He's too devoted to his business." Duhamel then notes that "Mr. Lytton lowered his eyes almost imperceptibly. It was nothing, and yet it was enough to make me understand that he drove very well, but that he did not like 'scenes.'"[73] This was in many respects an image familiar in representations of the United States in interwar France. Men, overly preoccupied with making money, lost control of their wives and other women who spent madly and drove automobiles. The

pillé!] . . . We fight waste everywhere in the factory. Are we going to accept it when it consists of human health? Certainly not!" The brochure points out that the law on accidents in the workplace merely demanded that the company continue to pay half of Ben Kacem's salary. Instead, "we got the whimsical idea a few days later to have him cared for in the best clinic of the city. However contrary that may be to established practice and however bold it may appear at first sight, it is perhaps not absurd." Then, to convince other employers, the company describes the bottom line; "the waste of Ben Kacem's leg cost 62.360 francs [a figure that included his half-salary, paid out in benefits]. How much would have cost the care for his broken leg in the best clinic? Total 6.374 [because the half-salary for no work could be avoided]." In a tone resembling the pieces on Taylorism, the brochure then notes that everyone's interest is served. In the interest of the employer, "Pension payments are rare and small, the medical leaves short. That more than pays for the increase in surgical costs. *As a whole, our costs have declined by 40%* [italicized in the original]." Workers, offered a choice of where to go for care, "want the clinic with the best reputation, the one that the employer himself would go to, with constant nursing care, a private room, two solid meals each day plus a choice of café au lait, hot chocolate or soup for breakfast." Moreover, it is good for the medical profession because it pays the best doctors and helps to undercut the mediocre, which is, Michelin notes, also in the interest of everyone.[80] Finally, Michelin writes that the resulting higher productivity of a healthier workforce is good for France as a whole.

Far from assuming that France needed to adopt American companies' or Taylor's own ideas about company welfare, Michelin coupled a productivist language with a nationalist one in order to argue that medical benefits, much like family allowances, were good for France. Michelin explicitly portrayed its own approach as superior to the German one. Michelin also assumed its policies were better than American ones. Taylor believed company welfare measures irrelevant. Ford himself concentrated on high wages, with his famous five dollars a day, although he combined it with a form of company welfare at least until the 1920s.[81] Michelin's own commitment to *oeuvres sociales,* as well as its public advocacy for them, also distinguished Michelin from American tire firms. Although American tire companies, most notably Goodyear and Firestone, also built housing for their workers (for purchase, not rent), set up sports leagues, and established medical and pension benefits, these efforts, taken largely to reduce turnover during labor shortages in the 1910s and 1920s, were generally introduced later and were less comprehensive than Michelin's own schemes.[82] Michelin did not copy its American competitors.

All told, Michelin's portrayal of France did not suggest that the country would be "Americanized" so much as it would simply be outfitted with automobiles. In a highly critical article on the company, its advertising, and its *oeuvres sociales* that appeared in the leftist review *Europe* in 1932, J. Lavaud accused Michelin of feigning French patriotism while trying to turn France into America, dominated by never-ending production and consumption, and creating a virtual *Etat Michelin* (Michelin state) not unlike the *Etat Ford* (Ford state). On one level, Edouard Michelin would surely have agreed: in regards to industrial production and markets for those goods, the company could not have been more clear or more vocal in attempting to create a society dependent on the production and consumption of automobiles. Yet on another level, in direct contrast to Lavaud's assertion that Michelin was "running after Uncle Sam," Michelin also distanced itself from "the American way of life," at least as described by critics like Duhamel and Lavaud (although perhaps not the real thing, which itself rested on more social hierarchies, including class but especially gender and race, than most Europeans thought).[83] Michelin wanted the domestic market for automobiles and thus tires that its American competitors enjoyed, but it did not argue for the wholesale "Americanization" of France any more than did any other interwar French observers, whether or not they admired certain American economic or cultural developments. In retrospect, the language of Michelin's brochures from the interwar years seems strangely prophetic, creating a discursive precedent for post–World War II developments; France needed to accept "American" methods of production in order to preserve France in the face of an American onslaught.

7

Defining France

Fusing Tourism, Regionalism, and Gastronomy in the Interwar Years

AFTER THE GREAT WAR, Michelin tapped into the widespread assumption in tourist circles that the development of French tourism could serve as the economic salvation of a cash-strapped France. In addition to undertaking yet another solution for the installation of road signs better suited to the changing expectations of tourists, Michelin adapted its guides to take into account the growing influence of gastronomic tourism. Increasingly, the French provinces, long held by historians to be in the process of being assimilated, were being reconsidered, even repackaged for tourists. The 1920s saw a veritable explosion of interest in the regions and in their distinct cuisines, while nevertheless placing those regions firmly into a national context, reconciling regional particularisms within a unified idea of the French nation. The Michelin guides reflected this development, as would the launching of the new Michelin regional guidebooks, the motorist's equivalent of the guides that Hachette had been producing for train travelers since the mid-nineteenth century. Michelin responded to the interest in cuisine by devising a system of stars for rating the "best" restaurants in France; the guide to *garagistes* became a guide to gastronomy.

In the process, Michelin not only witnessed the fusion of tourism, regionalism, and gastronomy that occurred in interwar France but also served as an accomplice. As Michelin's own efforts should make clear, France was increasingly defined not simply as the territory controlled by the French Third Republic and subject to its constitution but rather as a place where bourgeois French (and wealthy foreigners) ate and drank French food. That food came to include regional specialties, melding the various regional dishes, even (and especially) when they originated among ethnically distinct peoples across France into a broader notion of French cuisine, with all of the gastronomic fanfare that had long characterized notions of French *haute cuisine*. Promoters of tourism developed a form of regionalism, including the culinary manifestations of it, for the rest of the nation, particularly for Parisians, and even for foreigners. The renewed focus on food exemplifies a reformulation of French nationality that did not suppress regionalism so much as it appropriated the signs of difference, combined them for all French provinces, and thus used them to reinforce French unity. The presumed necessity of rating the restaurants of France, while itself fitting squarely into the hierarchical tradition of gastronomy, resulted in part from the new abundance of restaurants, particularly in the provinces, and the difficulty of knowing much about them as people traveled farther and farther afield. Michelin thus served in the interwar years to provide a clear hierarchy of French restaurants, incorporating those regional locales into a unified French whole. In the end, as the years after World War II ultimately showed, Michelin's packaging of the regions and its culinary stratification, both established in the interwar years, served as superb marketing devices for the tire company, not only by placing the Michelin name before a French audience but also by helping to equate France with the discovery, by automobile, of its regions and its food.

LA DOUCE FRANCE: TOURISM AS A NATIONAL INDUSTRY

Well before World War I, advocates of French tourism touted its importance as an "economic factor," providing the movement with a certain legitimacy as they argued for greater expenditure of state monies. Advocates encouraged tourists to see their travels as good for their country as well as themselves. As the *Revue mensuelle du Touring Club* addressed the association's members in 1912, "you [tourists] who are seeking only your pleasure . . . are at the moment one of the most important economic factors. . . . You are the cause and raison d'être of the hotel industry, which has become an element of primary importance, that is in constant progression, and that [supports] a thousand secondary industries." The *Revue* went so far as to claim that the growth of tourism had improved working-class salaries, local commerce, and even agricultural production, because peasants could save the cost of hauling their goods to market. As important for the TCF, tourists themselves "possess notions of culture, have traveled in various areas, and have learned to recognize their value," ultimately making tourists informed about the necessary components of "our national expansion."[1]

During the First World War, the language of the Touring Club changed in tenor and intensity; now French tourism became itself an "industry" capable of generating economic growth much as a factory would. The proclamations of the wonders of France increased, and promoters viewed tourism in France almost exclusively through the lens of French economic recovery. Tourism was, according to a book by Louis Forest, a journalist at *Le Matin*, an "industry: international tourism can . . . replace the export business." Forest proclaimed that there was little difference between exporting olive oil to Philadelphia or feeding it to tourists; the example seems silly given the limitations of how much olive oil one can actually ingest in a month's vacation, but the general argument itself reappeared frequently in the interwar years. According to the president of the Touring Club, Abel Ballif, increased tourism would be France's *revanche* against Germany, as France successfully drew in American dollars that would otherwise find their way to Germany. Ballif wrote in 1917 that the Germans had gone so far as to claim that Germany had more for tourists to see than did France. Of course, "this bluff, like so many others, will have no effect, and our France remains one hundred times more beautiful than . . . Germany."[2]

After World War I, the Office National du Tourisme of the Ministry of Public Works and the Touring Club de France defended tourism as a primary means for France to redress its trade imbalance after the war. It was an argument with some grounding in reality. Although the years just after the war, when the expected wave of Americans did not appear, were disappointing, the number of foreign visitors to France grew in number in the 1920s, reaching almost 2 million in 1926 before tapering off to under a million yearly in the early 1930s.[3] More important than the number of foreign tourists were the amounts that they spent in France in comparison with what French tourists spent outside France. Whereas the Americans spent some 868 million dollars in foreign travel in 1929 while foreigners spent 183 million in the United States, the French spent only 59 million dollars abroad while foreigners spent 392 million dollars in France.[4] Although there were many material reasons for France to have a favorable balance of tourist spending, including in the early 1920s the attraction of the weakened franc, propaganda within France also strongly encouraged the French to remain in France.

During and even after the war, the Touring Club called for the improvement of tourist facilities, namely hotels, to draw foreigners. At the same time, the group's *Revue* abandoned its earlier reports of foreign travel by French people. The TCF had always focused on tourism in France, but before the war articles on tourism to other countries regularly appeared. In 1905, Léon Auscher, whose trumpeting of the glories of France was so loud, even reported on his trip to the United States. Although insisting that France had a greater density of sights, his long-winded description of his transcontinental voyage was so rich in admiration that the article itself seems to advertise touring the United States.[5] By contrast, after the war, while tourism in the French colonies—itself a costly proposition—did receive regular attention in the pages of the *Revue,* there were no longer regular reports about foreign travel.

In the end, while there was a necessity of attracting foreign tourists, the TCF wanted the French themselves to stay within the French Empire. This seeming incongruity, driven by the assumption that tourism was a *national* industry, was now more than ever justified by a portrayal of a *douce* France (harkening back to the "Song of Roland") as so truly unique that the French would be crazy to ever leave, and it was only natural that foreigners would want to come. As Léon Auscher put it in a brochure published under the auspices of the Office National du Tourisme,

> France is one of the most beautiful countries in the world and, without contest, the most varied. It is perhaps the French who have been the last to realize that France is the prettiest of all, that her climate is mild [*doux*], that of all the qualities of the earth, however curious, wild, or delicate that one can imagine, are all brought together in her; . . . that her coasts on three seas are emerald, golden, and azur; that the highest mountain in Europe [Mont-Blanc] is hers; . . . that it is the territory of ancient monuments, traditional dress, old customs; that it is on her soil that art and science come to seek their consecration; that France alone, by the miracle of her

taste, knew how to make the bestiality of eating into a grand art; and finally that her inhabitants are welcoming and gay. And we call on all of those [French people] who have traveled abroad extensively to answer where one feels a more complete sense of liberty.[6]

France thus had it all and the best of it; there was no need for anyone to see any other country.

During and after the war, the leaders of the Touring Club thus attempted to mobilize public and private support for its initiatives by employing an extremely developed patriotic rhetoric. Their actions, however, were in other respects as practical as they had been before the war. The Touring Club prodded hotel keepers much as the TCF and the Michelin guides had before the war: hotel owners needed to keep clean hotels with adequate toilets, to serve good food, and to advertise a uniform set of prices (the perceived existence of separate, higher prices for Americans became, according to the TCF, a primary complaint of American tourists).[7] At the same time, the Touring Club worked closely with the Office National du Tourisme and the banking industry to create the Crédit National Hôtelier in 1925, a sort of cooperatively owned bank that provided low-interest loans to hotel owners desirous of making improvements or expanding.[8] At last French hotels were supposed to be able to compete successfully with the Swiss and the Germans. The Touring Club further hoped to encourage automobile tourism by founding a beautification competition, inspired by its *gares fleuries* (train stations beautified with flowers) contests established before the war. With an automobile and judges provided by Michelin, the Touring Club launched in 1921 the competition of *villages fleuris*, judging the beautification of well-kept villages adorned with flowers.[9] Presumably, the TCF could thus encourage villagers to make rural France prettier than other countrysides. Finally, as before the war, the Touring Club lobbied the government to improve the condition of French roads, frequently noting that they had once been the best in the world but were losing that position to the Americans and the Italians, whose new highways and autostrade supposedly put French roads to shame.

READYING THE ROUTES FOR TOURISM BY AUTOMOBILE

The vigorous prewar efforts by the Touring Club, the Automobile Club de France, the Office National du Tourisme, Michelin, and other advocates for automobile tourism had pushed the French government into maintaining better roads, providing a uniform system of numbering roads, and establishing road signs indicating towns, directions, and potential hazards on the roadway. Given the activity of these groups in the midst of the First World War and their high expectations for postwar tourism, it comes as little surprise that their efforts recommenced in earnest after the war.

After prodding from André Michelin and other representatives of French commerce and industry, in August 1919 the Ministries of Public Works and the Interior decreed for the first time that all French crossroads needed directional road signs that included road numbers. The decree also standardized the abbreviations *N, D, GC, IC,* and *VO* that Michelin had advocated before the war and noted that the letters on signs needed to be at least 15 centimeters in height, thus requiring much larger signs than had been the norm. Finally, the signs could not be affixed more than 2 meters above the pavement and needed to be situated so that headlights could make them legible after dark.[10]

Desirous of avoiding the delays that had characterized earlier such administrative decrees, the Touring Club offered to furnish the necessary road signs for the entirety of the French state, a proposal accepted by the government. The Touring Club undertook not only the production of the signs and the poles they would be placed on, but even packaged them up and delivered them to the train station at Bourget-Drancy. Only the transportation by rail and installation fell to government officials. By the summer of 1920 the TCF had persuaded donors to foot the bill for twenty-one thousand such signs (although one Michelin insider later claimed that Dunlop had offered a million francs' worth of signs before the TCF undertook the campaign).[11] In 1921, when the TCF ran out of willing donors after having provided about half of all of the signs necessary for the entire country, the French government took on the charge of paying for the rest of the forty-five thousand signs deemed necessary in 1920.[12]

The government not only accepted the Touring Club's donation of signs but the group also won the right to have *TCF* printed on the signs as well as the notation *Don de X* (the donor). The signs facilitated tourist travel and furthered the TCF's own efforts to build its membership, allowing it to wield yet more lobbying power in Paris. As interesting was the right granted the donors to have their generosity advertised on road signs that announced everything from directions and distances to the names of towns and *obstacles* on the road. French and British firms Dunlop, Citroën, Renault, Peugeot, Eco,

Hutchinson, and Rolls-Royce, anxious to keep their names before the motoring public, donated the vast majority of signs. Later signs even included companies' insignia. Several signs provided by Citroën were divided in half by the famous *deux chevrons;* those offered by Renault had its tell-tale diamond shape in the background; and the directional arrows on Peugeot signs featured its own symbol of a male lion standing atop the arrows.[13] French and British automobile companies could thus advertise among drivers, the people most likely to buy cars, at moments when the drivers needed direction. The companies could provide a service to potential clients not unlike Michelin's own efforts to aid tourists. As the management of Citroën, which began in 1922 to provide signs without working through the intermediary of the TCF, put it, "this publicity contributes to the popularity of [our] brand, thanks to the services rendered to drivers and the constant plugging inflicted [*matraquage infligé*] on the user each time he raises his eyes to find his way." Citroën had provided one hundred thousand such signs by 1926; in the 1930s the total number of Citroën signs approached one hundred sixty thousand scattered throughout the empire as well as the metropole.[14]

By far the largest donor of the TCF's road signs in 1920–22 was Dunlop, the British tire firm that built a French factory at Montluçon (in the Auvergnat department of Allier, adjacent to Michelin's own Puy-de-Dôme) in 1920. Of the initial 21,000 signs donated by various companies to the TCF, Dunlop pledged to pay for 20,000. By 1927, Dunlop had provided approximately 60,000 such road signs. In 1931 alone the company offered another 10,000.[15] By 1936, the firm had spent more than 6 million francs on the effort.[16] Without question, the company considered the signs excellent publicity, particularly for a French subsidiary of a British company; the signs' further association with the TCF, that bastion of patriotism, only helped. There was a certain psychological effect that allowed Dunlop to be considered a French company like any other; in 1928, the *Revue du Touring Club* referred to "Citroën, Dunlop, Eco, Michelin, Peugeot, [and] Renault" as "*our* major" industries. A further indication of the extent to which Dunlop considered the signs long-term publicity was its outrage in 1932 when government employees charged with maintaining the wooden signs were regularly painting over the portion that read "Don de Dunlop"; Dunlop enlisted the support of the Touring Club in protesting the practice.[17] The donations were obviously more than simple generosity.

Michelin, by contrast, stood aloof from the Touring Club's campaign. Dunlop's early willingness to support the TCF's effort afforded Michelin little possibility to play the role of dominant private sponsor of tourism in France, at least as far as signs were concerned. To place the Michelin name on a sign in the same way as Dunlop did might have equated the two companies in the minds of tire buyers. In fact, as was the case before the war, Michelin regularly chose an advertising strategy that set the Michelin name apart from those of its competitors; if even the TCF could consider Dunlop a French firm, Michelin clearly had good reason to avoid having its name placed in a similar position on the road signs.

Instead, Michelin offered its own *bornes d'angle* (crossroads markers). As early as April 1919, André Michelin tried to persuade the roads committee of the Touring Club not only that efforts needed to be concentrated at crossroads but also that the directions on the signs should be as legible after ten years as they were the first day. The *bornes* proposed and produced by Michelin were a far cry from the metal and wooden signs being so quickly erected in the 1920s by the Touring Club and the government. Measuring approximately 1 meter 75 in height, constructed out of concrete, and weighing around 380 to 500 kilograms, the bases of the *bornes* were literally indestructible. The signs on the *bornes* were equally durable. Modeled after a plaque on the monument to the Revolution of 1830 in nearby Riom (just outside Clermont), Michelin made the signs with enameled lava from the nearby mountains of Volvic and Puy-de-Dôme. After putting an enamel on the rock, the company fired the signs in kilns until the signs themselves were like rock. Inserted into reinforced concrete bases, the signs had the advantage of durability and the disadvantage of higher costs.[18]

In the early 1920s, the Michelin *borne* received little administrative attention. Beginning in 1920, Michelin had the authorization to place the *bornes* along local roads; in the department of Puy-de-Dôme alone, some 360 *bornes* were in place by 1927. The company experimented both with the shape, initially a pyramid but eventually a large circular post with a square top, and with the colors of the enamel on the signs in order to improve legibility, particularly after dark. In October 1927, the Touring Club and Michelin organized a tour in the Puy-de-Dôme that gave Michelin the opportunity to show the superiority of the *bornes* over traditional signs since the group could see not only many different examples but it could also examine *bornes* in place since 1920, revealing the durability of the Michelin alternative. The group making the trip included André Tardieu (minister of public works), Antoine Borrel (head of the tourism group within the Chamber of Deputies), a host of lesser officials, the president and vice president of the Touring Club as well as the editor of the *Revue*, Louis Baudry de Saunier, and representatives of the press, including the head of *L'Illustration*. They met André and Marcel Michelin, local road engineers, and the prefect of Puy-de-Dôme in Clermont. The format was classic for the Touring Club. The men drove around the Puy-de-Dôme examining the *bornes* before they had an "excellent dinner [at noon] at the Paradis restaurant in Royat," during which André Michelin explained the advantages and costs of the *bornes,* reissuing his offer to the administration for Michelin to provide free of charge the *bornes* for every crossroads on the *route nationale* 7, the main route from Paris to the French Riviera. The group then adjourned for a reception (*vin d'honneur*) atop the Puy-de-Dôme.[19]

Ultimately the trip and Michelin's lobbying had the desired effects. Tardieu asked both the president of the TCF and the head of the Office National du Tourisme to provide reports of the various participants' impressions. They were overwhelmingly positive. In February, a committee of the Touring Club announced its public support for the *bornes*. By 1930 some 376 Michelin *bornes* had been placed on *route nationale* 7. As a result of the TCF's support and that of Baudry de Saunier in particular, in February 1931 Gaston Gérard, the minister of public works, and Pierre Laval, the minister of the interior as well as premier, decreed that the Michelin *bornes* could be used at all of the crossroads on all categories of roads (from *voies ordinaires* to *routes nationales*) across France. During the 1930s Michelin signs came to include not only the *bornes* but a whole series of directional and other road signs, all characterized by the nearly indestructible enameled volcanic rock. By the early 1940s, some forty-five thousand *bornes* had been placed at crossroads while there were seventy thousand Michelin signs along the sides of roads.[20] The advertising potential of the signs was not overlooked by fellow advocates for increased tourism, although their own patriotic discourse saw such advertising as different, more patriotic than other forms. As Baudry de Saunier wrote, "[if] the Michelin company, having spent several dozens of millions in the obvious interest of the public, [also] finds supplementary pleasure in the publicity that the company gets from it, what honest man would blame them? Nevertheless, I know my friends [the Michelin brothers], and I know that the love of France and of tourism is here the motivation for their action."[21]

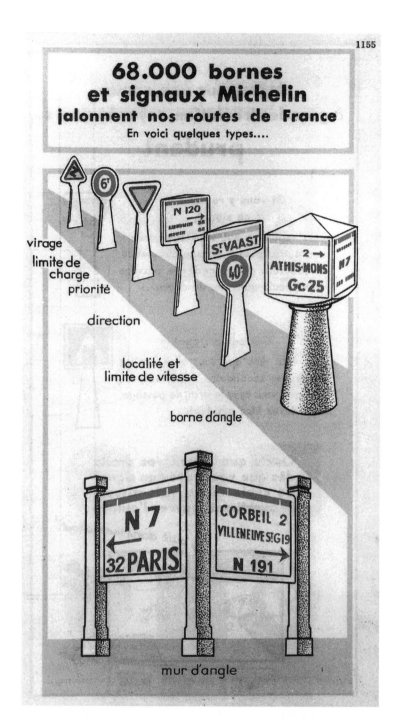

27. "68,000 Michelin milestones and signs mark our French roads," advertised Michelin in the red guide. Although Michelin deleted the customary "don de Michelin [donated by Michelin]" from the signs pictured, the company did show the diversity of the concrete and enameled signs as they existed in the interwar years. *Guide Michelin France*, 1938, p. 1155.

By the eve of the Second World War, France had not only well-paved roads, as a result of the asphalting of the *routes nationales,* but also well-marked ones. After the war, the Michelin *bornes* were in widespread use as were signs not of enameled lava but that were nevertheless anchored in reinforced concrete. By the late twentieth century, of course, most road signs had been installed at government expense and did not include the subtle advertising that became so frequent in the interwar era. In fact, the veritable absence of the formerly well-known *don de X* expression on French roads tends to obscure the centrality of efforts by the TCF and the array of manufacturers, including Michelin, that had such a crucial role in marking French roads in the first place. Now that automobile tourism is so well-entrenched, it is ironic that there are no longer many physical reminders of what private interests stood to gain from the expansion of tourism in provincial France in the first place. In their attempts to turn a profit, manufacturers combined forces with nonprofit advocates for tourism, helping to create the assumption that tourists in automobiles should be able to travel quickly across France, day or night, in quest of regional sights to see and regional dishes to eat.

REGIONAL GASTRONOMY

Well before the First World War, travel to and within France was for many a culinary experience. When foreigners visited France in the late nineteenth century, they recorded their meals in letters and diaries.[22] When early cyclists of the TCF took day-trips, they set off early in the morning and met their less athletic brethren who came by train for a huge noon meal; members' reports of these trips often included a notation of where and what the group ate and drank. With the advent of the automobile, tourists increasingly sought opportunities for fine dining at inns and restaurants in out-of-the-way places as well as in provincial cities. After World War I, as the network of roads improved and as the number of automobiles increased, gastronomy came to include fine dining in the provinces to a far greater extent than it had before the war. The interwar years brought the development of French regional gastronomy, a pairing of the appreciation of French *haute cuisine* and of traditional regional dishes. Michelin did not invent regional tourism or regional cuisine, but the company managed in the interwar years to capitalize on the newly expanded interest in regional gastronomy in order to identify the company name with both regional tourism and French gastronomy at a time when both "French" food and "French" regions increasingly came to define what it meant to be French, at least in the eyes of advocates of tourism.

French gastronomy, seen by the interwar years as a timeless characteristic of France, was itself largely a creation of the early nineteenth century. Despite later claims that French cuisine had been at the pinnacle of fine dining at least since the Gauls (just as the idea of France itself was so often anachronistically projected back in time), there is no historic evidence that either the French or foreigners viewed French cuisine as superior before the late seventeenth century. During and after the reign of Louis XIV, however, when France itself was the predominant military and political power in Europe, "French" cuisine—as eaten by royalty and the aristocracy in France and elsewhere—set a standard for elite consumption in Old Regime Europe. After the first restaurants emerged in the late eighteenth and early nineteenth centuries, in part as a result of the dispersion of French chefs during the noble emigration of the French Revolution, the cuisine originating in the kitchens of the nobility was increasingly construed as a national cuisine. Simultaneous with the founding of the first restaurants came the first French food critics, or *gastronomes*, the best known of whom were Alexandre Grimod de la Reynière and Anthelme Brillat-Savarin. What distinguished *gastronomes* from mere diners was that they wrote about the food they had eaten, critiquing its quality as well as its presentation, while also setting rules for how excellent cuisine and wine should be appreciated by patrons. The term *gastronomie* as it appeared in early-nineteenth-century France thus referred not only to the preparation and consumption of good food and wine but also the critical appreciation of it, often in written form.[23]

French cuisine was by the nineteenth century a symbol of French greatness, a cultural manifestation of French political and military influence in Europe. French food was at the top of a perceived culinary hierarchy in the eyes of elite diners in France as well as in Britain, the German states, and the United States. Yet as French economic and military might seemed to decline vis-à-vis that of Britain and Germany, French cuisine continued to set the standard for fine dining into the twentieth century. Anne-Marie Thiesse has argued that the loss of the Franco-Prussian War seemed to bring, in the eyes of many French commentators, a renewed focus on the livability of France, including its cuisine. Its climate was *doux*, or mild, not harsh like that of countries less subject to the Gulf Stream currents of the Atlantic. The micro-climates and

superb soil of France could supposedly enable the growing of superior produce. And most important, its richly diverse regional cuisines and wines set
France apart. Thiesse persuasively asserts that French nationalism came increasingly to be construed in cultural, defined as daily practices such as eating
and drinking, as well as in political terms (especially on the Right, where most
gastronomes usually placed their political allegiance, even though the notion
that French food symbolized all that was excellent in France was hardly unique
to the Right). In short, as the disaster of 1870 seemed to make clear, France
may not have been able to dominate Europe militarily or politically, but it was
the best place to live. The pioneering geographical work of Vidal de la Blache
called attention to the wonders of French geographical diversity, embodied in
the subsequent cultural traditions that had emerged in the different "soils" of
France. To oversimplify this face-saving nationalist discourse, France did not
need any more territory beyond the return of Alsace and Lorraine because
France already had it all: a mild climate and terrific food and wine.[24] The
French supposedly had a *joie de vivre* and a *savoir-vivre* that were the envy of
the world; the centrality of France as an elite tourist destination in the late
nineteenth and early twentieth centuries appeared to reinforce this position.

The superiority of France as a place to live and eat rested on the idea that
France had a unique composition of geographical, historical, and cultural diversity. Such a formulation of French nationalism maintained an important
role for the various French regions. While historians, most notably Eugen
Weber, have focused on the very real weakening of regional linguistic differences within France after 1870, that process of acculturation was accompanied
by another simultaneous development generally ignored by historians: increasingly, the historic French regions, divided into departments during the
Revolution, came to be seen as constituent parts of the French nation as a
whole.[25] For the entirety of France, regional products, particularly culinary
ones, supposedly made France unique among nations. While the centralized
French political state avoided any devolution of political power to the regions,
French cultural discourse provided a place for the regions, albeit one in which
differences were lined up, homogenized, and packaged for Parisians, other
French, and even foreigners to experience.

This homogenization of avowedly different French regions was embodied
in the most popular school reader in Third Republic France, *Le tour de la
France par deux enfants,* by G. Bruno (Mme. Augustine Fouillée), a book first

appearing in 1877 that sold some 6 million copies by 1901. *Le tour de la France* reinforced a national perspective that united French provinces while simultaneously pointing out their distinctive qualities, particularly any local specialties. *Le tour de la France* told the story of two orphaned brothers, André and Julien, who left their native Lorraine for France just after the Franco-Prussian War. They traveled throughout the French countryside, learning about the various customs and specialties of each French region. A child studying *Le tour de la France* could not miss coverage of his or her home region, but to follow the story line, that same child learned about other regions of France. Ultimately, a child should have learned a little about all of the different French regions.[26] *Le tour de la France* forms a marked contrast with contemporary readers in imperial Germany that focused on a single state and allowed a child to learn a great deal about his or her home region, or *Heimat*, but little about anyone else's. Within France, *Le tour de la France* thus features regional specialties, but it juxtaposes them and makes them equal, so that the regions fit neatly together to make up France. Customs and cuisine even from peoples as ethnically and linguistically distinct as Alsatians, Basques, Bretons, Corsicans, and Flemings could be seen as constituting regional diversity within France rather than as separate traditions distinct from those of the "original" *franciliens*.

The advent of automobile tourism for the wealthy, combined with the longstanding notion that regional differences constituted part of the greatness of modern France, led in the early twentieth century to the development of regional gastronomy. Even before the First World War, the interest in regional specialties—as distinct from the *haute cuisine* of Parisian and Lyonnais restaurants—began to grow. In 1911, the Touring Club solicited readers of its *Revue* to send in any information about regional specialties across France. In 1912, the *Revue* published a five-page list of the products from various French regions. In introducing the need for the list, the *Revue* claimed that great regional cuisine had flourished alongside Parisian fare during the supposedly golden age of restaurant-based gastronomy of early-nineteenth-century Paris, creating what was in fact a mythical history of regional specialties. "In producing delicacies, the provinces did not fall behind [Paris]. They carefully nursed the sacred fire in the stoves of their inns and in the rotisseries at the hearths of the leading kitchens of the city. Every canton of our country of France possessed its specialty and the secret of some tasty morsel [*friande gueulardise*], devotedly, delicately, and lovingly prepared, that one could not find anywhere else, and that it was even sometimes necessary to travel there to savor [*déguster*]."

The Touring Club thus implicitly denied that the new regional specialties were being created or altered for urban palates—which is precisely what occurred in the early twentieth century—by pointing out their nineteenth-century heritage. The tourist could thus experience the "genuine" culinary differences that made France such a wonderful place for eating and drinking: "In passing through the provinces, the tourist will soon be able to enjoy, as in the good old days, a delicious meal, made up of well-known specialties and respected dishes, all washed down with the best vintages of the area; it will be the end of the eternal menu, always the same, served in most restaurants here and there on the roads from Nancy to Pau, from Grenoble to Nantes, or from Calais to Marseilles."[27] Automobile tourists, bored with the standard fare that hotel owners had adopted to please them, were thus supposed to be seeking out the "authentic" regional dishes that hotel and inn owners now needed to adapt and even create for these well-heeled travelers, all the while "maintaining" the presumed "genuine" quality of the food.

The results of the TCF survey found their way into the *annuaire* of the Touring Club, which listed under each city or town any noteworthy local specialties, be they food or wine. At the end of the volume, the TCF grouped the various local specialties under one broad rubric. The specialties were becoming quite specific, tied to individual towns or villages as much as to the region as a whole. In most cases, as in the following entry for Normandy, the place names as well as the special names for the dishes make the list virtually untranslatable:

Normandie (Seine-Inférieure, Eure, Calvados, Manche, Orne)
Mets: Truites de la Bresles, de la rivière d'Arques et de la Durdent; petits maquereaux et soles de Dieppe; harengs saurs à l'huile de Fécamp; caneton rouennais; canard à la Duclair; oies grasses d'Alençon; fromages de Camembert, Neufchâtel-en-Bray, Livarot, Pont-L'Evêque; beurre de Gournay et d'Isigny; andouilles de Vire; volailles de Carentan, Evreux, Thévray et Crèvecoeur; galets du Havre; agneaux du Mont-Saint-Michel; tripes à la mode de Caen; lamproies d'Harfleur; poulets et canards de la Vallée d'Auge; huîtres de Saint-Waast-la-Hougue; crevettes: bouquet de Chaussey, de Senneville, et demoiselles de Cherbourg.[28]

Similarly, the pre–World War I Michelin guides briefly noted under the listing of a particular city what the primary local specialty was. The Michelin guide also offered a map of France entitled "La France gastronomique," which featured the historic regions of France and the primary regional foods and wines of each. The notations were quite general when compared with the Touring Club's specificity; Auvergne, was, according to Michelin, a center of "fruits, preserves [*confits*], and fromages."[29] The notion that touring rural France consisted of sampling local wines, cheeses, pastries, pâtés, and regional dishes was thus already established before the war. Automobiles, owned and operated by the same wealthy folk most likely to be interested in French gastronomy in Paris or Lyon, made the quest for regional specialties possible. Michelin, in the footsteps of the Touring Club, could promote tourism by fostering regional gastronomy.

After the First World War, tourist literature uniting fine dining and regional specialties literally exploded. The Touring Club, bemoaning the vogue of "exotic, foreign" cuisine in Paris, called for a "return" to French food embodied in French regional cuisine.[30] In 1920, under the direction of *Revue* editor Louis Baudry de Saunier, that longtime champion of automobile tourism, the Touring Club launched the "inventory of our gastronomic riches." Baudry de Saunier, claiming once again the uniqueness of French food and fitting regional gastronomy into the broader postwar discourse of tourism as economic recovery, called on members of the group to help in the creation of a master list for each French region. "One of the major fortunes of France is gastronomy. This richness is caused simultaneously by the rare gifts of our soil, the fertility of our land, and the elegant, balanced, and refined art that our race [*notre race*] has developed in exploiting them. No corner of our country is without its culinary traditions, its special dishes, its gastronomic technique. . . . But it is necessary to recognize that this richness, more or less modified by the war, today requires an inventory. . . . The general question of French gastronomy is of major importance, not only for our well-being and our reputation but also for our financial recovery." Readers were supposed to send in any information about regional dishes, wines, and recipes as well as about traditional pottery, wine bottles or glasses, and even the regional decor and dress, "that still occasionally exist in the country and that should be resurrected in our tourist hotels."[31] Here Baudry de Saunier implicitly admitted the extent to which the "regional traditions" were being re-created for tourists' consumption.

The *Revue* then featured each designated region in alphabetical order, beginning with the department of Ain. The TCF worked with the departmental Syndicat d'Initiative, which mailed a detailed questionnaire to 1,200 people in Ain, including 457 mayors and 150 hotel owners, but also chambers of commerce, small-town syndicats d'initiative, and even "winemakers, fishermen, hunters, botanists, and mycologists [specialists in fungi, especially mushrooms]."[32] The results were then published in the *Revue*. Subsequent issues took up other regions, but as Baudry de Saunier suggested at the outset, each region was fit neatly under the broader rubric of "traditional French cuisine." In the case of Alsace, whose regional specialties closely resembled those across the Rhine, the *Revue* denied that its regional cuisine was anything but French: "Sauerkraut [Choucroute] [is an] eminently [French] national, and contrary to common belief, not a German dish. In Germany, they don't know how to prepare it; it is insipid and bad, and it is not eaten much except as a hors d'oeuvre."[33] Of course, sauerkraut in Germany differed little from sauerkraut in Alsace, in the method of production, in its accompanying pork products, or in its place at the center of the meal. This French national appropriation of regional cuisine, even when the latter resembled other nations' cuisines, seems most ridiculous in the case of Alsace, but the overall procedure was the same for Brittany and Provence, where the regional specialties varied widely from each other and from the norms of Ile-de-France.

In its focus on regional cuisine, the Touring Club was by no means alone. In fact, the best-known *gastronome* of the interwar years, Curnonsky, advocated "good, old French cuisine," which he believed consisted of regional as much as Parisian cuisine. Maurice-Edmond Sailland, an aspiring writer who served before the war as a ghostwriter for Willy, the first husband of Colette, and claimed to have written some of the "Lundis" for Michelin, took the name "Why not sky" (Cur non sky) in the pro-Russian fervor that surrounded the formation of the Franco-Russian alliance in the 1890s. By the interwar period, Curnonsky's informal, humorous accounts of his gastronomic adventures had a distinct following, allowing him to be elected "Prince des gastronomes," in a poll of the magazine *Le bon gîte et la bonne table* in 1927. A native of Anjou, where he learned to appreciate good food prepared by the female family cook, Curnonsky claimed that regional cuisine cooked by women, as opposed to the *haute cuisine* that was the preserve of male chefs, could embody the very essence of French cuisine.[34]

In the 1920s, Curnonsky and Marcel Rouff launched a series of gastronomic guides to the French provinces. Before Rouff's death, twenty-eight of the projected thirty-two individual guides to various regions had appeared. Here as elsewhere, the differences of the regions blended together to form a larger French whole, a fact reinforced by the existence of two volumes on the restaurants of Paris. Curnonsky and Rouff attempted to equalize all French regions; again, the example of Alsace reveals the extent of their efforts. According to the authors, despite the German "occupation," "Alsace has been able to preserve all of the great culinary and gastronomic traditions by which she is attached to the country that knew how to elevate cuisine to the dignity of a grand art—that is to France! Like Bresse, Périgord or Franche-Comté, Alsace is one of those blessed lands where nature laid out all of her gifts." The link between Alsatian specialties and those of other French regions was made in the specific notations of regional foods and the locales where they were consumed: in Alsace, "as in Bresse and Anjou, pâtisseries are the meeting places of pretty mondaines." The use of herbs in Alsatian cooking, by contrast, resembled Provence: Alsace "is not afraid of mustard or even the garlic that Provence did not monopolize for herself!" With an illustration of a woman force feeding a goose, Curnonsky and Rouff equated even Alsatian pâté de fois gras with that of Périgord: "in Alsace, as in Périgord, one finds confit d'oie and especially this incredible treasure (as is sung in "Carmen"!), the divine, the incomparable pâté de fois gras, a gourmet's delight." [35] The subtext is obvious: Alsace is a French region different yet like all the others, a fact made clear by its fundamentally "French" culinary tradition.

The volume on Alsace also offered Curnonsky and Rouff several opportunities to define French food not only as what it was supposed to be, that is, a compilation of regional cuisines, but also as what it was not: foreign. Like the notion of national identity to which it became so closely tied in this period, French food was defined in opposition to that of other countries, particularly Germany and the United States, the two countries that in very different ways posed the primary threats to French political sovereignty, economic independence, and cultural nationalism in the interwar years. According to Curnonsky and Rouff, the Germans were ignorant of gastronomy and fine wine. The Germans had supposedly attempted to copy and appropriate Alsatian wines just as they had French architecture and champagne:

The Boches are incomparable aces at counterfeiting. They invented *ersatz* and put it everywhere. From their famous cathedral at Cologne (this

Gothic church completed around 1860!) to our champagne corks; from our fine champagne to our articles de Paris, they have imitated everything, copied everything, disfigured everything, and adulterated everything. You would expect that under their domination the fine wines of Alsace suffered from the laws of this general falsification. Just as their Rhineland wines— with the exception of three exceptional vintages—are especially distinguished by their flavor of cooled pee [*pipi refroidi*] and dirty water, the Boches generously mixed the wines of the plain with those of Alsatian hills, or more simply contented themselves with decking out the wines of Alsace with a German etiquette and baptized the Riesling, Traminer, and Cleyner with pretentious and baroque names like "Milk of the Loved Woman" [presumably *Liebfraumilch; Lait de la femme aimée* in the original] and "Saliva of the Last Born [*Salive du dernier né*]!"

The authors also claim that sauerkraut was specifically Alsatian, despite German attempts to appropriate and "dishonor" it, much as they had taken the supposedly virgin Alsace, in one of their stranger uses of gender. "Sauerkraut is . . . quite specifically Alsatian and only Pan-Germanists attempted to incorporate it into Germany to dishonor it. We have eaten alongside the Ill [the river running through Strasbourg] this tender and young fresh sauerkraut that is to the old, well-marinated kind of winter what a timid and pale virgin is to a forty-year-old woman." It is noteworthy in passing that Curnonsky and Rouff also claimed Gothic architecture as solely French, that they condemned German copying of French champagne when it was German entrepreneurs in Champagne who had proven so successful in developing champagne in nineteenth-century France, and that here, as in the case of the Touring Club's framing of regions as women, the two men portrayed as women the foods to be consumed by men.[36]

Once gastronomy, including regional gastronomy, became a marker of what it meant to be French, French food, like France itself, was thus defined in contrast with the "other." Germany, the recent wartime enemy, thus represented the worst in eating, even when the Germans ate food and drank wine that closely resembled Alsatian or even Parisian fare. The United States, which represented an economic threat of "modernization" and "Americanization" in the interwar period, was also held up as an example of a place where one could not possibly expect to eat well. A critique of American food, particularly the industrial-style production symbolized by the Chicago slaughter-

DEFINING FRANCE

houses, was frequent among interwar French critics of "Americanization"; the slaughterhouse visit reported in Georges Duhamel's popular *America the Menace* is the best-known example.[37] Similarly, Curnonsky and Rouff established their credibility on the first page of the English-language volume that combined their smaller volumes on Paris and Normandy for British and American travelers by announcing that they were experienced travelers who had covered some two-thirds of the globe: "we have eaten all kinds of cooking, from the admirable and learned Chinese cooking to the dreadful thermochemical and doctored food served in the American Palace-Hotels and caravansaries [sic, meaning trailers that served food] which brought us, together with the best epicureans of all times and all countries, to the conclusion that one can be fed almost anywhere but that one really eats in France only."[38] For Curnonsky and Rouff, one of the worst insults to be leveled at a bland sauce was to refer to it as *américaine*.[39] Clearly, notions of gastronomy were cultural constructions, and in France those constructions carried the weight of a certain idea of French national identity. By contrast, the Americans and Germans were lesser peoples because their food was inferior.

Like Curnonsky and Rouff, other interwar *gastronomes* focused on regional cuisine as well as the Parisian norms of *haute cuisine* in their descriptions of the superiority of French food and drink. J.-A.-P. Cousin wrote three guides in the 1920s in a series called *Voyages gastronomiques au pays de France*. A reminder of the importance of automobile tourism, the guides centered on the region of Lyon to Marseilles and Nice; on southwestern France from Nîmes to Bordeaux, and the Paris region (including seven itineraries for the trip from Paris to Nice).[40] Edouard Dulac published a guide to regional gastronomy for all of France.[41] Austin de Croze compiled a cookbook of regional recipes from across France in 1928, and coauthored another list of specialties and a recipe book with Curnonsky in 1933 called *Les trésors gastronomiques de la France*.[42] Significantly, this short list does not include works on gastronomy or tourism per se, which themselves increasingly included discussions of the importance of French regional cuisine as sources for innovation in *haute cuisine* generally.[43]

In the form of food and drink, French advocates of tourism, including most notably the Touring Club, accepted a construction of French national identity buttressed by regionalism. Their actions fit within a larger reconsideration of the regions revealed in the 1937 World's Fair in Paris.[44] Like Curnonsky and company, the TCF articulated the importance of regional specialties in setting

France apart from other countries, in even constituting an idea of French greatness in the early twentieth century. Although the Michelin guides did not include the rhetorical flourishes common among promoters of gastronomy and regionalism, their evolution in the interwar years must be placed fully in the context of the fusion of tourism, regionalism, and gastronomy, therefore into the widespread definition of modern France as a superior place for eating and drinking. The works by Michelin's "fellow travelers" made it possible for Michelin guides to be the simple, practical, no-nonsense guides that they appear to be, yet at the same time a means of further associating the company's name with France.

MICHELIN RED GUIDES: FROM GARAGES TO GASTRONOMY

Before the First World War, Michelin guides had been eminently practical from a prewar tourist's perspective. They provided the necessary information about how to change a tire and offered a comprehensive list of Michelin dealers, or *stockistes*. They offered detailed information about hotel accommodations, including acceptable places to eat. The guides even went so far as to list and guarantee prices of those hotels that chose to submit prices to Michelin. In the interwar years, Michelin guides remained the practical, and as the titles put it, much heralded vade mecums of automobile tourism. But just as improved roads and road signs made provincial France more accessible by car, there was a simultaneous expansion in the number of hotels, inns, and restaurants catering to automobile tourists. In addition, tires themselves required less specialized knowledge on the part of the driver because the number of *stockistes* who could repair tires increased, tires lasted longer, and they were easier to change. As a result, the prewar guide to hotels and *stockistes* quickly became a guide to hotels and restaurants. The company continued to claim that it was at the service of the client, but these elite clients' perceived needs changed considerably after World War I. Not surprisingly, given the interest in fine dining in provincial France after the war, gastronomy soon replaced technical information as a noteworthy raison d'être of the Michelin guide.

Since their debut in 1900, the Michelin guides had been divided into three main sections: technical information about tires; a list of cities and towns with the names of hotels, *stockistes*, and other vital information; and practical information for traveling inside and outside France, including rules of the road, customs, and city entrance taxes (*octroi*). After the war, the section on tires became both much less detailed and much shorter. In 1900, the guide contained 399 pages, 36 of them devoted to tires. In 1912, the guide had over 600 pages, 62 of which concerned tires. By 1927, however, the first section of the guide devoted to changing tires included only 5 pages, out of 990 total.

Despite the reduction of pages devoted to narrow, technical information, the size of the guides grew considerably, from 399 pages in 1900, to 774 in 1922, to 1,022 (not including maps) in 1929, to 1,107 in 1939. The numbers of *stockistes* and garages grew. More towns received a listing, and more of those with listings had a map. The number of hotels and later restaurants grew as did prefatory information about how to use the complicated abbreviations of the guide, designed to save space. At the same time, advertising, except that for Michelin tires, tire changing equipment, guides to the battlefields, regional guidebooks, and maps disappeared entirely.

In the meantime, the company began charging 7 francs for the guides to France in 1920, a price that grew to 10 francs in 1925 (about twice the price of a decent hotel room, equaling approximately 5 hours of work of a provincial worker earning the average wage),[45] 20 francs in 1928, 25 francs in 1933, and 30 francs in 1939. Although Alain Jemain has reported the company's version of the pricing strategy to be the result of André Michelin's trips to garages, where he found Michelin guides used to prop up a work bench and André's assertion that people only respect what they pay for, the new pricing coincided with the guide's increased focus on hotels and restaurants, rather than information about Michelin tires.[46] Initially, people appear to have bought fewer guides than they had accepted gratis. Whereas Michelin had printed 75,000 in 1919 and 90,000 in 1920, the company only printed 60,000 in 1922 (no guide appeared in 1921 because of the firm's preoccupation with guides to the battlefields).[47] The number printed climbed in the late 1920s and 1930s. Between 1926 and 1940, Michelin sold approximately 1,340,000 guides to France, just under 100,000 yearly on average since no guide appeared in 1940.[48] Although these numbers were quite high compared with the press runs of moderately popular novels, for which the average printing was about 15,000 copies in the early 1920s, they paled in comparison with those of Michelin's own interwar pamphlets, which ranged in several cases from 500,000 to a million copies, and with the sales of maps, which totaled 33,300,000 from 1926 to 1945.[49]

The major innovation of interwar guidebooks was their inclusion of restaurants. Before the war, Michelin included the price of a hotel's board, as well as that of the room, but in a period before there were many restaurants in provincial France, the guide had no category for restaurants per se. In 1923, the Michelin guide began to include restaurants. Three gradations of modest, average, and first-class restaurants received one to three stars.[50] Michelin thus began ranking restaurants at the same time that it began to list them, much as it already listed hotels by level of comfort and amenities. From the outset the company thus took for granted a notion widespread in interwar France: that there was a clear hierarchy of restaurants in France and elsewhere, one that could be discerned and reported to the traveling public. It was not, however, clear in 1923 what the relative weights of the surroundings versus the quality of the food were in the determination of the rankings. As interesting, while soliciting information from guide users to perfect the list, Michelin distanced itself from these initial rankings with the passive construction, *qui nous ont été signalé* (that have been pointed out to us), implying that Michelin was doing little more than repeating reports it had received.

The Michelin system evolved rapidly in the interwar years. In 1925, the company instituted five categories for restaurants, which corresponded closely with the gradations established for hotels. The introduction to the rankings was the same as in 1923, including that passive expression "qui nous ont été signalés comme faisant de la bonne cuisine":

. . .   Restaurants de tout premier ordre—grand luxe
         [First-class restaurants—real luxury]

. .   Restaurants de très belle apparence—cuisine recherchée
         [Well-appointed restaurants—meticulous cuisine]

.   Restaurants renommés pour leur table
         [Restaurants renowned for their food]

**   Restaurants moyens
         [Average restaurants]

*   Restaurants simples, mais bien tenus
         [Simple but well-maintained restaurants]

In the meantime, hotels "possédant une table renommée" (possessing renowned cuisine) received a single star or, in 1927, a diamond.[51] That is, Michelin ranked hotels for their level of comfort, but then noted places with particularly good food. In 1927 rankings of restaurants, Michelin replaced the star/period combination with small diamonds, ranging from five diamonds to one, but the five categories remained the same. In 1929, the tiny stars and periods returned, and Michelin maintained the same five categories.[52]

The 1930s brought even more changes, most notably distinctions between the physical plant of restaurants and the quality of the food. In 1931, a hotel or restaurant received two distinct ratings. In the case of restaurants, there were five gradations:

Restaurants de tout premier ordre—grand luxe
   [First-class restaurants—real luxury]
Restaurants de grand confort moderne
   [Very comfortable, modern restaurants]
Restaurants très confortables
   [Very comfortable restaurants]
Restaurants moyens
   [Average restaurants]
Restaurants simples
   [Simple restaurants]

At the same time, Michelin introduced its system of stars for both restaurants and hotels, a ranking based on the quality of their food:

*   Cuisine de très bonne qualité
   [Cuisine of very good quality]
**   Cuisine d'excellente qualité
   [Cuisine of excellent quality]
***   Cuisine fine et justement renommée
   [Fine and justly renowned cuisine][53]

Henceforth, Michelin also listed the specialty of the restaurant or hotel, which sometimes coincided with the very regional specialties being inventoried by the Touring Club and Curnonsky among others, under the institution's entry in the guide. A sign of the extent to which the Michelin guide was designed for Parisians visiting the provinces was the fact that it was only in 1933, after provincial establishments had been ranked, that Michelin subjected the cuisine of Parisian institutions (and not just their level of creature comforts) to the star system.

By 1933, Michelin's system of stars for restaurants and hotels serving food was in place. In the course of the 1930s, further changes established the broad contours of Michelin's rating system. The guide of 1939, the last before the Second World War, maintained the separate ratings for the comforts of hotels and restaurants and the assessment of the quality of their food. Moreover, the three stars remained, although the explanations stressed that the tourist was traveling for the sake of food, a fundamental part of seeing the regions of France by the interwar years:

***   Une des meilleures tables de France; vaut le voyage
          [Some of the best cuisine in France; worth the trip]
**   Table excellente; mérite un détour
          [Excellent cuisine; worth a detour]
*   Une bonne table dans la localité
          [Good cuisine in the area]

In the 1930s, the company vouched for its ratings, no longer attributing them to others' assessments. Michelin did, however, qualify the awarding of stars, pointing out the conditions under which the ratings should be used. For establishments with * and **, "We have first taken into account the **price of the meal** [bold in the original]." That is, the stars indicated a better meal for the price than one would pay at another local restaurant. Accepting widespread early-twentieth-century French notions that some regions simply had "better" food than others, the guide noted that "Certain regions, . . . such as the Lyonnais, are traditionally regions of fine dining. Wherever the motorist stops, he is more or less sure of finding a good meal. The stars thus indicate 'what is better among the good.' Other regions are less well provided for: a meal at any old place [*au petit bonheur*] risks being mediocre. In an establishment with a star, one has a better chance of 'eating better.' One can even eat very well there. For the same price, **one must not then do a comparison except among establishments of the same region** [bold in the original]." Only the three-star restaurants escaped this relative assessment; here "it is a matter of cuisine 'without rival,' [they are] the flower of French cuisine [Il s'agit de tables 'hors classe,' la fleur de la cuisine française]. Whatever the region, everything must be perfect: food, wine, service. There is no longer a question of price."[54]

Michelin clearly assumed and defined a hierarchy of cuisine both among restaurants and among French regions. By the late 1930s, Michelin devoted several introductory pages in the guide to gastronomic maps of France in which towns with a three-star restaurant were in bold, capital letters, those with a two-star restaurant in large font, and those with a one-star restaurant in small font. The maps allowed a tourist to plan an itinerary, a veritable *tour de la France gastronomique*, around the meals that might be consumed. For the very well-heeled *gastronomes*, the long-standing importance of the gastronomic voyage from Paris to the French Riviera was confirmed: in 1939, with the exception of Bordeaux and Annecy, every town or city with a three-star restaurant was on or near the axis from Paris to the Riviera. Regions well-represented with one- and two-star restaurants included mostly the usual tourist destinations: Alsace, Brittany, Normandy, the Loire valley, the Lyonnais, the Pyrenees, and the French Riviera.[55] Large sections of northern and south-central France were apparently bereft of any fine French cuisine.

Under the individual restaurants and hotels, Michelin recommended the special dishes and wines at some restaurants receiving at least one star, and very frequently at those receiving three stars. At the pinnacle of French cuisine were those restaurants with three stars, which served traditional French, but not usually distinctive regional, dishes. While restaurants recommended for their regional dishes, such as those in Brittany and Alsace, could make it into the ranks of one-star restaurants, in 1939 not one had three stars and only a couple had two stars. That is, regional gastronomy was a fundamental part of French gastronomy more generally, but it had little hope of reaching the top, where the restaurants of Paris, and to a lesser extent Lyon, largely served the classic dishes from the nineteenth century and before. In the provinces, the one- and sometimes two-star restaurants usually served the regional specialties inventoried by the Touring Club and Curnonsky in the interwar years. In 1939, all but three of the twelve one-star restaurants (none received two or three stars) in Nice received mention for their bouillabaisse, two for their raviolis à la niçoise.[56] In Strasbourg that same year, only five restaurants received one star (none received two or three stars), and in all but one case the establishments were noted for their choucroute garnie or their choucroute à l'alsacienne.[57] Whereas it is true that Michelin recognized such regional fare by giving it a single star, it is equally true that the hierarchy of gastronomy remained dominated by Paris, where six of the fourteen three-star restaurants in France could be found in 1939. To use the guidebook's own language, the so-called flower of French cuisine clearly grew best in Ile-de-France.

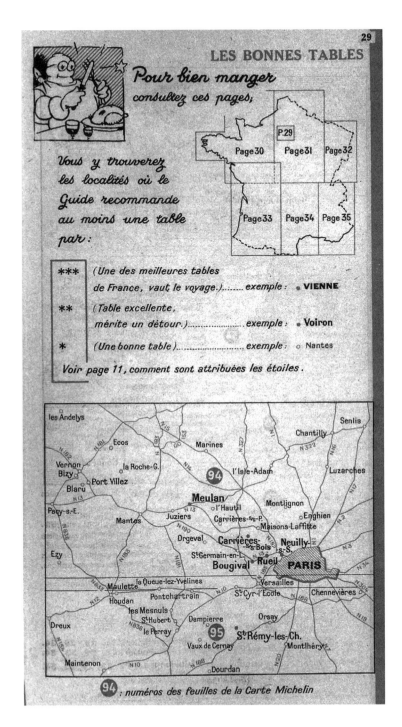

LES BONNES TABLES

Pour bien manger

consultez ces pages;

Vous y trouverez les localités où le Guide recommande au moins une table par :

P.29

Page 30 Page 31 Page 32

Page 33 Page 34 Page 35

*** (Une des meilleures tables de France, vaut le voyage.)........ exemple : ● **VIENNE**

** (Table excellente, mérite un détour.).................. exemple : ● **Voiron**

* (Une bonne table)..................... exemple : ○ Nantes

Voir page 11, comment sont attribuées les étoiles.

les Andelys

Ecos

Vernon
Bizy

Port Villez

Blaru

Pacy-s.-E.

Ezy

Marines

la Roche-G.

Meulan

l'Hautil

Juziers

Mantes

Orgeval

St Germain-en-L.

Bougival ● Rueil

Maulette

la Queue-lez-Yvelines

Houdan

Pontchartrain

les Mesnuls

St Hubert

Dreux

le Perray

Maintenon

Dourdan

Senlis

Chantilly

l'Isle-Adam

Luzarches

Montlignon

Carrières-s.-P.

Enghien

Maisons-Laffitte

Carrières-s.-Bois

Neuilly-s.-S.

PARIS

Versailles

St Cyr-l'École

Chennevières

Dampierre

Orsay

St Rémy-les-Ch.

Vaux de Cernay

Montlhéry

94 : *numéros des feuilles de la Carte Michelin*

28. Michelin's star system soon found its expression in the form of maps. At the beginning of the red guides, Michelin explained the geographical hierarchy of French gastronomy, featuring several pages of *les bonnes tables* across France. At a glance, the tourist could locate a city with restaurants earning one to three stars. *Guide Michelin France*, 1938, p. 29.

Consistent with the generalized association of French cuisine with French national identity in early-twentieth-century France, the Michelin guides largely ignored foreign food. The foreign restaurants of Paris did not receive any stars. In the guide to Belgium, Luxembourg, and southern Holland, that French-speaking city dominated by French culinary norms, Brussels, did have a three-star restaurant, and the seaside city of Ostende as well as Anvers and Bruges had two-star restaurants. But only a handful of other cities and towns in Flemish-speaking Belgium had even one-star restaurants, whereas Wallonia, or French-speaking Belgium, had a concentration that matched several tourist regions within France. Within Belgium, more telling is the utter absence of regional specialties besides Waterzoie. The fine beers of Belgium receive no similar attention, nor do chocolates. No restaurant in Holland received a ranking of any kind.[58] The guidebook to Switzerland, the Tyrol, and northern Italy similarly ignored most regional specialties. There was little special to be found in Geneva, Milan, Neuchâtel, Salzburg, Zürich, or even Venice; the regional specialties noted in Switzerland are the unspecified "wines" of Maienfeld and the "biscuits 'Ours de Berne'" in Bern.[59]

In France's own empire, the Michelin guide also assumed that the French ate French food. In 1930, Michelin issued a new guide to Morocco, Algeria, and Tunisia to celebrate the hundredth anniversary of French intervention in Algeria. It replaced the earlier Pays du Soleil guide that grouped the Riviera, Italy, Greece, Egypt, Algeria, and the Mediterranean. In the new guide, Michelin advised French tourists not to drink local water in favor of bottled mineral water. They needed to avoid raw vegetables and any fruits that could not be peeled.[60] More telling because it was less related to health concerns resulting from bacteria in the water, the company took for granted that French tourists to the empire would be looking at sights rather than experiencing the cuisine, a marked contrast to interwar norms of touring in the metropole. The guide assumed that French tourists would not be seeking couscous or other North African specialties, so it offered no such lists of local cuisine. Recommended restaurants and hotels clearly served European, and especially French, food. As was the case in Indochina at the time, French colonists themselves ate French food, which symbolized French civilization and marked the French and other Europeans as superior to the *indigènes*.[61] Tourists were hardly supposed to be different. The exceptions to this norm are rare: whereas there are no regional specialties listed for Algiers, Casablanca, or Tangiers, Fez is unique in receiving the short notation, "gâteaux arabes (kaabrezel)."[62]

Without question, Michelin reflected preexisting French and often European notions of gastronomy and codified them. In creating a hierarchy of restaurants the firm revealed one of the ironies of its own and the TCF's efforts to encourage the development of hotels and restaurants to serve tourists. Just as many elite tourists since the nineteenth century attempted to make a distinction between themselves as "travelers" and the less well-heeled as "tourists," self-proclaimed *gastronomes* viewed the expansion of the number of restaurants with alarm. As more and more middle-class people went to restaurants while touring, those diners who saw themselves as preserving the tradition of fine dining set themselves apart from the new hordes, including those from outside France expecting their own versions of "French" food, by founding exclusive gastronomic clubs. These organizations, which rarely admitted women, met periodically at restaurants in Paris and the provinces in order to sample what they considered to be the finest cuisine in France. In 1939, R. Bodet estimated that there were some twelve hundred such clubs in France.[63] They included Curnonsky's own Académie des gastronomes, modeled on the Académie Française with forty members and founded in 1928, as well as one that played on the title of Brillat-Savarin's *Physiologie du goût.*[64]

One of the oldest and best-known of such clubs was the Club des Cent, founded in 1912 by Louis Forest, journalist at *Le Matin.* Forest expected the exclusive Club des Cent to include one hundred *gastronomes* who met together at leading restaurants. Forest wrote in 1912 that the "mission" of the group was "to defend in France the taste of our old national cuisine, threatened by chemical formulas imported from countries where they have never even known how to prepare chicken stew." Members of the group traveled frequently to the provinces and then reported in detail on the fine meals they had eaten in the pages of group's private publication, signing their membership numbers rather than their names. In 1921, Louis Forest and Emile Lamberjack sponsored André Michelin's entry into the "Cent." Michelin soon began to supply a gastronomic map of France to members of the Cent along with a special edition of the Michelin guide.[65] In return, Michelin had access to members' recommendations to restaurants across France. Both the notion of a hierarchy of French cuisine as well as specific details about individual restaurants were obviously well-established before the Michelin guide introduced its stars.

Although Michelin may today be the single organization most associated with a system of stars, the firm was not the first to use stars or asterisks to rank either tourist attractions or hotels. John Murray's guides, which began to appear in 1836, had used stars. Since the middle of the nineteenth century, Baedeker guides, including those to France, included an asterisk next to a hotel that was particularly recommended and eventually one or two asterisks for a noteworthy tourist sight.[66] When Hachette launched, under the direction of Marcel Monmarché, the new Guides Bleus (blue guides) to provincial France after World War I, one asterisk noted an especially good hotel, one or two indicated an important tourist sight.[67] The Touring Club's *annuaire* had also used asterisks to denote the price ranges of hotels.[68] Michelin's innovation was not in using stars or asterisks but in doing so systematically to recommend places for fine dining. To a greater extent than any interwar guidebook or any interwar writer on gastronomy, Michelin provided an inventory of French hotels and restaurants ranking their fare. In essence, the company took the Touring Club's *concours de la bonne cuisine* (good cooking competition), which focused on the restaurants and inns of a few departments at a time, to include the entirety of France.[69] By the 1930s, Michelin had managed to represent the best in French gastronomy, at the same time that gastronomy itself was becoming an important part of French tourism.

A seemingly Taylorized brevity distinguished the new Michelin ratings. The stars became the sole indicator of relative quality of restaurants. The long descriptions of meals that were so important in the writings of Curnonsky, Touring Club members, and various *gastronomes,* had no counterpart chez Michelin. In gastronomic circles, writing, reading, and talking about meals was as much a part of the process as eating itself, hence the irony that Michelin, as identified with gastronomy as any French institution in the late twentieth century, provided no commentary whatsoever with its ratings. Restaurants received no more detail in the guide than did tire dealers. The very brevity of Michelin entries added over time to the mystery surrounding the Michelin rankings. Although the company has made periodic references to its inspectors, their absolute anonymity, their procedures, and even at times their number,[70] Michelin has carefully cultivated a notion of secrecy that garners yet more attention than outspoken clarity of criteria could ever offer. Nevertheless, the ratings were reliable enough that the company's recommendations, however disputed from time to time, appear to have been generally trusted. In the interwar years, Michelin also encouraged trust in the ratings by constantly reminding readers that Michelin accepted no payments, advertisements, or other influence from restaurant or hotel owners.

What clearly improved the quality of Michelin's assessments of restaurants' cuisine was the information the company received from tourists. Since 1900, in both the red guides and the "Lundis," the company solicited letters from its readers in much the same manner that the Touring Club solicited travelers' accounts. So at the same time that Michelin offered little information about its own process in determining ratings, it continued to appeal constantly to readers to provide any and all information about their own experiences, providing a handy form to be filled out and mailed back to the company. As in the prewar years, Michelin prided itself on "service to the client." By reinforcing the link between the company and readers of the guide, Michelin could not only serve clients but could use their comments and complaints to further perfect the guide.

Michelin thus managed a considerable feat in interwar France. Although "service to the client" had been a preoccupation of prewar businesspeople, such as champagne makers, after the First World War markets were becoming bigger and more anonymous.[71] Michelin maintained the image of service to the individual client while simultaneously profiting from the growth in the number of tourists in the interwar years and doing everything in its power to foster that expansion. Clearly, Bibendum's recommendations could replace the word-of-mouth recommendations of restaurants' clients to each other; Bibendum himself became a friend in the know. James Buzard has asserted that part of the success of Thomas Cook and Karl Baedeker in the nineteenth century

Pour être inscrit au Guide Michelin: pas de piston, pas de pot de vin!

29. In the 1930s, the red guides regularly noted that "to be listed in the Michelin guide [there could be] no string-pulling and no bribes." The French original uses both alliteration and plays on words: the term *piston* has the same automotive technical meaning as in English; and *pot de vin* means bribing, pictured here as an actual carafe of wine. The reputation of the Michelin guide, and thus of the tire company generally, rested on the credible impartiality of its ratings. After the First World War, the company reinforced its independence from influence by refusing to include ads for any products besides its own. *Guide Michelin France*, 1938, p. 14.

resulted from the ways that companies used the men themselves, even after their deaths, as images of personal service for tourists in unfamiliar environs. Cook and Baedeker guaranteed their information, serving as knowledgeable personal contacts.[72] Michelin, represented by Bibendum, provided a similar service for early-twentieth-century motorists.

The fact that most people, even wealthy interwar tourists, could not afford three-star restaurants did not keep the guides from working as a marketing device. For the well-off, Michelin offered the ratings of the restaurants and hotels. For the less fortunate, Michelin offered a reminder of the glories of French cuisine that was not necessarily less pervasive among the French bourgeoisie for being hierarchical. By providing the list, Michelin made clear that the exclusive nectar of the "flower of French cuisine" was open to all, allowing those who could afford it to distinguish themselves from others and those who wished they could afford it to dream of the possibilities if only they had the cash. In either case, Michelin could proclaim that its sole objective was to serve the client while tying itself quite closely to the very essence of what many *gastronomes* thought it meant to be French: to appreciate good, old French food (*la vieille cuisine française*), much of it to be found in the provinces. The question was how to get there.

MICHELIN REGIONAL GUIDES

Before the Great War, the Michelin tourist office and the Michelin guide offered an obvious service for motorists otherwise dependent on guidebooks designed for the railroad. The Michelin tourist office provided itineraries that listed the roads and crossroads with notations of sights to see. On the eve of the war, Michelin guides also provided several overviews of possible itineraries. In either case, motorists could use the Michelin itineraries to reach a destination where they could then grab a Guide Joanne or a Baedeker, look up the sight, and read the requisite pages. Offering free itineraries seems to have worked reasonably well for the company with limited costs; constant exhortations to use the service indicate that Michelin was by no means overburdened. Given the paucity of prewar tourists by automobile, the personalized itineraries gave Michelin the opportunity to cater individually to any and all motorists wanting assistance.

In the 1920s, however, the total number of automobiles increased dramatically in France, resulting in greater demand for automobile itineraries. There were 3,000 registered automobiles in France in 1900 (when the first Michelin guide appeared), 53,669 in 1910 (just 2 years after the establishment of the tourist office), and 107,535 in 1914. By 1926 (the year of the first regional guide) there were 836,449, a number that would grow to 1,544,057 by 1930.[73] Admittedly, most of these automobiles were not touring cars, but some certainly were, as the increased demand for itineraries from the Michelin tourist office revealed. There were 18,925 requests in 1921 but 155,267 in 1925. Whereas one employee had filled all requests in 1908, the company employed 120 in 1925.[74] The notion of individual service to each motorist was getting expensive.

In the meantime, changes in the railway guidebooks did little to resolve the problem of how to determine the best road for provincial automobile travel. Baedeker was essentially hounded out of the French market altogether in the 1920s; during and after the war, the TCF had repeatedly attacked the supposedly biased nature of the Baedeker guides.[75] Hachette's new Guides Bleus series, designed to replace the Baedekers,[76] made an effort to accommodate automobile tourists by including brief descriptions of the roads one might take along the railway itinerary, an approach embodied in several cases on the covers, which showed a train and a car traveling side by side. Of course, the flexibility inherent in having an automobile, such as taking an alternate scenic route, was lost if one traveled more or less along rail lines; these were guides for train travelers above all. The Guides Bleus, like the earlier Guides Joanne and the Baedekers, provided abundant detail about various tourist sights, worrying comparatively little about the means of transportation. This was not problematic for train travelers, but before French roads were entirely passable by automobile or were well-marked, the guides worked less well for motorists than for train passengers. The Guides Thiolier, which contained much less detail than the Guides Bleus, had even less space to accommodate the specific information needed by motorists.

In 1926, Michelin launched a series of guidebooks to various French regions, beginning with one on Brittany. In-octavo, measuring approximately 12.5 x 20 centimeters and nicely bound, the guides roughly resembled the Guides Bleus except in their slightly larger format and bright red cloth cover. By 1933, ten such guides had appeared, mirroring rather closely the tourist regions defined by both Hachette Guides Bleus and the earlier Guides Joanne: regions included the Alpes (1927), the Pyrenees (1928), castles of the Loire

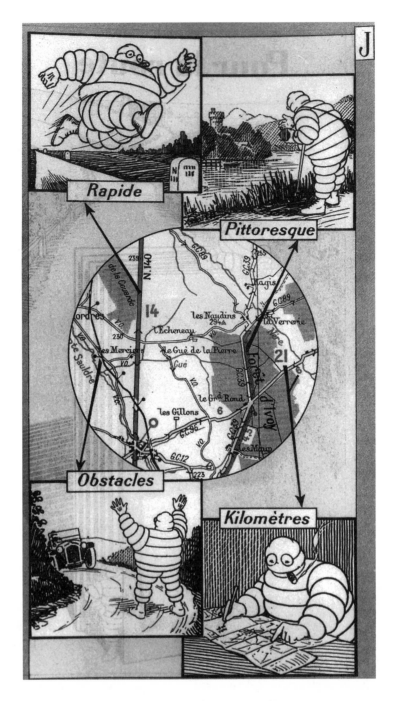

30. Bibendum educates users of the Michelin maps to France, illustrating "rapid" or "picturesque" routes as well as "obstacles" and "kilometers" between localities. With four colors and precision in the detail, Michelin maps began to set the standard for road maps in France after their introduction in 1910. Although Bibendum had by the 1920s lost most indicators of his own social class, he remained the male embodiment of counselor to tourists. *Guide Michelin France*, 1929, back matter.

valley (1928), Auvergne (1929), Gorges du Tarn (1929), the French Riviera (1930), the Vosges (1930), Provence (1931), and Normandy (1933).[77]

Each guide contained two or three parts. Each one included a brief preface, introducing the geography, history, and architectural, artistic, and folk traditions of the region. The second section, by far the largest in each guide, consisted of the individual itineraries. Some guides, such as those devoted to Brittany and Normandy, also provided a third section that listed alphabetically the leading cities of a region, then described their individual sights in some detail. Since these cities were often in different places in the itineraries, serving as starting and end points for several excursions, it was more logical to list the detailed information about them in a single location.[78]

By far the most important part of all of the early Michelin regional guides were the itineraries. For example, the guide to Brittany in 1926 contained some 430 pages, of which 346 were devoted to the itineraries and 44 to the descriptions of the individual cities. In choosing roads instead of rail lines, the guides were an interesting hybrid of the information to tourist sights that one could find in a Guide Bleu and the automobile itineraries that had earlier been offered by the Michelin tourist office. In the guide to Brittany, Michelin laid out the justification for a new guidebook, noting that its "originality resides in the sketches that illustrate it. Drawn from the [actual] terrain, they represent in a precise fashion the forks in the road or the obstacle that the traveler should expect. However, for tourists who find the use of such sketches tedious, we have translated our designs into a clear language in the text."[79] Closely resembling the illustrations used in the guides to the battlefields to help tourists find their way at crossroads, the sketches and accompanying text offered the clarity of the earlier typewritten itineraries with the visual reinforcement that served much like the Michelin maps. At the same time, one could stick to a single volume while traveling, using Michelin's description of tourist sights, just as earlier travelers had been able to use a Guide Joanne or Baedeker while in the train.

The itineraries themselves reveal much about interwar automobile travel. In most cases, early itineraries ran in both directions because each crossroads would look different, and *obstacles*, such as hills, provided very different challenges depending on whether one was climbing or descending. Michelin's attempt to expand the possible places a motorist might go meant not the abandonment of itineraries, at least not until improvement of roads and signs by the late 1930s, but their accretion, so that the primary difference between the guide to Brittany in 1926 and the second edition in 1928 was the inclusion of ninety-five rather than thirty-eight itineraries. Although the guides occasionally noted that one need not be slavish in following itineraries, Michelin also reminded ambitious tourists of the perils they faced. Michelin calibrated each itinerary to take a certain number of days, and various parts fit neatly into each day's travel. As the 1926 guide to Brittany announced, Michelin had designed each itinerary so that "the midday stops are situated in a locality possessing hotels or restaurants about which we have specific information. . . . For the final stop in the evening, our preferred choices were towns in the interior rather than along the coast. Coastal hotels frequently have no vacancies in the high season, and the traveler risks not finding a place."[80] The itineraries thus reflected the paradox that characterized early automobile tourism: on the one hand, the automobile offered freedom from the train's itinerary; but on the other, the limits of the infrastructure and tourists' own expectations of good food and accommodations could limit that "liberty" to little more than was possible by train. Moreover, in laying out the sights along the itinerary, the Michelin guides further suggested very precisely what one needed to see, whatever the presumed potential of the automobile to get off the beaten track.

In the content of its introductions and descriptions of tourist sights, Michelin offered little that was terribly new for longtime users of the Guides Bleus. The introductory material in Michelin regional guides, like that of the Guides Bleus, considers the region's geology, geography, history, particular customs and dress, cuisine and gastronomy, economic life, art, and architecture, while also providing a short bibliography for those interested in further reading. Although the Michelin guides offer only about half the number of introductory pages of the Guides Bleus, the content differs remarkably little. All the guidebooks accepted a generally center-right political consensus that avoided controversy: high points of French history include the time of the Gauls and Romans (before there was a "France"), the Middle Ages (always with a focus on Joan of Arc if she passed anywhere near the area), and the Old Regime, but the Revolution and the nineteenth century receive little attention. For example, the Michelin guide to Auvergne contains five full pages devoted to history. That newfound Gallic hero Vercingétorix receives nearly a full page, and the history of Auvergne up to the Renaissance occupies four of the five pages. Three-quarters of the final page considers the Old Regime, the account of which notes the economic growth occurring under the watch of the *intendants* before noting briefly that "The Revolution interrupted this economic revival." The guide then jumps into the nineteenth century with a paragraph on the "era of the railroad" dedicated solely to the "economic revolution" that ensued.[81] As in the Guides Bleus and some nineteenth-century French schoolbooks, politics essentially ended in 1789; it was in the absence of detail that controversy could be avoided.

What distinguished the Michelin guides from the Guides Bleus, aside from the former's brevity, were Michelin's careful hand-drawn illustrations. The Michelin guides were considerably more accessible, particularly in their coverage of geology and architecture, where Michelin explained specialized terms in detail. Here illustrations serve an obvious didactic function, explaining material taken for granted in the Guides Bleus. Clearly, Hachette destined the Guides Bleus for a would-be erudite bourgeoisie that already knew a specialized terminology whereas the Michelin regional guides read more like today's introductory college textbooks. Michelin regional guides are explicitly and unself-consciously pedagogical, attempting to initiate the uninitiated, presuming only that the user of the guide could read and look at pictures.

The most obvious innovation of the Michelin guides was their use of a system of three stars for rating tourist sights, or *curiosités*. Since the 1830s, the Murray guidebooks had used an asterisk or star to denote a particularly interesting sight. In 1844, Baedeker adopted the same practice for an especially good sight.[82] By the end of the nineteenth century, Baedeker had adopted a system of two asterisks; special sights, such as the Louvre, received two asterisks instead of one, whereas the modern Eiffel Tower merited only one.[83] The Guides Bleus also used a system with two asterisks. In the guide to the Vosges and Alsace, Monmarché awarded the cathedral of Strasbourg two asterisks but the nearby Palais des Rohans only one.[84] Michelin took the procedure a step farther, offering one to three stars representing

***   itinéraire ou curiosité de très grand intérêt
        [very interesting itinerary or sight]

**   itinéraire ou curiosité de grand intérêt
        [interesting itinerary or sight]

*   itinéraire ou curiosité d'un certain intérêt
        [somewhat interesting itinerary or sight]

Because the Michelin regional guide was a road guide, the company also rated the various itineraries as well as the individual sights, a practice that grew out of the demarcations of especially picturesque routes in the pages of the earlier Michelin red guides. As in the Guides Bleus, Michelin printed information about tourist sights in two font sizes. As Michelin put it in the introductions to the guides, "The visit to the towns is described in complete detail. But the important sights [*curiosités*] are always indicated in bold print; thus the tourist in a hurry will have no difficulty in passing quickly from one to the other while neglecting everything in small print ([such as] history, anecdotes, and secondary sights)."[85] In short, Michelin helped complete the triage between the sights that needed to be seen and those that could be seen if time and will permitted.

Since the mid-nineteenth century, the Murray, Joanne, and Baedeker guidebooks had all listed what tourists "needed to see." Across the board, they either claimed or implied an objectivity in judging what tourist sights were worthy, eliminating the more subjective, romantic, and personalized observations that had characterized many earlier travelers' accounts. At the same time, the guides employed an increasingly utilitarian style that became downright terse in the hands of both Baedeker and Michelin in the interwar years, a style that left little room for any information explaining why editors made certain choices. In short, the Michelin regional guides, like the Baedekers and to a greater extent than the Guides Bleus, implied that sightseeing was becoming increasingly divorced from any ongoing engagement on the part of tourists as to why they were touring in the first place.[86]

The most important difference between the Michelin regional guides and the comparable Guides Bleus was Michelin's careful calibrations of time. The itineraries themselves took a specific number of days, with lunches and dinners already timed. Michelin frequently offered two timelines for visiting a specific city, labeled a "detailed visit" and a "rapid visit." In Rouen, a tourist could choose between the "rapid visit," taking 3 hours and 30 minutes, a time that "has been counted without the visit of the Trésor—15 min.—and without the ascent of the cathedral spire—1 h.)" or a "detailed visit" taking one and a half days.[87] A Taylorist preoccupation with efficient use of time is one of the defining characteristics of the Michelin guides; one can almost imagine Michelin employees carefully timing tourist visits much as the stopwatch reorganized the production of Michelin tires in the 1920s. Quite without judging the individual tourist, Michelin facilitated the development of the notion that automobile tourists could swoop in, see the necessary sights without worrying much about history or the broader context, and speed off in time to catch the next sight. Interestingly, to the extent that tourists actually followed Michelin's rapid visits, this supposed defilement of European culture by speed, so derided by contemporaries and more recently by Paul Fussell, was at the hands of wealthy bourgeois who could afford cars. Harvey Levenstein perceptively claims that the well-off "travelers" were as much "tourists" as the lower-middle-class hordes they so often resented.[88]

As was the case with Michelin's earlier tourist initiatives, the regional guides reveal a preoccupation on the part of the company with soliciting readers' responses. Michelin seemed desperate for any information, including criticism, in its efforts to make further innovations in the guides. In the nineteenth century, the Guides Joanne and Baedekers had also called on readers to write in, but Michelin was yet more insistent and printed questionnaires that could simply be filled out and mailed back.[89] Combined with the tourist office's continued offering of itineraries for routes not covered by the regional guides, Michelin's own knowledge of which Michelin maps sold best, and information gathered from users of the older red guides, Michelin could attempt to keep abreast, and even ahead of, the expectations of tourists in the 1930s.

The evolution of regional guides reveals where Michelin thought the tourist market was headed. In 1935, Michelin began to revamp the regional guides. Information about hotels and restaurants disappeared, with readers sent off to consult the annual red guide. The physical dimensions of the guide changed to an oblong format measuring first 25.3 × 12 centimeters and ultimately 26 × 12 centimeters, with a thickness of less than 1 centimeter.[90] The regional guide went from looking like a Guide Bleu with a red cover to that of a fat, bound Michelin map, capable of being slipped into the suit pocket of an itinerant tourist. With an accessible layout and short, pithy descriptions of sights, they were in essence what became the postwar Michelin green guides. At exactly the same time that Michelin, in control of Citroën, worked to develop a small automobile for the masses that ultimately became the Deux Chevaux, the company introduced guidebooks designed to reach an ever larger readership of potential tourists.

Michelin's pricing strategy revealed an attempt to sell more guidebooks. Since the guidebooks were for Michelin as much a form of advertising and "service" as a profit-generating product, the company did not have to break even on the guides themselves. Initially, Michelin regional guides, despite obvious start-up costs, had already undercut the Guides Bleus. In 1926, the Michelin guide to Brittany sold for 15 francs while the Guide Bleu sold for 25 francs. In the early 1930s, Michelin guides to the Pyrenees and Auvergne cost roughly half of the price of the comparable Guides Bleus, which were approximately the same size. The new Michelin format further lowered the price so that no guide in the new format cost more than 12 francs in the late 1930s while the two to Brittany cost as little as 7 francs each and that to Gorges du Tarn 5 francs.[91] The lower prices eventually allowed Michelin to compete with the cheaper Guides Thiolier, which cost around 5 francs. But whereas the Michelin guide usually offered several hundred pages with illustrations and abundant detail, the Guide Thiolier looked more like a pamphlet, and had around 100 pages. Designed for less well-heeled tourists in trains and especially buses, the Thiolier claimed in its own advertising a total circulation of 550,000 guides in 1938, whereas Michelin claimed to have sold only a total of 653,000 regional guides during all of the years 1926 to 1945.[92]

The new regional guidebooks, like the postwar green guides, reduced the space given to itineraries.[93] By the eve of the Second World War, the *routes nationales* and other major roads had asphalt surfaces and were well-marked. The exponential growth in the number of hotels and restaurants for tourists, so obvious if one compares the sparse list of the first red guide of 1900 to the fat volume of 1939, also made it increasingly possible to tour without following a specific prepackaged itinerary before leaving. Instead, cities and villages appeared, ironically enough, in alphabetical order, much as they had in some guidebooks before the advent of the railway in the first place.[94] Although it was still only the French bourgeoisie that could afford to tour by automobile, Michelin was clearly laying the foundation with regional guides, reinforced with maps and red guides, for the expansion of automobile tourism. Its efforts attempted to create an ever-expanding market for tires while at the same time that they reflected changes in tourism already under way. Michelin's place as the foremost private promoter of tourism in twentieth-century France had never been more secure.

PROVINCIAL TOURISM AND MODERN FRANCE

A fundamental irony characterized interwar French tourism. At the same time that the automobile—the symbol and reality of modern transportation—

MARKETING MICHELIN

264

made possible the penetration of small towns only marginally accessible by train, tourists increasingly sought what they considered to be an authentic, tradition-bound provincial life replete with local culinary specialties, held up as the greatness of France. In several respects, the new fusion of French gastronomy and regionalism was what Dean MacCannell has called "staged authenticity."[95] Urban, bourgeois tourists came to the countryside in search of traditional dress, wine festivals, and the Breton *pardons*, descriptions and calendars of which occupied a full three introductory pages of the Michelin guide to Brittany, one more than Michelin devoted to the entire history of the region.[96] Guidebooks and the Touring Club's *Revue* directed tourists to places where they might find "authentic" regional dishes, meals that were in fact considerably altered to fit an urban palate: "the potato and bacon pies [*pâtés*], buckwheat pancakes [*galettes*], and boiled chestnuts could not be served without alteration to tourists."[97] True, the freshness of local products was emphasized, but the dishes that became regional specialties were prepared differently, cooked differently, and consisted of foods rarely eaten by the peasantry.

Experiencing authenticity was the act of an outsider looking in; it was like the 1937 Paris World's Fair exhibition of the regions, except that one could find an even more "genuine" version in the provinces.[98] One could ogle at traditional dress (worn often, primarily by women, for the sake of outsiders), appreciate local architecture (beautified with flowers as encouraged by the TCF), and eat (adapted) regional cuisine, but there was never a question of shedding one's own city clothes, working in the fields, sleeping in cottages with dirt floors on straw beds with bedbugs, eating what the overwhelming majority of peasants actually ate on a daily basis, or traveling on foot or in a horse-drawn cart. Within tourist literature, even locals themselves, preferably in traditional dress while serving tourists, were part of the landscape dominated by local sights and quaint villages. The automobile, improved roads, and tourist guides thus made possible the rapid, modern, time-conscious travel to the French countryside, where one could experience the presumed timeless essence of a provincial France of *pardons*, peasant dress, and particularly regional cuisine, that ultimate manifestation of *la vieille France*.

Curnonsky best articulated this contradiction also embedded in Michelin's own efforts to promote the automobile by facilitating visits of the urban bourgeoisie to experience regional cuisine prepared and served by locals. In his autobiography, Curnonsky recounted his travels in Languedoc in 1927. After noting the "powerful Bugatti" automobile they were driving, he described his attempts to get a Mme. Adolphine, a seventy-two-year-old woman known locally for preparing "the cassoulet of the gods." In Castelnaudary, the cassoulet of "La mère Adolphine" was so much more precious for being difficult to get. As Curnonsky learned in a nearby town, Mme. Adolphine had a "very modest restaurant where she only works on the days and the hours that she wants, and where she only cooks by special order for a few elite guests who have made a cult to her—and who know how to appreciate her talent and her incomparable 'manner.' She cannot stand foreigners, from whatever country, tourists, motorists, or any one else passing through." In order to gain entrée, Curnonsky had the chauffeur hide the Bugatti once in Castelnaudary and came equipped with a letter of introduction. He timidly called on the grumpy old woman, who carefully checked his credentials. Although she claimed to have heard the name Curnonsky (with its foreign ring), he quickly added that his real name was Sailland and that he was from Anjou. After nearly receiving

31. Curnonsky wrote the "Art of Good Eating in Alsace" for distribution by the French government. Illustrated by Dorette Muller, the drawing on the back cover embodied much of interwar automobile tourism. Wealthy, fashionable, and even modern-looking tourists could be served by a young woman in traditional dress, the sort of feminine representation of the region that had emerged in French tourist circles by the late nineteenth century. The modern urban bourgeoisie could thus go to the villages of Alsace for authentic rural Alsatian food and decor. Other, apparently less well-heeled tourists arriving by autocar, admire the scene. By the early twentieth century, the wooden chairs with hearts, the pottery, the architecture, and the geraniums had all become the signs of "traditional Alsace." "L'art du bien manger en Alsace," n.p., n.d. [1920s].

a refusal to be served for implying that he, three friends, and their chauffeur would like cassoulet that same day, Curnonsky ultimately extracted a meal for the following day. Mme. Adolphine then set about cooking for fifteen hours, waking six times in the night to check on the slowly cooking cassoulet. The following day the guests ate not just a cassoulet, but "*the Cassoulet,* with a capital C [his emphasis]." As the guests complimented their chef, she noted that it was the beans and water of Castelnaudary that were essential for a good cassoulet, and all agreed that "good cooking takes time."[99]

Admittedly, Curnonsky was a humorous *gastronome* specialist in the extreme, but he nevertheless vocalized clearly the themes implicit in much of the interwar focus on traveling to the provinces to eat regional specialties. In being served by a true local, embodied in the person of a woman, hence without the taint of professional training that a professional male chef might have, Curnonsky and his urban fellows could experience the truly authentic cassoulet, prepared in Castelnaudary as it could be prepared nowhere else, by a seventy-two-year-old woman with interrupted sleep, a grouchy manner, and eccentricities in abundance, who had at her fingertips the appropriate water and beans that only Castelnaudary could provide. In an age when the regions were made accessible by the automobile, in this case a fancy Bugatti that along with the chauffeur reinforced the social standing of the travelers, the foods themselves could only be made more authentic, more local, and more refined by creating an aura of inaccessibility, as Curnonsky's Mme. Adolphine had presumably done—at least in his imagination.

Michelin never made similar claims about cassoulet or any regional dishes, preferring instead crisp, efficient lists of regional specialties and where they might be found. The company's style in the guidebooks was concise and seemingly all the more objective as a result. It is nevertheless true that the company's own innovations in the red guides and the introduction of the regional guides were concurrent with the explosion of interest in the regions and in gastronomy in the interwar years. In short, Michelin did not need to make the connection between its guides and the focus on regionalism, gastronomy, or French nationalism so often suggested by other interwar advocates of automobile tourism. Michelin's own actions, however less obvious, reveal the same irony of nineteenth- and twentieth-century tourism: technological changes facilitated tourism by improving mobility and increasing leisure time, including the time well-off people were willing to devote to touring. The advent of the railroad and particularly the automobile, that "modern" invention, certainly coincided and sometimes helped to spawn attempts to "preserve" the "genuine" and the "authentic" nature of rural France, transforming it in the process. This leading French industrial firm, and one of France's foremost advocates of Taylor's and Ford's notions of both mass production and mass consumption, worked to create a mass market for the tire by promoting good, old *French* gastronomy and an appreciation for regional differences in France while simultaneously taking the more obvious tack of providing road signs. Associating the company with interwar notions of French gastronomy and regionalism so steeped in French patriotism, articulating the very hierarchy of French gastronomic tourism, Michelin placed itself in the service of the French nation in the early twentieth century.

CONCLUSION

AFTER 1940, the general contours of Michelin's advertising within France changed little from those established by André Michelin before 1931. Guidebooks and maps remained central to the company's efforts to promote itself. After the war, Bibendum continued to adorn the company's posters, the guides themselves, and the caravans to the French countryside, particularly during the Tour de France. The company worked much less hard to make the link between Michelin and France, in part because Michelin was already widely recognized as a French firm, the legacy of the company's earlier efforts. This short conclusion will trace the expansion of Michelin after the Second World War with an interest in related changes in its marketing. In several respects, Michelin's postwar advertising and developments within France generally place in even greater relief the company's attempts to create a mental association of the Michelin name with French national identity in the early twentieth century.

In 1940, some nine years after the death of his older brother, Edouard Michelin himself died. His sons, groomed to succeed their father at the head of the business, had both died accidentally in the 1930s: Etienne in an airplane wreck atop Puy-de-Dôme in 1932 and Pierre when driving from Paris to Clermont in 1937. As a result, Edouard entrusted Pierre Boulanger, who was not of the family, and Robert Puiseux, the husband of Edouard's daughter Anne, to take over the company until Edouard's grandson François came of age. The immediate and extended Michelin family continued to hold the active shares in the company, as it does today. François joined his uncle Robert Puiseux at the head of the company in 1955 and took full control in 1959.

The year 1940 also brought the defeat of France. In June, Philippe Pétain signed the Armistice with Germany and established a new French government at the Auvergnat spa town of Vichy, just north of Clermont-Ferrand. Michelin, reputedly desirous of maintaining its workforce, manufactured tires for the German army. While Michelin initially had difficulty locating rubber because the British had cut off shipping into France, by 1942 the Germans had begun to supply their own synthetic rubber, the *Buna*, so that Michelin could recommence production. By 1944, approximately 80 percent of Michelin's production was for Germany. Although Robert Puiseux is reputed to have resisted various German demands on the company's production, the Michelin plant was deemed sufficiently important for the German military that the British bombed the Cataroux plant in Clermont in 1944, almost completely idling it.[1]

Vichy presented Michelin with a considerable challenge. Always on the political right, the family found itself perilously close to the array of antirepublican, fascist, and paramilitary leagues that sprang up in the politically radicalized 1930s. Unlike André, who had taken such a public role in nationalist, proaviation, and pronatalist circles in the 1920s, Edouard remained generally silent in the 1930s. Michelin was nevertheless accused by contemporaries and at least one historian (without solid evidence) of subsidizing the activities of Colonel de la Rocque, whose Croix de Feu and Parti Social Français became essentially a French variant of fascism.[2] Pierre Michelin, Edouard's son, was also accused (without the allegation having ever been proven) of giving one million francs to the Comité Secret d'Action Révolutionnaire, or Cagoule, a far-right group that made attacks that could then be attributed to the Communists and used for repression.[3] The only clear link between Michelin and the

Cagoule was the police's discovery of the participation of several engineers employed by Michelin in the bombing of the Confédération Générale du Patronat Français in 1937. Michelin's reluctance to immediately fire the engineers raised suspicions about the Michelin family's loyalty to the Third Republic and predictably infuriated the communist CGT.[4]

Like all right-wing nationalists in 1940, including of course Charles De-Gaulle, members of the Michelin family had to decide whether to support the Vichy regime, which pursued conservative social policies but was directly subject to the Germans. Despite the factory's contribution to the German war effort, as individuals several of the Michelins chose literally to fight Vichy. Marcel Michelin, son of André, organized the resistance in the Puy-de-Dôme, was caught, deported, and then died in a concentration camp. Of his sons, one served as an Allied pilot, a second was also deported to Germany, and a third died in the liberation of Corsica. Marguerite Michelin, the wife of André's second son Jean, was deported to Germany for hiding resisting priests from the Gestapo. Their son, Jean-Luc, narrowly escaped execution by the Germans for attempting to send radio signals aiding Allied pilots over Germany.[5] In part as a result of the patriotism and the contribution to the Resistance of various family members, not to mention Puiseux's efforts to limit the scope of German demands, the Michelin company avoided postwar nationalization, the fate of Louis Renault's firm.[6]

After 1945, Michelin's postwar prosperity resulted directly from its investments in research and development of the interwar years. Immediately after the end of the war, Michelin filed for a patent on technology for its new "pneu x," a tire that had in various ways been in development since the late 1930s. This new tire was quite literally "radial"; whereas bias-ply tires had been built by placing interior cords onto a tire mold in a diagonal fashion, the cords of the radial tire were at a 90 degree angle from the direction of the tire. The new tire also included a thin steel belt that ran in the direction of the tire between the base made of fabric and the sculpted tread of the tire. It was a technological breakthrough in tire building, arguably surpassing any of the American innovations of the 1910s and 1920s. Although more expensive to build and thus priced higher, the radial could generally last twice as many kilometers as a conventional bias-ply tire.[7] Moreover, in creating less resistance to the road, the radial saved fuel. Without internal sources of petroleum, European countries, including France, continued after the Second World War to tax gasoline considerably more than did the United States, making fuel economy an important factor in tire purchases throughout the postwar era.

Although it had lost market share to Dunlop within France in the 1930s, Michelin remained dominant in France and was a strong competitor throughout Europe on the eve of the war. Most of its factories eventually reopened for civilian production after the war, and the company catered to clients long familiar with the Michelin name. After the war, peddling a technically far superior product at a time when automobiles were for the first time being produced for a mass European market, Michelin prospered. Throughout the 1950s and early 1960s, Michelin could hardly keep pace with consumer demand, despite the new construction of factories across Western Europe. In the years 1947 to 1969, Michelin built some fifteen factories in Europe, as well as one in Algeria, one in Vietnam, and a third in Nigeria.[8]

In the 1960s, François Michelin set his sights on the United States, which had the largest national market for tires in the world. Postwar economic expansion had also brought prosperity to the American tire companies, but they did not initially make an effort to develop their own radial tires as Michelin's European competitors had. Price, above all else, drove the market for tires in the United States. Detroit auto makers demanded the lowest possible price from their suppliers, offering little incentive for vast improvement of the tire itself. Price was also crucial in the replacement market, where chain stores such as Sears dominated. Moreover, most American tires were produced in Akron, Ohio, which had been a veritable rubber city since the turn of the century. In Akron, antiquated plants, companies' sense of loyalty to workforces, strong labor unions, and management's proud complacency in the face of a much smaller European rival led American firms to underestimate the importance of the radial tire.[9]

Michelin's primary problem in the United States was the absence of a distribution network. It was a hard sell to convince independent tire dealers, already a tiny minority when compared to the Goodyear and Firestone tire centers and the major retailers, why they should risk the transportation and communication problems of dealing with a foreign company whose products were at least 20 percent more expensive. In 1965, Michelin circumvented the problem by signing a contract with Sears, whereby Sears would market Michelin tires under the Sears Allstate label. Since Michelin had no real name recognition in the American market anyway, it was not an impossible sacrifice.[10]

Outside factors quickly provided unforeseen opportunities for the company. In the mid-1960s Michelin soon got a boost from Ralph Nader's publicity against highway fatalities in America, many of which were caused by blowouts. Radials, in contrast with traditional tires, rarely had blowouts. More important, the oil embargo resulting from the Yom Kippur War in 1973 led to a quick hike in American gasoline prices, placing a premium on greater fuel efficiency. The radial's higher cost was now an investment not only in security but in long-term fuel savings. American companies' general denial that radial technology was indeed a threat to their markets allowed Michelin to gain a foothold.[11] After establishing two plants in Canada in 1971, the company built two others in South Carolina in 1975. Four more appeared in North America by 1991.[12] Michelin's place in the North American market grew even faster when Michelin bought Uniroyal-Goodrich in 1990. In 1998, only 10 percent of Michelin's sales were in France, the company had seventy-seven different factories in eighteen countries, and Michelin had a world market share of 18.6 percent. Michelin, Goodyear, and Bridgestone-Firestone, created when the Japanese firm Bridgestone bought Firestone, were the three major players in the world tire market at the end of the 1990s.[13] As this book went to press in 2000, it was obvious that Michelin and Goodyear would be the beneficiaries of the massive recall of Firestone tires. Implicitly, the quality of Michelin's own tires was again confirmed.

MARKETING AT HOME AND ABROAD

Michelin's industrial expansion in Europe and North America offered challenges for a company whose marketing was initially so closely tied to the expectations of bourgeois French car owners. In France, and to some extent in Europe generally, Michelin already had terrific name recognition. After the war, the company watched France realize the dream of so many of Michelin's interwar brochures; at last France had a truly mass market for automobiles. Yet while Bibendum and the guidebooks, already omnipresent in France before World War II, seem to have held the course, changes in the form of Bibendum and in the form and content of the guidebooks show how the company recast its symbols in order to reach out to an ever larger audience of potential tire buyers. In the process, Michelin clearly hoped to offer its services to a new class of consumers, educating and helping the masses as it had the bourgeoisie.

Bibendum's transformations signaled the firm's changing assumptions about potential buyers. That *bon vivant* of the Belle Epoque, a white man adorned with cigar, champagne, and women, had already mutated into a potential father figure during the interwar years, dumping the champagne for armloads of children. He lost the pince-nez and, by the 1930s, his cigar. After the Second World War, when Michelin advertising avoided comment on any social policy in France, Bibendum even lost the children that had appeared in the advertisements for the guides to the battlefields and in the pronatalist pamphlets. Over time, he even looked remarkably little like the stack of tires that he had once been. As the market for tires expanded, the clear references to class disappeared altogether. Given the generally inclusive approach of advertising, it is not surprising that Michelin needed to avoid alienating any potential buyers by leaving Bibendum bourgeois or by making him working-class. Although he has remained white, except for one very interesting poster portraying him as black, other references to race have disappeared.[14] The use of the English Michelin Man notwithstanding, Bibendum has appeared without any commentary reminding the viewer that he is definitely male. He became almost a puffy little doll until the company reduced his weight in the 1990s. His rather commanding prewar poses are gone. Except in his international presence and greater postwar sobriety, he has come increasingly to resemble the Pillsbury Dough Boy (in contrast with Aunt Jemima), welcoming all and offending none.[15]

The centerpiece of Michelin advertising in France, and to some extent Europe, has remained the red guides to hotels and restaurants. Although ratings of hotels, names of tire and automobile dealers, and a host of other practical bits of information continued to fill the guides after 1945, French gastronomy took pride of place. In 1951, when Michelin assured readers that the restaurants of France had finally recovered from the war, the Michelin guide to France again began to award one to three stars to the best restaurants in the country.[16] In a mass tourist market, the red guide is a curious work. It assumes a hierarchy of foods with French cuisine at the summit, and with yet another clear hierarchy of the restaurants serving that food; only in 1999 did the first Chinese restaurant in Paris manage to get a star, creating quite a stir because of the traditional franco-centric nature of the ratings. But the costs of such restaurants, particularly in the case of those with three stars, are prohibitive for the vast majority of French people, except in the case of a meal of a lifetime. Presumably some readers merely like to dream of the possibilities, and

32. One of the few examples of a black Bibendum ever to appear, this poster advertises the new Michelin Xa radial tire. The radial, first introduced by Michelin after World War II, carried the name "X." Before the First World War, ironically enough, Michelin had regularly used the term "pneu x" to denote competitors, particularly Continental. Poster by Raymond Savignac, 1965.

the presence of the stars reassures them of the accuracy of Michelin's other ratings of hotels and restaurants, not to mention the quality of Michelin tires.

At the same time, several innovations in the red guide, including notations of cheaper yet very good restaurants, have also served to reach out to the budget-conscious. The red guide remains one of the company's best marketing tools, published in some six hundred thousand copies each year.[17] Its appearance every spring, offering new ratings of restaurants, receives mass media attention in France and is covered in leading national newspapers abroad. Other competing guides have appeared, but they have hardly detracted from the attention received by the Michelin guide. The company's cultivated secrecy in not divulging how rankings are arrived at or even how many inspectors exist only whets the appetite of the media. As had been the case even before the First World War, Michelin has covered the rest of Western Europe with its guides as well; in 1998, the company published red guides to the Benelux countries, Germany, Spain and Portugal, Great Britain and Ireland, Italy, and Switzerland. A separate guide covers the major cities of Europe, with the same French-oriented assumptions about what constitutes good cuisine. With multilingual introductions and elaborate symbols, the guides are accessible in several different European languages.[18]

The company's green tourist guides, the successors of the interwar Michelin regional guides, have come to complement the red guides by providing a wealth of easily accessible information about all French regions. Until the slight shrinkage of the past couple of years, Michelin maintained the oblong format (12 × 24 centimeters), introduced just before the Second World War, and has continued to use a layout, including abundant illustrations, aimed increasingly at a mass audience. The content and tone are deliberately pedagogical, distinguishing the green guides from the more detailed and more erudite blue guides of Hachette. With the exception of sights tied to modern industry and those of World War II battles, the recommended tourist sights included in the green guides remain in large part those of the nineteenth- and early-twentieth-century Guides Joanne, Baedeker guides, and blue guides.[19] The green guides have also evolved from covering only the traditional tourist regions to include all of France in some twenty-four different volumes, not including the one to Disneyland Paris.[20] As the guide to EuroDisneyland itself reveals, and as Marc Francon has shown, in its treatment of sights Michelin has managed to keep abreast of the changing expectations of tourists.[21] In the year 2000, the green guide metamorphosed into a mix of the earlier guide and the Guide de Routard, now including practical information about shopping, entertainment, public transportation, and parking. Some twenty-four additional guides to other European countries and cities further link the Michelin name with European tourism generally. By 1998, most of the guides also appeared in English, and some appeared in German, Spanish, Italian, Japanese, Dutch, and Portuguese editions. Ranging in length from 180 to 500 pages, the green guides have a circulation of 40,000 to 100,000, depending on the subject of the volume. To accompany the green guides, Michelin has also continued to publish maps of France and Europe; there were by 1998 Michelin maps for virtually all of Europe, save the countries of the former Soviet Union.[22] In all, the various tourist aids represented approximately 2 percent of Michelin's total sales in the late 1990s.[23]

Aside from Bibendum, maps, and guides, Michelin's earlier initiatives found no counterpart after the Second World War. The company offered no guides to the battlefields of World War II, although the green guides quickly incorporated information about the war. The Resistance received pride of place as did the Normandy landings, and Michelin was fond of noting that the Allies copied a 1939 Michelin guide to France and issued it to invading troops and that Michelin supplied the Allies arriving in Paris with some 4 million maps of northeastern France and western Germany.[24] After the war, the company also prepared a commemorative map to the D-Day landings, reedited in 1994 for the fiftieth anniversary. The company could imply, as it had said explicitly with the guides to the World War I battlefields, that Michelin was a patriotic French (and Allied) firm, although the ambiguities of French loyalties during the war would have seriously complicated any similar effort to produce a series of battlefield guides. After 1945, Michelin also avoided any renewed public campaign in favor of aviation, pronatalism, or American production methods within France.

Although Michelin has not tied its presentations of Bibendum or the guidebooks as closely to the French nation as it did in the early twentieth century, the identification of Michelin, a multinational firm moving its production across the globe, as "French" is probably made by most French consumers, a group far more aware of the national provenance of their consumer products and foodstuffs than most Americans are. The movement to European unity also benefits Michelin in Europe, allowing what was once a firm with an essentially French persona to be seen as a European company. Michelin's maps and guides to various European countries further solidify recognition of Michelin as European. Moreover, the company's decision in 1963 to begin supplying tires for racing allowed the company both to make claims of superiority about its radial—as it had about its tires during races before 1912—and to reinforce an association of Michelin with European racing.[25]

In contrast, Michelin's postwar expansion into the American market has underlined just how thoroughly Michelin's pre–World War II marketing in France relied on the company's ability to cast an image of itself as quintessentially French. The company's long-term marketing strategy focused on tourism, so successful in France and even Europe, has not really been implemented in the United States out of the realization that it would not work.[26] In the realm of maps, Rand McNally already had in the United States roughly the same brand loyalty that Michelin possessed in France. Perhaps most significantly, cultural differences in both touring and eating abounded between America and France. The notion of experiencing a region, its customs, and its cuisine never developed in the United States as in France, making the very concept of the green guides a much harder sell. The existence of both hotel and food chains in postwar America, not to mention real differences in culinary assumptions, further distinguished the Americans from the French and other Europeans. As Alain Arnaud, the spokesperson for Michelin's tourist service, has recently reported, Michelin's market surveys indicate that there is little interest within the United States for a red guide.[27] There are but a handful of Michelin maps to various parts of the United States, and the green guides cover only Canada, Quebec, New England, Florida, California, New York City, Chicago, San Francisco, and Washington, D.C.[28]

Ironically, Michelin's French origin, the very thing that makes the Michelin name so familiar in France and even in Europe, is as much a stumbling block as an advantage in the United States. It is altogether unclear whether average American tire buyers—save those who studied or toured abroad—even know that Michelin is a French company.[29] The priorities embedded in the Michelin guides and so closely associated with cultural construction of France that Michelin helped to build, notably regional touring and gastronomy, are not broached in Michelin advertisements in the United States. In fact, Michelin ads even avoid the suggestion that Michelin is a French-based company. While Americans may associate France with quality perfumes, champagne and other fine wines, silks, and designer dresses, giving France a veritable cachet in the realm of luxury goods, French manufacturers do not have a widespread reputation for high-quality products within the United States. As a result, Michelin ads reinforce the quality of their tires and the security they provide, without tying the products to the company's national origin. The contrast with German automobile manufacturers, who can assume that Americans associate technical prowess with Germany, is telling. In the 1980s, Volkswagen aggressively promoted its products with a single German word *Fahrvergnügen* (joy of driving) unfamiliar to the overwhelming majority of Americans. The point was obvious: Volkswagens are German, making them ipso facto excellent cars, just as higher-priced BMWs and Mercedes are.

A CERTAIN IDEA OF FRANCE

Without question, Michelin helped to define tourism in twentieth-century France, and the company continues to play that part. In the process, Michelin, like other advocates of tourism, has in some respects helped to define France itself. Michelin framed an idea of French national identity that included certain notions of gastronomy, of diverse regions that together make up a strong unity, and of the superiority of France as a tourist destination. Although today largely forgotten, Michelin's other early-twentieth-century efforts to keep its name before the French public also had a place in reflecting and conditioning how French people have seen France.

Since the Second World War, France's system of family allowances has become the most developed in the world. As Hervé Le Bras has shown with no small amount of controversy, there is a direct line of continuity in personnel and ideology from the Alliance Nationale de l'Accroissement de la Population Française and other interwar pronatalists to Vichy's Fondation Alexis Carrel, and to the foundation of the state-funded Institut National d'Etudes Démographiques after the war. Within French government and within French society as a whole, there has remained a strong consensus that the French state has a role in promoting a higher birthrate among native French by paying parents. As Le Bras himself has suggested, a politician opposing the allowances would appear to be opposing France itself.[30] While Michelin never itself advocated state involvement, its significant financial support of the Alliance Nationale and Michelin's own promotion of family allowances encouraged the development of an interwar consensus that has since become an institutionalized entitlement guaranteed by the French state.

Although Michelin ended its support for French aviation in the 1930s, the early-twentieth-century blending of aviation, modernity, and French national-ism has had an obvious legacy in postwar France. Governmental support for aviation, most striking in the development of the Concorde and in fostering Airbus, generally resulted from a perceived consensus in France that the exis-tence of European, and particularly French, aviation was an issue of interna-tional prestige.[31] When reading the French press, it is at first astonishing to an American to realize just how important the international competition between Airbus and Boeing is assumed to be, as if French national identity hangs in the balance. Whereas the American press does not generally report Boeing's vicissitudes as national triumphs and losses, that is the norm for Airbus in France. Of course, Michelin is today silent on the issue, but the company did as much as any company, including airplane manufacturers, to associate the future of French aviation with the future of France itself in the early twenti-eth century.

In one final respect, Michelin's interwar actions seem to have presaged post-war developments. Michelin publicly maintained an ostensibly ambiguous po-sition on what has often been called the Americanization of France. On the one hand, America presented an obvious international economic threat. On the other, Michelin admired the very success that made American business so threatening. That apparent ambiguity continues to characterize French ap-proaches to American society after the Second World War. So at the same time that McDonald's outlets can do a fine business of selling hamburgers to people of all ages in French cities, it can be difficult to find anyone besides teenagers who will admit to ever eating there. And at the same time that adver-tisers aggressively use English in their attempts to identify their products with youth, with the future, and with the United States because it works in selling more goods, there also exists in France a sufficient consensus among the elec-torate for stringent laws to limit the use of English in advertising. Of course, every society and every individual in any society is riddled with contradic-tions; in that France is not unique. But in the particular way that France focuses on its own industrial modernization to defend itself against the United States while simultaneously accusing the United States of threatening a French way of life is a trope that has been around since the 1920s.[32] Michelin itself resolved that contradiction by implying that France could have economic moderniza-tion in the form of automobiles without accepting wholesale some "American way of life." In the end, despite a discourse that has changed remarkably little since the interwar years, the selective adoption of global changes usually deemed "American" is exactly what has happened in France, not because Michelin said so, but because it was the only option.

Clearly, Michelin had an influence, however unquantifiable, in creating the cultural assumptions, rather widespread in France, that France is *the* place to eat, drink and tour, and a place of modern innovation. Each summer French newspapers remind the reading public that France receives more foreign tourists than any other country in the world, making tourism an "industry" selling glimpses at French cultural "traditions."[33] The notion that France can be both "modern" and "traditional," that it can have the world's best and fastest civilian aviation and high-speed trains but also remain the most beautiful, nostalgia-laden countryside for touring, is in part a legacy bequeathed by the various advocates of early-twentieth-century tourism, notably the Touring Club de France and Michelin. Michelin further shared in defining notions of social solidarity in the form of family allowances and in offering a selectively ambiguous image of the United States. In the end, the company tapped into earlier images about what constituted France, but in so doing participated fully in rearticulating them, helping to give them a cultural life of their own. While by no means acting alone, there seems little question that Michelin helped to formulate a certain idea of France. It is easy to forget, particularly after suffering through an entire book on the subject, that all this began as an effort to sell more tires.

NOTES

Introduction

1. Jean Arren, *Sa majesté la publicité* (Tours: Mame, 1914), especially pp. 5–6, 74–85.

2. Ibid., pp. 292–300.

3. On Torrilhon, the pamphlet "Manufacture Générale de Caoutchouc" (Clermont-Ferrand: n.p., 1908), in the Archives Départementales du Puy-de-Dôme (hereafter ADPD) 8 BIB 2310. On Bergougnan, "Les poilus Bergougnan pendant la grande guerre (1914–1919)" (Clermont-Ferrand: Bergougnan, 1919).

4. "Michelin, André, 31 May 1897, Propositions individuelles, Mia-Mig, for the Légion d'honneur," Archives Nationales (hereafter AN) F/12/5211.

5. René Miquel, *Dynastie Michelin* (Paris: Table Ronde, 1962); Herbert Lottman, *Michelin: 100 ans d'aventures* (Paris: Flammarion, 1998), pp. 13–35; Lionel Dumond, "L'arrière plan technique et commercial," in *Les hommes du pneu: Les ouvriers Michelin, à Clermont-Ferrand, de 1889–1940*, ed. André Gueslin (Paris: Editions de l'Atelier/Editions Ouvrières, 1993), pp. 9–11.

6. Lionel Dumond, "L'industrie française du caoutchouc, 1828–1938: Analyse d'un secteur de production" (Doctorat Nouveau Régime, Université de Paris VII, 1996), pp. 559–60.

7. Annie Moulin-Bourret, *Guerre et industrie: Clermont-Ferrand, 1912–1922; La victoire du pneu*, 2 vols. (Clermont-Ferrand: Publications de l'Institut d'Etudes du Massif Central, 1992).

8. Antoine Champeaux, "Michelin et l'aviation, 1896–1936" (Mémoire de diplôme d'études approfondies, Universités de Montpellier III et Paris I, 1988); "Michelin et l'aviation: De l'aéro-cible au Breguet-Michelin XVV BZ, l'avion de la victoire." *Revue historique des armées* (1995): 33–44; Moulin-Bourret, *Guerre et industrie*.

9. Mansel G. Blackford and K. Austin Kerr, *BFGoodrich: Tradition and Transformation, 1870–1995* (Columbus: Ohio State University Press, 1996), pp. 60–61.

10. Continental's operations were confiscated by the government at the outbreak of the war. Continental's factory at Clichy was sold in 1921. Daniel Bordet et al., *Pneu Continental: Le temps des pionniers, 1890–1920* (Paris: Somogy, 1996), p. 48.

11. Dumond, "L'industrie française," pp. 560–61.

12. Blackford and Kerr, *BFGoodrich*, pp. 77–110; Michael J. French, *The U.S. Tire Industry: A History* (New York: Twayne, 1990).

13. Dumond, "L'industrie française," pp. 558–59.

14. "Le Lundi de Michelin," *Le Journal*, 19 January 1914, p. 5.

15. On the preference for "quality" over cheap quantities on the part of the French bourgeoisie, see Whitney Walton, *France at the Crystal Palace: Bourgeois Taste and Artisan Manufacture in the Nineteenth Century* (Berkeley: University of California Press, 1992); Leora Auslander, *Taste and Power: Furnishing Modern France* (Berkeley: University of California Press, 1996); and Rosalind Williams, *Dream Worlds: Mass Consumption in Late Nineteenth-Century France* (Berkeley: University of California Press, 1982).

16. Emmanuel Chadeau, *Louis Renault* (Paris: Plon, 1998); Patrick Fridenson, *Histoire des usines Renault: Naissance de la grande entreprise, 1898–1939* (Paris: Seuil, 1972).

17. Jean-Louis Loubet, "La société André Citroën (1924–1968)" (Thèse de 3e cycle, Université de Paris X-Nanterre, 1979). For the comparative context of the creation of brand names, see Richard S. Tedlow, *New and Improved: The Story of Mass Marketing in America* (New York: Basic Books, 1990).

18. David S. Landes, "French Business and the Businessman: A Social and Cultural Analysis," in *Modern France*, ed. E. M. Earle (Princeton: Princeton University Press, 1951), pp. 334–53.

19. Michelin, like many French and British firms, thus did not meet the single most important criterion to be considered successful by leading business historian Alfred D. Chandler (*Scale and Scope: The Dynamics of Industrial Capitalism* [Cambridge, Mass.: Harvard University Press, 1990]); other historians have attempted to place France, which unlike Britain, Germany, and the United States, was not included in Chandler's study, into a Chandlerian framework: Maurice Lévy-Leboyer, "The Large Corporation in Modern France," in *Managerial Hierarchies: Comparative Perspectives on the Rise of the Modern Industrial Enterprise*, ed. Alfred D. Chandler and Herman Daems (Cambridge, Mass.: Harvard University Press, 1980), pp. 117–60; Patrick Fridenson, "France: The Relatively Slow Development of Big Business in the Twentieth Century," in *Big Business and the Wealth of Nations*, ed. Alfred D. Chandler, Franco Amatori, and Takashi Hikino (Cambridge: Cambridge University Press, 1997), pp. 207–45; and Michael S. Smith, "Putting France in the Chandlerian

Framework: France's 100 Largest Industrial Firms in 1913," *Business History Review* 72 (Spring 1998): 46–85.

20. Emmanuel Chadeau, "The Large Family Firm in Twentieth-Century France," *Business History* 35 (1993): 184–205. The international importance of family firms, understated from the Chandlerian perspective, is the subject of Roy Church, "The Family Firm in Industrial Capitalism: International Perspectives on Hypotheses and History," *Business History* 35 (1993): 17–43. Youssef Cassis, although less concerned with family ownership per se, considers European big business without focusing on managerial capitalism or other aspects of the Chandlerian framework: Youssef Cassis, *Big Business: The European Experience in the Twentieth Century* (Oxford: Oxford University Press, 1997).

21. Landes, "French Business."

22. William H. Sewell Jr., "The Concept(s) of Culture," in *Beyond the Cultural Turn: New Directions in the Study of Society and Culture*, ed. Victoria E. Bonnell and Lynn Hunt (Berkeley: University of California Press, 1999), pp. 35–61.

23. Claude Lévi-Strauss, *The Savage Mind* (Chicago: University of Chicago Press, 1966); Clifford Geertz, *The Interpretation of Cultures: Selected Essays* (New York: Basic Books, 1973).

24. Roland Barthes, *Mythologies*, trans. Annette Lavers (New York: Hill and Wang, 1972); Jacques Derrida, *Of Grammatology*, trans. Gayatri Chakravorty Spivak (Baltimore: Johns Hopkins University Press, 1976); Michel Foucault, *The Order of Things: An Archaeology of the Human Sciences* (New York: Vintage, 1970), and *The Archaeology of Knowledge and the Discourse on Language*, trans. A. M. Sheridan Smith (New York: Pantheon, 1972).

25. Sewell, "The Concept of Culture," p. 50.

26. This is particularly true of Michel Foucault, who has been the most influential among historians of modern France: *Discipline and Punish: The Birth of the Prison*, trans. Alan Sheridan (New York: Pantheon, 1977).

27. Jackson Lears, *Fables of Abundance: A Cultural History of Advertising in America* (New York: Basic Books, 1994); Roland Marchand, *Advertising the American Dream: Making Way for Modernity, 1920–1940* (Berkeley: University of California Press, 1985); idem, *Creating the Corporate Soul: The Rise of Public Relations and Corporate Imagery in American Big Business* (Berkeley: University of California Press, 1998); Ellen Garvey, *The Adman in the Parlor: Magazines and the Gendering of Consumer Culture, 1880s–1910s* (Cambridge: Cambridge University Press, 1996); Pamela Walker Laird, *Advertising Progress: American Business and the Rise of Consumer Marketing* (Baltimore: Johns Hopkins University Press, 1998).

28. Exceptions include Ellen Furlough, "Packaging Pleasures: Club Méditerranée

and French Consumer Culture, 1950–1968," *French Historical Studies* 18, no. 1 (Spring 1993): 65–81; and Kolleen M. Guy, "'Oiling the Wheels of Social Life': Myths and Marketing in Champagne during the Belle Epoque," *French Historical Studies* 22, no. 2 (Spring 1999): 211–39. Historians of France have begun to consider advertising as a profession and as an aspect of "Americanization" without yet concentrating on the content of that advertising. See, for example, Victoria de Grazia, "The Arts of Purchase: How American Publicity Subverted the American Poster, 1920–1940," in *Remaking History*, ed. B. Kruger and P. Mariani (Seattle: Bay Press, 1989), pp. 221–57; idem, "Changing Consumption Regimes in Europe, 1930–1970: Comparative Perspectives on the Distribution Problem," in *Getting and Spending: European and American Consumer Societies in the Twentieth Century*, ed. Susan Strasser, Charles McGovern, and Matthias Judt (Washington, D.C.: German Historical Institute; Cambridge: Cambridge University Press, 1998), pp. 59–83; Marie-Emmanuelle Chessel, *La publicité: Naissance d'une profession, 1900–1940* (Paris: CNRS, 1998).

Chapter 1. The Making of the Michelin Man

1. Jean Arren, *Sa majesté la publicité* (Tours: Mame, 1914), pp. 292–305; Alain Weill, *Nectar comme Nicolas* (Paris: Herscher, 1986), p. 1. Marc Martin claims that Nectar had first been drawn in 1914 but that Dransy redrew him in 1921: *Trois siècles de publicité en France* (Paris: Odile Jacob, 1992), p. 203. Benjamin Rabier created the creamery Bel's "Vâche qui rit" in 1924: F. Ghozland, *Un siècle de réclames alimentaires* (Tournai: Editions Milan, 1984), plates 135–44. Goodyear's less well-known Wingfoot did not appear until the turn of the century. The Wingfoot, which was obviously not anthropomorphic, did not regularly substitute for Goodyear, instead appearing along with the company's name.

2. I am basing my assessment of most people's idea of Bibendum on my discussions with both French and Americans (admittedly, primarily academics) as I described my research. There are only two historical works on Bibendum, both by popular historians/journalists: Pierre Gonzalez, *Bibendum: Publicité et objets Michelin* (Paris: Editions du Collectionneur, 1995); and Olivier Darmon, *Le grand siècle de Bibendum* (Paris: Hoëbeke, 1997). More general works take up the creation of Bibendum, but they too ignore the extent to which the figure might represent larger cultural issues or social hierarchies: Herbert Lottman, *Michelin: 100 ans d'aventures* (Paris: Flammarion, 1998), pp. 63–66; and Marie-Emmanuelle Chessel, *La publicité: Naissance d'une profession, 1900–1940* (Paris: CNRS, 1998), p. 178.

3. Martin, *Trois siècles*, p. 17. Recent work on the poster and other modes of advertising has focused on the profession of advertising, the extent of American influence,

and local adaptations of American methods of advertising, particularly in interwar France, without focusing on its content: Chessel, *La publicité;* Daniel Pope, "French Advertising Men and the American 'Promised Land,'" *Historical Reflections* 5, no. 1 (1978): 117–39; Victoria de Grazia, "Changing Consumption Regimes in Europe, 1930–1970: Comparative Perspectives on the Distribution Problem," in *Getting and Spending: European and American Consumer Societies in the Twentieth Century,* ed. Susan Strasser, Charles McGovern, and Matthias Judt (Washington, D.C.: German Historical Institute; Cambridge: Cambridge University Press, 1998), pp. 59–83; idem, "The Arts of Purchase: How American Publicity Subverted the American Poster, 1920–1940," in *Remaking History,* ed. B. Kruger and P. Mariani (Seattle: Bay Press, 1989), pp. 221–57; on the advertising profession and its place in the evolution of French regional and national identities before the First World War, see Aaron Jeffrey Segal, "The Republic of Goods: Advertising and National Identity in France, 1875–1918" (Ph.D. diss., University of California at Los Angeles, 1995); and for survey-style overviews of French advertising, see Martin, *Trois siècles* and the older Philippe Schuwer, *Histoire de la publicité* (Geneva: Rencontre, 1965).

4. On Dunlop, see James M. Laux, *The European Automobile Industry* (New York: Twayne, 1992), pp. 8–9. The Michelin company's account can be found in "Le Lundi de Michelin," *Le Journal,* 22 November 1909, p. 5; "Le Grand Pierre et le petit père des démontables," *L'Illustration,* 3 July 1920, advertising section. Succeeding works have essentially recapitulated the company's account. See Lionel Dumond, "L'arrière plan technique et commerciale," in *Michelin, Les hommes du pneu: Les ouvriers Michelin, à Clermont-Ferrand, de 1889 à 1940,* ed. André Gueslin (Paris: Editions de l'Atelier/Editions Ouvrières, 1993), p. 16; Lottman, *Michelin,* pp. 37–38, 42–43.

5. Claude Bellanger, ed., *Histoire générale de la presse française,* vol. 3, *1871–1940* (Paris: Presses Universitaires de France, 1972), p. 299.

6. According to René Miquel, the "Eclair" was a refashioned Peugeot and the "Hirondelle" a refashioned Benz. René Miquel, *Dynastie Michelin* (Paris: La Table Ronde, 1962), p. 376.

7. James M. Laux, *In First Gear: The French Automobile Industry to 1914* (Montreal: McGill-Queen's University Press, 1976), p. 23.

8. As later recounted in quasi-legendary terms in "Le Lundi de Michelin," *Le Journal,* 24 June 1912, p. 5.

9. Dumond, "L'arrière plan technique," pp. 19–23.

10. "Le Lundi de Michelin," *L'Auto,* 10 October 1904, p. 7; "Bibendum en Amérique," *L'Auto,* 6 March 1905, p. 8.

11. On James Gordon-Bennett and the *New York Herald,* see Bellanger et al., *Histoire de la presse,* vol. 3., pp. 71, 382–83. Michelin ads trumpeted victories in the Gordon-Bennett and other contests: "Nous avons eu la bonne fortune," *L'Auto,* 28

November 1904, p. 8; "Bibendum à la Coupe Gordon-Bennett 1905," *L'Auto*, 31 July 1905, p. 2.

12. Publicly announced in short advertisements such as that in *Le Journal*, 22 July 1912, p. 7; as early as 1905, André Michelin successfully advocated that the Automobile Club de France stop sponsoring the international Coupe Gordon-Bennett by claiming that it merely provided an occasion for American and British auto and tire makers to learn of the advances being made by the French. André Michelin, "A propos de la Coupe Gordon Bennett," *Revue mensuelle du Touring Club* (July 1905): 291–92.

13. Handbill in the Michelin archives, facsimile reproduced in Darmon, *Le grand siècle*, p. 9.

14. Darmon reprints the texts of these advertisements in *Le grand siècle*, p. 11.

15. The number of shares owned by the top four investors in *L'Auto-Vélo* in 1900 were the Société de Dion-Bouton with 192 shares worth 96,000 francs, Adolphe Clément with 50 shares worth 25,000 francs, Michelin with 20 shares worth 10,000 francs, and the Baron de Zuylen with 20 shares worth 10,000 francs. Christopher S. Thompson, "The Third Republic on Wheels: A Social, Cultural, and Political History of Bicycling in France from the Nineteenth Century to World War II" (Ph.D. diss., New York University, 1997), p. 487.

16. Bellanger, *Histoire de la presse française*, vol. 3, pp. 383–84; Eugen Weber, *France: Fin de Siècle* (Cambridge, Mass.: Harvard University Press, 1986), pp. 207–9; Thompson, "The Third Republic on Wheels," pp. 245–46. Thompson also notes that Giffard's paper had been seen by Dion and Clément as a monopoly with high advertising rates and a bias in favor of Darracq, one of Dion and Clément's competing automobile manufacturers.

17. Thompson, "The Third Republic on Wheels," p. 246.

18. In 1908, however, the "Lundis" began appearing in *Le Journal* instead of *L'Auto*, although *Les Sports* carried a more technical version. Louis Baudry de Saunier's *Omnia* reprinted the "Lundis" of *Le Journal*.

19. Thompson, "The Third Republic on Wheels," pp. 287–89. *Le Vélo* failed in 1904. Bellanger, *Histoire de la presse*, vol. 3, pp. 383–84.

20. On the price of the tire, "Le Lundi de Michelin," *L'Auto-Vélo*, 1 April 1901, p. 2; on the laborers' wages, Jean Fourastié, with Jacqueline Fourastié, *Pouvoir d'achat, prix et salaires* (Paris: Gallimard, 1977), especially pp. 65–66.

21. "Le Lundi de Michelin," *L'Auto-Vélo*, 11 March 1901, p. 3.

22. "Le Lundi de Michelin," *L'Auto-Vélo*, 18 March 1901, p. 2.

23. "Le Lundi de Michelin," *L'Auto-Vélo*, 8 April 1901, p. 2; "Le Lundi de Michelin," *L'Auto-Vélo*, 15 April 1901, p. 2; "Le Lundi de Michelin," *L'Auto-Vélo*, 6

May 1901, p. 2, and 13 May 1901, p. 2; "Le Lundi de Michelin," *L'Auto-Vélo,* 9 September 1901, p. 2; "Le Lundi de Michelin," *L'Auto-Vélo,* 16 September 1901, p. 2; "Le Lundi de Michelin," *L'Auto-Vélo,* 30 September 1901, p. 2; "Le Lundi de Michelin," *L'Auto-Vélo,* 14 October 1901, p. 2.

24. Laux, *In First Gear,* p. 32.

25. Ibid.; on salons, see also Rosalind Williams, *Dream Worlds: Mass Consumption in Late Nineteenth-Century France* (Berkeley: University of California Press, 1982), pp. 87–90.

26. "Le Lundi de Michelin," *L'Auto-Vélo,* 8 December 1902, p. 7, and 15 December 1902, p. 7.

27. "Le Lundi de Michelin," *Le Journal,* 23 December 1912, p. 5.

28. The illustration of Bibendum began appearing regularly after the "Lundis" primary venue was changed from *L'Auto* to *Le Journal* in 1908.

29. Alain Weill, *The Poster: A Worldwide Survey and History* (Boston: G. K. Hall, 1985), pp. 24–28.

30. Ibid., pp. 31–53.

31. Martin, *Trois siècles,* pp. 109–20. On posters, see also Miriam R. Levin, "Democratic Vistas—Democratic Media: Defining a Role for Printed Images in Industrializing France," *French Historical Studies* 18, no. 1 (Spring 1993): 82–108.

32. Arren, *Sa majesté la publicité,* pp. 298–300. Although O'Galop also claimed credit for Bibendum, the company generally recapitulated André Michelin's account, as have historians and journalists: Darmon, *Le grand siècle,* pp. 22–23; Lottman, *Michelin,* pp. 63–68; Miquel, *Dynastie Michelin,* p. 379. In the pages of *L'Illustration* Bibendum later recounted his naming by Théry: "J'existais. Mais je n'étais pas encore baptisé. Pour que le gérondif latin devint un nom propre français, il fallait un parrain qui ne comprît pas la langue d'Horace. Théry a vu mon papa Michelin et dit: voilà Bibendum, vive Bibendum." "Le Samedi de Michelin," *L'Illustration,* 4 June 1920, advertising page in back.

33. Ads for potions to make one stronger abounded. For one that could make buxom women as well as bigger men for only 2 francs 50, see "Laissez-nous vous faire engraisser" [Let us fatten you up], *Le Journal,* 18 February 1912, p. 10. On the reversal of body size as an indicator of class since the nineteenth century, see Pierre Bourdieu, *Distinctions,* trans. Richard Nice (Cambridge, Mass.: Harvard University Press, 1984).

34. Kolleen M. Guy, "'Oiling the Wheels of Social Life': Myths and Marketing in Champagne during the Belle Epoque," *French Historical Studies* 22, no. 2 (Spring 1999): 211- 39.

35. Darmon, *Le grand siècle,* pp. 22–23.

36. On class differentiation between race organizers and racers in the Tour de France, see Thompson, "The Third Republic on Wheels."

37. He sports a fur coat in the ad "Le pneu Michelin," *L'Auto,* 7 January 1907, p. 8.

38. "Les Lundis de Michelin," *L'Auto-Vélo,* 17 November 1902, p. 5, and *L'Auto,* 17 August 1903, p. 7.

39. "Le Lundi de Michelin," *L'Auto-Vélo,* 25 November 1901, p. 2.

40. Poster collection, Michelin archives; also reprinted in Darmon, *Le grand siècle,* p. 32.

41. Poster collection, Michelin archives, reprinted in Darmon, *Le grand siècle,* p. 40.

42. "Le Lundi de Michelin," *L'Auto,* 23 October 1905, p. 7.

43. "Le Lundi de Michelin," *L'Auto,* 9 October 1905, p. 8.

44. Michelin, "Le pneumatique et l'automobile, 1894–1914: Influence du perfectionnement du pneumatique sur le développement de l'industrie automobile" (Clermont-Ferrand: Michelin, 1904), p. 27.

45. "Le Lundi de Michelin," *Le Journal,* 25 November 1912, p. 5.

46. The "Lundis" constantly explained that Michelin tires were more expensive because they were of higher quality, a claim certainly perceived to be true before the war because Michelin was able to control the price structure in France. Lionel Dumond, "L'industrie française du caoutchouc, 1828–1938: Analyse d'un secteur de production" (Doctorat Nouveau Régime, Université de Paris VII, 1996), pp. 559–60.

47. Jürgen Kocka, "The Middle Classes in Europe," *Journal of Modern History* 67 (December 1995): 783–806; and the introduction to Victoria E. Bonnell and Lynn Hunt, eds., *Beyond the Cultural Turn: New Directions in the Study of Society and Culture* (Berkeley: University of California Press, 1999).

48. Whitney Walton, *France at the Crystal Palace: Bourgeois Taste and Artisan Manufacture in the Nineteenth Century* (Berkeley: University of California Press, 1992); and Leora Auslander, *Taste and Power: Furnishing Modern France* (Berkeley: University of California Press, 1996).

49. This is based on images of Nectar in Weill, *Nectar comme Nicolas.*

50. Other tire manufacturers also used the category of class to sell their products. An advertisement for Goodrich, which established a factory in France, shows a man in suit and top hat painfully walking with bare feet. As the caption reads, "Une automobile sans pneus Goodrich, c'est un gentleman qui va nu-pieds!" [An automobile without Goodrich tires is a gentleman going barefoot!]. *Le Journal,* 26 July 1912, p. 5.

51. "Bibendum à la Coupe Gordon Bennett," *L'Auto,* 31 July 1905, p. 2.

52. "Le Lundi de Michelin," *Le Journal*, 12 December 1914, p. 5.

53. Undated (although obviously prewar) postcard by O'Galop, postcard collection, Michelin archives.

54. "Un mot historique" and "Ein historisches Wort," postcard collection, Michelin archives.

55. Undated drawing, reprinted in Darmon, *Le grand siècle*, p. 88. Within France, Bibendum was also associated with the "great emperor," as in 1921 when he dressed as Napoleon: "Le Samedi de Michelin," *L'Illustration*, 8 January 1921, inside back cover.

56. "Michelin Tyres," reprinted in Darmon, *Le grand siècle*, p. 95.

57. "Théâtre illustré du pneu," 2nd series, prologue, Michelin archives.

58. Bellanger, *Histoire de la presse*, vol. 3, pp. 297, 314–16. Fernand Gillet claimed that Michelin pulled the "Lundis" from *L'Auto* because the newspaper, like French auto makers, blamed Michelin's new detachable rim for the French loss of automobile races in 1908: "Cent ans d'industrie: Histoire anecdotique de la Maison Michelin," vol. 2, p. 32, Michelin archives.

59. Although some of the *Journal* "Lundis" were simply reprinted in *Les Sports*, the articles in the "Lundis" in the latter newspaper were generally longer and much more detailed, designed for a more specialized readership.

60. "Le Lundi de Michelin," *Le Journal*, 21 March 1910, p. 5.

61. "Le Lundi de Michelin," *Le Journal*, 20 March 1911, p. 5.

62. Ibid.

63. "Le Lundi de Michelin," *Le Journal*, 10 February 1908, p. 5.

64. Arren, *Sa majesté la publicité*, p. 299.

65. "Le Lundi de Michelin," *Le Journal*, 13 November 1911, p. 5; "Le Lundi de Michelin," *Le Journal*, 5 January 1914, p. 5; the brochure "Les joyeux de Bibendum," probably from the late 1920s, Michelin archives.

66. For a brilliant analysis of the role of sex and gender relations in the private versus the public domain, see Edward Berenson, *The Trial of Madame Caillaux* (Berkeley: University of California Press, 1992).

67. Michelin employee Fernand Gillet did note the resemblance of Bibendum to Gargantua: "Cent ans," vol. 2, pp. 30–31.

68. The "Commandments de Bibendum," Michelin archives. A reprint of the version of the Ten Commandments translated here can be found in Darmon, *Le grand siècle*, p. 39. The others are also in the "Théâtre illustré du pneu," epilogue, Michelin archives, and in "Le Lundi de Michelin," *L'Auto*, 14 January, 1907, p. 8.

69. This comment about the marketing by religious groups is based on Suzanne K. Kaufman, "Sacred Goods and Mass-Produced Miracles: Commercialization and

Materiality at the Lourdes Sanctuary," and Aaron J. Segal, "Between the Sacred and the Profane: The Afterlife of Religious Symbols in Fin-de-Siècle Advertising," papers presented at the Society for French Historical Studies meeting, Phoenix, Ariz., 30 March–1 April 2000.

70. "Théâtre illustré du pneu," prologue, Michelin archives.

71. *Illustration théâtrale*, 11 February 1911, last page.

72. *Illustration théâtrale*, 29 April 1911, back page, and 3 June 1911, back page; *Illustration théâtrale*, 27 May 1911, back page; *Illustration théâtrale*, 25 March 1911, back page; *Illustration théâtrale*, 2 September 1911, back page.

73. "Théâtre illustré du pneu," last page, Michelin archives.

74. *Illustration théâtrale*, 25 November 1911, back page.

75. M. Didier Derbal of Michelin pointed out this fact, individual conversation, July 1997.

76. One pamphlet featured a black Bibendum as a jazz player ("Les joyeux de Bibendum" in the Michelin archives); a drawing of a caricatured samurai Bibendum committing harikari (*Illustration théâtrale*, 17 June 1911, back page); a "Lundi" that describes "a Bibendum-Rouge [a red Bibendum, presumably Native American], a Bibendum-Nègre, and a Bibendum-Chinois" at the Salon de l'Aviation of 1913 ("Le Lundi de Michelin," 8 December 1913, p. 5); and in 1965 Raymond Savignac drew a black Bibendum in an advertisement for the XA tire.

77. The term *commodity racism* comes from Anne McClintock, *Imperial Leather: Race, Gender, and Sexuality in the Colonial Contest* (New York: Routledge, 1995), chapter 5. For the broad critique of how the Western scholars defined the West itself in opposition to an amalgamated, stereotypical East, see Edward Said, *Orientalism* (New York: Pantheon, 1978).

78. Segal, "The Republic of Goods," p. 196.

79. "Le supplice de la roue," by O'Galop; reproduction in Darmon, *Le grand siècle de Bibendum*, p. 31; Segal, "The Republic of Goods," pp. 195–96. Michelin employed the usual stereotypes of the Ottomans both in this *image* and also in an advertisement in *Le Rire* in which Bibendum is caught in the harem of a sultan with three harem women, including two who are topless. They are, as is usual in the portrayal of women, about one-fifth his size. The guards wield scissors to cut, that is, puncture him. Bibendum is then pictured walking away unscathed with the women while the guards' scissors have been worn down. The motif of harem girls and vicious guards, like that of horsemen, was a frequent advertising image in turn-of-the-century France. G. Hauser's sketch, *Le Rire*, 20 December 1913, p. 2. Patrick Young has made the connection between tourists' images of North Africa and the commercial stereotypes of harem girls and Arab horsemen ("The Consumer as Na-

tional Subject: Bourgeois Tourism in the French Third Republic, 1890–1914" [Ph.D. diss., Columbia University, 1999]).

80. "Vie et aventures du célèbre Bibendum: Vélautobiographie transcrite par O'Galop," poster collection, Michelin archives.

81. O'Galop, "La bicyclette à travers les âges," poster collection, Michelin archives.

82. "Le Lundi de Michelin," *Le Journal*, 13 January 1913, p. 5.

83. Gaugé, postcard collection, Michelin archives.

84. Philippe Perrot has shown with subtlety how the bourgeoisie used clothing to define an identity as bourgeois in nineteenth-century France: *Fashioning the Bourgeoisie: A History of Clothing in the Nineteenth Century*, trans. Richard Bienvenu (Princeton: Princeton University Press, 1994).

85. On popular images of Africans in turn-of-the-century France, see William H. Schneider, *An Empire for the Masses: The French Popular Image of Africa, 1870–1900* (Westport, Conn.: Greenwood, 1982). For advertisements of food products, see Ghozland, *Un siècle de réclames*.

86. "La Végétaline," Collection Bibliothèque Forney; Ghozland, *Un siècle de réclames*, plate 90.

87. A. Gallice, "Old Jamaïca," 1893, Bibliothèque Forney. Images of the *tirailleur Sénégalais* in Banania advertisements were somethat less stereotypical: Ghozland, *Un siècle de réclames*, plates 238–43.

88. McClintock, *Imperial Leather*, chapter 5; see also Thomas Richards, *The Commodity Culture of Victorian Britain: Advertising and Spectacle, 1851–1914* (London: Verso, 1990).

89. It is true that tire changing itself required a fair amount of strength in the early days of the automobile, but Michelin itself trumpeted its own detachable rims as making it possible for children to change a tire: leaflet by Poulbot, dated 1913, reprinted in Darmon, *Le grand siècle*, p. 72. Although not directed at the French experience, recent work has focused on the gendering of the automobile and of technology generally: Virginia Scharff, *Taking the Wheel: Women and the Coming of the Motor Age* (New York: The Free Press, 1991); Ruth Oldenziel, "Boys and Their Toys: The Fisher Body Craftsman's Guild, 1930–1968 and the Making of a Male Technical Domain," *Technology and Culture* 38 (1997): 60–96; idem, *Making Technology Masculine: Men, Women, and Modern Machines in America, 1870–1945* (Amsterdam: Amsterdam University Press, 1999); Nina E. Lerman, Arwen Palmer Mohun, and Ruth Oldenziel, "The Shoulders We Stand On and the View from Here: Historiography and Directions for Research," *Technology and Culture* 38 (1997): 9–30; idem, "Versatile Tools: Gender Analysis and the History of Technology," *Technology*

and Culture 38 (1997): 1–8; Barbara Orland, "Geschlecht als Kategorie in der Technikhistoriographie," in *Geschlechterverhältnisse in Medizin, Naturwissenschaft und Technik* (Bassum: Verlag für Geschichte der Naturwissenschaft und der Technik, 1996), pp. 30–42.

90. In a very nuanced essay, taking a page from Michel de Certeau who has claimed that no fundamental distinction could be made between consumption and production, Leora Auslander has argued that the "production/consumption dichotomy is misleading" because "the bourgeoisie of both genders were cast as consumers, albeit consuming to different ends." Auslander, "The Gendering of Consumer Practices in Nineteenth-Century France," in *The Sex of Things: Gender and Consumption in Historical Perspective*, ed. Victoria de Grazia with Ellen Furlough (Berkeley: University of California Press, 1996), p. 79; and Michel de Certeau, *The Practice of Everyday Life*, trans. Steven F. Rendall (Berkeley: University of California Press, 1984). For the traditional notion of separate spheres in France, see James MacMillan, *Housewife or Harlot: The Place of Women in French Society* (New York: St. Martin's, 1981); Bonnie Smith, *Ladies of the Leisure Class: The Bourgeoises of Northern France in the Nineteenth Century* (Princeton: Princeton University Press, 1981); and Louise A. Tilly and Joan Scott, *Women, Work, and Family* (New York: Methuen, 1987).

91. The cultural link between women and cars, so obvious for artists, has not been explored by historians of the automobile. On the portrayal of women in advertising for cars and in art generally, see Gilles Néret and Hervé Poulain, *L'art, la femme et l'automobile* (Paris: E.P.A., 1989).

92. Among the exceptions, the Italian subsidiary portrayed him as Primavera, modeled after Botticelli, on the cover of the in-house publication in 1926, reprinted in Darmon, *Le grand siècle*, p. 88. A drawing of a female Bibendum watched her *jumelé* (twin) tire children at play ("Vernissage d'Automne," *Le Journal*, 27 November 1911, p. 5); a female Bibendum is pictured next to a male one, "Solide au poste," *Le Journal*, 25 September 1911, p. 5; and a "jolie Bibende," is referred to having been present at the booth of the Salon de la Locomotion aérienne, "Notre Bibendum National," *Le Journal*, 4 October 1909, p. 5.

93. Louis Hindre, "Le pneu vélo Michelin: Le meilleur, le moins cher," 1911, poster collection, Michelin archives.

94. *L'Auto*, 16 September 1907, p. 8. A German version appeared in the form of a postcard distributed by Michelin's German subsidiary, "Bibendum's Eroberungen, 1907," postcard collection, Michelin archives.

95. *L'Auto*, 23 September 1907, p. 8.

96. "Le Lundi de Michelin," *Le Journal*, 6 April 1914, p. 5.

97. "Le Lundi de Michelin," *Le Journal*, 19 June 1911, p. 5.

98. "Le Lundi de Michelin," *Le Journal*, 27 November 1911, p. 5.

NOTES TO PAGES 46–50

294

99. "Le Lundi de Michelin," *Le Journal*, 15 June 1908, p. 5.

100. "Le Lundi de Michelin," *Le Journal*, 26 October 1908, p. 5.

101. "Le Lundi de Michelin," *Le Journal*, 18 March 1907, p. 7.

102. For a subtle analysis of the place of gender in advertisements for bicycles, see Thompson, "The Third Republic on Wheels," pp. 175–215.

103. Poster for semelles Michelin, poster collection, Michelin archives.

104. "Le Lundi de Michelin," *Le Journal*, 25 April 1910, p. 5.

105. "Le Lundi de Michelin," *Le Journal*, 24 January 1910, p. 5.

106. Front cover of *Le Rire*, 17 November 1906.

107. Arren, *Sa majesté la publicité*, p. 299.

Chapter 2. Finding France

1. Catherine Bertho Lavenir, *La roue et le stylo: Comment nous sommes devenus touristes* (Paris: Odile Jacob, 1999), p. 97; anonymous history of the Touring Club de France in its files in Archives Nationales (hereafter AN) 53/AS/163.

2. The Touring Club de France maintained an outwardly cooperative relationship with its potential rivals, recognizing that the groups had the same goals. In 1895, the review of the TCF announced the foundation of the Automobile Club de France, noting that virtually all of its members were also members of the TCF. The two groups would work together on a whole series of projects, although the TCF maintained an interest in other forms of touring whereas the ACF focused on travel by automobile. "L'Automobile Club de France," *Revue mensuelle du Touring Club de France*, December 1895, p. 699. The Michelin brothers were both listed as members of the ACF (*Annuaire de l'Automobile Club de France* [Paris, 1907, 1929–30]); the size of the TCF precluded a comprehensive list of its members in the *annuaire* after the early 1890s.

3. Marina Duhamel, *Un demi-siècle de signalisation routière: Naissance et évolution du panneau de signalisation routière en France, 1894–1946* (Paris: Presses des Ponts et Chaussées, 1994), p. 18. *Guide Michelin*, 1905, p. 751; "Le Lundi de Michelin," *L'Auto*, 8 June 1903, p. 7.

4. Bertho Lavenir, *La roue et le stylo*, pp. 96–99.

5. The Touring Club's political neutrality was laid out in 1891 and 1892, as well as intermittently thereafter. "A nos lecteurs," *Revue mensuelle du Touring Club de France*, January 1892, p. 2; Bertho Lavenir, *La roue et le stylo*, pp. 99–100.

6. *Guide Michelin*, 1900, p. 5.

7. Ibid., pp. 9, 11–13. There were only two foreign dealers, in Brussels and Vienna.

8. Ibid., pp. 54–56; Marc Francon, "Le guide vert Michelin: L'invention du

tourisme culturel populaire" (Thèse de doctorat du Nouveau Régime, Univeristé de Paris VII, 1998), p. 15.

9. Lists of appropriate hotels had long appeared in earlier guidebooks.

10. Touring Club de France, *Annuaire*, 1891; *Guide Michelin*, 1900, p. 56.

11. *Guide Michelin*, 1900, pp. 54–55.

12. Ibid., p. 6.

13. Ibid., p. 5.

14. Ibid., pp. 5, 57–58.

15. "Le Lundi de Michelin," *L'Auto-Vélo*, 26 August 1901, p. 2; *Guide Michelin*, 1902, p. 78.

16. "Le Lundi de Michelin," *L'Auto*, 17 August 1903, p. 7; "Le Lundi de Michelin," *L'Auto*, 5 February 1905, p. 7.

17. "Le Lundi de Michelin," *Le Journal*, 27 July 1908, p. 5; "Le Lundi de Michelin," *Le Journal*, 3 August 1908, p. 5; "Le Lundi de Michelin," *Le Journal*, 8 March 1909, p. 5.

18. *Guide Michelin*, 1902, p. 618; *Guide Michelin*, 1906, n.p.

19. Emile Gautier, "Une croisade," *Revue mensuelle du Touring Club de France*, February 1897, pp. 43–44.

20. "Note," *Revue mensuelle du Touring Club de France*, April 1899, pp. 142–43; "W-C," *Revue mensuelle du Touring Club de France*, June 1899, p. 229.

21. Dr. Léon-Petit, "Hôtel recommandé," *Revue mensuelle du Touring Club de France*, December 1899, pp. 507–8.

22. "La défense des intérêts nationaux," *Revue mensuelle du Touring Club de France*, May 1901, p. 197.

23. H. Berthe, "Patriotisme et profits," *Revue mensuelle du Touring Club de France*, August 1903, pp. 390–91.

24. *Guide Michelin*, 1908; "Le Lundi de Michelin," *Le Journal*, 20 April 1908, p. 5.

25. *Guide Michelin*, 1908; "Le Lundi de Michelin," *Le Journal*, 20 April 1908, p. 5.

26. "Le Lundi de Michelin," *L'Auto-Vélo*, 19 May 1902, p. 5; *Guide Michelin*, 1911; *Guide Michelin*, 1912; "Le Lundi de Michelin," *L'Auto-Vélo*, 19 May 1902, p. 5; *Guide Michelin*, 1913.

27. "Ce que Michelin a fait pour le tourisme," 1912.

28. "Le Lundi de Michelin," *Le Journal*, 6 July 1908, p. 5.

29. "Le tourisme: Important facteur économique," *Revue mensuelle du Touring Club de France*, July 1912, p. 305.

30. "Le Lundi de Michelin," *Le Journal*, 24 February 1908, p. 5.

31. "Le Lundi de Michelin," *Le Journal*, 20 April 1908, p. 5.

32. Michelin attacked Dunlop's influence in protecting the British market from

Michelin products. "Le Lundi de Michelin," *L'Auto-Vélo,* 27 May 1901, p. 2; "Le Lundi de Michelin," *Le Journal,* 22 November 1909, p. 5.

33. In 1913, France produced 45,000 automobiles and had about 125,000 in operation in the country whereas Germany produced 23,000 and had 70,615 in operation. James M. Laux, *In First Gear: The French Automobile Industry to 1914* (Montreal: McGill-Queen's University Press, 1976), p. 196.

34. Daniel Bordet et al., *Pneu Continental: Le temps des pionniers, 1890–1920* (Paris: Somogy, 1996), p. 24; Continental, *Guide routier et aérien: France, Corse, Algérie, Tunisie* (n.p., 1912).

35. In 1912, in the wake of a border incident known as the Graffenstaden Affair, the French press undertook a fervent campaign against German firms in which Continental was the preferred target. German diplomats believed that *Le Matin* led the charge against foreign firms and that it had received payments—a widespread practice in prewar journalism—from Michelin for the campaign. Raymond Poidevin, *Les relations économiques et financières entre la France et l'Allemagne de 1898 à 1914* (Paris: Armand Colin, 1969), pp. 767–69.

36. "Le Lundi de Michelin," *Le Journal,* 6 April 1914, p. 5; "Le Lundi de Michelin," *Le Journal,* 13 July 1914, p. 5. For reproductions of the pamphlets, see Bardou et al., *Le Pneu Continental.*

37. "Le Lundi de Michelin," *Le Journal,* 16 June 1913, p. 5.

38. Daniel Nordman, "Les Guides-Joanne," in *Lieux de mémoire,* vol. 2, *La Nation,* ed. Pierre Nora (Paris: Gallimard, 1997), p. 1048; Adolphe Joanne, *Itinéraire général de la France: Réseau de Paris à Lyon et à la Méditerranée* (Paris: Hachette, 1861); P. Joanne, *Normandie,* Collection des Guides Joanne—Guides Diamant (Paris: Hachette, 1882). Even the later guides aimed primarily for cyclists took the train lines as a fundamental principle of organization: P. Joanne, *Normandie 1901: Les routes les plus fréquentées,* Collection des Guides Joanne—Guides Diamant (Paris: Hachette, 1901). On Hachette bookstores in train stations, see Eileen Sposato Demarco, "Reading and Riding: Hachette's Railroad Bookstore Network in Nineteenth-Century France" (Ph.D. diss., University of California at San Diego, 1996).

39. Karl Baedeker, *Le Nord-Est de la France: De Paris aux Ardennes, aux Vosges, et au Rhône: Manuel du voyageur* (Leipzig and Paris: Baedeker, 1903); and idem, *Le Sud-Ouest de la France: De la Loire à la frontière d'Espagne: Manuel du voyageur* (Leipzig and Paris: Baedeker, 1906).

40. Touring Club de France, *Excursions et voyages (France-Algérie-Tunisie)* (Paris, n.d.); notices of different editions of these guides began to appear in the *Revue mensuelle du Touring Club de France* in the 1890s.

41. Touring Club de France, *Guide routier: Le Nord* (1897); idem, *Excursions et*

voyages (Paris, n.d.). The combination of practical information and sights was frequent in the TCF's members' submissions describing their trips. See, for example, J. Demartial, "Une excursion dans les Vosges," *Revue mensuelle du Touring Club de France,* October 1898, pp. 388–93.

42. Michelin published a list of the different itineraries in the *Guide Michelin,* 1905, p. 753.

43. *Guide Michelin,* 1900, p. 230; *Guide Michelin,* 1912; Marc Francon has pointed out that the tourist areas for which excursions were provided (Brittany, Normandy, the Châteaux de la Loire, the area around Paris, the Ardennes, the Vosges, the Jura, Savoy, Dauphiné, Provence, Côte d'Azur, Cévennes, Cantal, Auvergne, Dordogne, and the Pyrenees) closely resembled the areas eventually covered by the interwar regional touring guides published by Michelin. Francon, "Le guide vert Michelin," p. 24.

44. It was reported in the 19 June 1907 issue of *La vie au grand air* that a "sumptuous edifice" designed to pamper customers included a "bureau de tourisme." Cited by Bardou et al., *Le Pneu Continental,* p. 32; "Le Lundi de Michelin," *L'Auto,* 10 June 1907, p. 7.

45. "Le Lundi de Michelin," *Le Journal,* 14 March 1910, p. 5; "Le Lundi de Michelin," *Le Journal,* 21 March 1910, p. 5.

46. "Le Lundi de Michelin," *Le Journal,* 14 March 1910, p. 5; "Le Lundi de Michelin," *Le Journal,* 17 June 1912, p. 5.

47. "Le Samedi de Michelin," *L'Illustration,* 17 July 1920, advertising section.

48. J. B. Harley articulates very well this point: "Deconstructing the Map," in *Writing Worlds: Discourse, Text, and Metaphor in the Representation of Landscape,* ed. Trevor J. Barnes and James S. Duncan (London: Routledge, 1992), pp. 231–47. The influence of politics per se in cartography is clear in Jeremy Black, *Maps and History: Constucting Images of the Past* (New Haven: Yale University Press, 1997); Mark Monmonier, *How to Lie with Maps* (Chicago: University of Chicago Press, 1991); and idem, *Drawing the Line: Tales of Maps and Cartocontroversy* (New York: Henry Holt, 1995).

49. Marc Duranthon, *La carte de France, son histoire* (Paris: Solar, 1978), pp. 44–47; Georges Reverdy, *Atlas historique des routes de France* (Paris: Presses des Ponts et Chaussées, 1986), pp. 130–31; *Guide Michelin,* 1900, p. 340; Léon Auscher, *Le tourisme en automobile* (Paris: Dunod, 1904), p. 49.

50. *Guide Michelin,* 1900, pp. 340–41. Auscher also described the maps' advantages and disadvantages in *Le tourisme automobile,* p. 49.

51. The actual work on the roads was done by localities who assessed the service or *prestation,* which could be commuted to a cash payment, required from their in-

habitants. In the course of the nineteenth and twentieth centuries, people increasingly chose the cash payment, essentially a tax. *Les routes de France depuis les origines à nos jours* (Paris: Association pour la Diffusion de la Pensée Française, 1959), p. 140.

52. *Guide Michelin,* 1900, p. 340; Duranthon, *La carte,* p. 49.

53. "Cartes topographiques," *Revue mensuelle du Touring Club de France,* July 1892, p. 141; "Carte vélocipédique des environs de Paris dressée avec le concours du TCF," Bibliothèque Nationale de France (hereafter BN) G CC 428, and "Carte touristique de France dressée avec le concours du TCF," BN G C 379. André Berthelot, "Les cartes du Touring Club," *Revue mensuelle du Touring Club,* October 1897, pp. 403–8; advertisements for the maps, "Carte touriste," *Revue mensuelle,* March 1905, p. ix; and "Carte," *Revue mensuelle,* July 1905, p. ix; Auscher, *Le tourisme en automobile,* p. 50. Maps by Taride, Forest, and Campbell also appeared at approximately the same time, although they do not appear to have captured much of a market: Reverdy, *Atlas,* p. 161.

54. *Guide Michelin,* 1900, p. 341. Another in the scale of 1/200.000e covered the area around Paris. "Carte Routière de Dion-Bouton: Environs de Paris," BN Ge C 3174.

55. Michelin's file for the award of Légion d'honneur, 1897, in AN F/12/5211 Légion d'honneur, Propositions individuelles, "Mia-Mig."

56. *Guide Michelin,* 1900, pp. 340–41. "Circuit d'Auvergne, Eliminatoire française, Coupe Gordon-Bennett, 1905." I have used a facsimile edition done by Michelin for the ninetieth anniversary in May 1995. Michelin explained the map and smaller versions of the racing area appearing in the 1905 red guide in "Le Lundi de Michelin," *L'Auto,* 22 May 1905, p. 7.

57. Louis Baudry de Saunier's article in *Omnia* of 14 April 1906, reprinted in "Le Lundi de Michelin," *L'Auto,* 30 April 1906, p. 7; "Ce que seront les nouveaux Guides-Atlas Michelin de 1907," in *Guide Michelin France,* 1906, p. 2.

58. Baudry de Saunier's article in *Omnia* of 14 April 1906, reprinted in "Le Lundi de Michelin," *L'Auto,* 30 April 1906, p. 7; "Ce que seront les nouveaux Guides-Atlas Michelin de 1907," in *Guide Michelin France,* 1906, p. 2.

59. *Guide Michelin France,* 1907, p. 2.

60. "Le Lundi de Michelin," *Le Journal,* 10 August 1908, p. 5. In 1909, the Michelin guide to Switzerland also included a map in 1/1,000,000. "Le Lundi de Michelin," 16 August 1909, p. 5.

61. "Le Lundi de Michelin," *L'Auto,* 15 July 1907, p. 7; "Le Lundi de Michelin," *Le Journal,* 10 August 1908, p. 5.

62. On the survey, see Ribeill, "Du pneumatique," p. 194.

63. *Guide Michelin France,* 1907; and Francon, "Le guide vert Michelin," p. 43.

64. In addition to the maps themselves, "Le Lundi de Michelin," *Le Journal*, 10 January 1910, p. 5; "Le Lundi de Michelin," *Le Journal*, 16 May 1910, p. 5; "Le Lundi de Michelin," *Le Journal*, 31 July 1911, p. 5; "La carte Michelin," *Le Journal*, 26 August 1912, p. 5.

65. "Le Lundi de Michelin," *Le Journal*, 10 January 1910, p. 5.

66. Francon, "Le guide vert Michelin," p. 45. On the tarring of roads and other interwar improvements, see Reverdy, *L'atlas*, p. 162; and *Routes de France*, p. 135.

67. "Le Lundi de Michelin," *Le Journal*, 25 December 1911, p. 5; Duhamel, *Un demi-siècle*, p. 33; Ribeill, *Du pneumatique*, p. 197.

68. "Le Lundi de Michelin," *Le Journal*, 25 December 1911, p. 5.

69. Léon Auscher compared tourism to an art, which can only be appreciated slowly. It was absolutely essential that drivers not "release the anger of peasants and citydwellers against the innocent [drivers] who will pass after you; don't run over other people's things, whether they are in the form of calves or chickens . . . *and do not scare people*." Auscher, *Le tourisme en Automobile* (Paris: Dunod, 1904), pp. ix, 18.

70. "Le Lundi de Michelin," *Le Journal*, 25 December 1911, p. 5; Duhamel, *Un demi-siècle*, p. 33.

71. Duhamel, *Un demi-siècle*, pp. 25, 27.

72. Ribeill, *Du pneumatique*, p. 197.

73. "Le Lundi de Michelin," *Le Journal*, 9 September 1912, p. 5. The article in *L'Illustration* by F. Honoré had appeared in 8 June 1912. The first "Lundi" calling for the numbering of all roads appeared on 2 September and "Lundis" on the subject continued for the rest of the year.

74. As cited in Ribeill, *Du pneumatique*, p. 198.

75. Because of its interest in aviation, including most obviously the Prix Michelin, the company regularly had a stand at the salon. Assuming that the constituencies for airplanes and automobiles overlapped considerably, Michelin had earlier displayed its signs for town entries (the *plaques merci*) at the 1911 salon.

76. "Le Lundi de Michelin," *Le Journal*, 28 October 1912; "Le Lundi de Michelin," *Le Journal*, 4 November 1912, p. 5; "Le Lundi de Michelin," *Le Journal*, 11 November 1912, p. 5. It has frequently been asserted that Fallières did not know what he was signing and that the wily André Michelin thus tricked him. Given that the entire Michelin booth was covered with information about the numbering system and that a visitor's book to the salon would not be placed in the Michelin booth, the assertions seem to be an example of mythmaking. Fallières was probably operating in the long tradition of official patronage of tourist associations.

77. Clippings pasted in a volume marked "Publicité—Voiture," Michelin archives.

78. Laux, *In First Gear*, p. 196.

79. The original petition can be found in AN F/14/17461. Several other notables, Michelin's illustrator O'Galop, and about twenty-five thousand other people signed the petition at the salon itself; "Le Lundi de Michelin," *Le Journal*, 16 December 1912, p. 5; Ribeill, *Du pneumatique*, p. 198; Francon, "Le guide vert Michelin," p. 47.

80. "Le Lundi de Michelin: L'Angleterre attend que notre voirie fasse son devoir," *Le Journal*, 14 October 1912, p. 5; *Le Plein Air*, 8 November 1912, unnumbered advertisement.

81. "Le Lundi de Michelin," *Le Journal*, 16 September 1912, p. 5. Eugen Weber's point that French peasants and colonized peoples were often compared seems vindicated by this kind of use of the term *indigène*. See *Peasants into Frenchmen: The Modernization of Rural France, 1870–1914* (Stanford: Stanford University Press, 1976).

82. "Théâtre illustré du pneu," 2nd series, last page; smaller version in "Le Lundi de Michelin," *Le Journal*, 19 May 1913, p. 5.

83. Decrees from the minister of public works to the prefects, 17 March 1913, and from the minister of the interior to the prefects, 12 April 1913, both reprinted in *Revue mensuelle du Touring Club*, May 1913, pp. 195–97; "Le Lundi de Michelin," *Le Journal*, 28 April 1914, p. 5.

84. "Le Lundi de Michelin," *Le Journal*, 20 April 1908, p. 5.

85. "Le Lundi de Michelin," *Le Journal*, 24 February 1908, p. 5.

86. Ibid.

87. "Le Lundi de Michelin," *Le Journal*, 28 October 1912, p. 5.

Chapter 3. Touring the Trenches

1. Before the war, the Touring Club de France advocated use of Esperanto, and Michelin even ran advertisements in Esperanto when the international meetings took place in Paris. After World War I, neither organization encouraged use of the international language, and the TCF specifically abandoned the idea in "Un langage international par les chiffres," *Revue mensuelle du Touring Club*, February 1920, p. 4. For a history of Esperanto, including its sizable following in France and the effect of the World Wars on support for it, see Peter G. Forster, *The Esperanto Movement* (The Hague: Mouton, 1982).

2. On the importance of motorized vehicles to the war effort, see André Doumenc, "Les transports automobiles pendant la guerre de 1914–1918," in *Les fronts invisibles: Nourir, fournir, soigner*, ed. Gérard Canini (Nancy: Presses Universitaires de Nancy, 1984), pp. 371–80.

3. Annie Moulin-Bourret, *Guerre et industrie: Clermont-Ferrand, 1912–1922: La victoire du pneu*, vol. 1 (Clermont-Ferrand: Publications de l'Institut d'Etudes du Massif Central, 1997), pp. 172–73, 199–201.

4. Ibid., pp. 192, 202–4; Antoine Champeaux, "Michelin et l'aviation, 1896–1936" (Mémoire de diplôme d'études approfondies, Universités de Montpellier III et Paris I, 1988), pp. 167–232; Lionel Dumond, "L'arrière plan technique et commercial," in *Michelin, Les hommes du pneu: Les ouvriers Michelin, à Clermont-Ferrand, de 1889 à 1940*, ed. André Gueslin (Paris: Editions de l'Atelier/Editions Ouvrières, 1993), pp. 52–54; Michelin, "Michelin pendant la guerre, 1914–1917" (Clermont-Ferrand: Michelin, 1917); Alain Jemain, *Un siècle de secrets* (Paris: Calmann-Lévy, 1982), pp. 77–79; Bergougnan, "Les poilus Bergougnan pendant la grande guerre (1914–1919)," n. p., n.d., in Archives Départementales du Puy-de-Dôme (hereafter ADPD) 6 BIB 3142.

5. Moulin-Bourret, *Guerre et industrie*, vol. 1, pp. 276–86, 303–9; Dumond, "L'arrière plan," pp. 52–54.

6. Moulin-Bourret, *Guerre et industrie*, vol. 2, p. 442; *Bulletin Intérieur Michelin*, 20 June 1946, p. 2; Antoine Champeaux, "Les guides illustrés Michelin des champs de bataille, 1914–1918" (master's thesis, Université de Paris IV, 1984), p. 24.

7. Henry Defert, "Guides à proscrire," *Revue mensuelle du Touring Club*, August 1914–April 1915, p. 28; Henry Defert, "Guides Plagiaires," *Revue mensuelle*, November–December 1917, p. 128; Henry Defert, "L'hôtellerie française aux Français," *Revue mensuelle*, May 1919, pp. 97–98; and Abel Ballif, "Pour 'notre Revanche,'" *Revue mensuelle*, May–June 1916, pp. 81- 85.

8. Léon Auscher, "L'aurore du tourisme nouveau," *Revue mensuelle du Touring Club*, September–October 1916, pp. 132–33.

9. The term comes from Herman Lebovics, *True France: The Wars over Cultural Identity, 1900–1945* (Ithaca, N.Y.: Cornell University Press, 1992).

10. Abel Ballif, "Pour 'notre Revanche,'" *Revue mensuelle du Touring Club*, May–June 1916, pp. 81–85.

11. Pierre Chabert, *Le tourisme américain et ses enseignements pour la France* (Paris: Hachette, 1918), pp. 126–27.

12. Ibid., pp. 127–30.

13. Ibid., pp. 130–31.

14. Antoine Prost, "Les monuments aux morts: Culte républicain? Culte civique? Culte patriotique?" in *Les lieux de mémoire*, ed. Pierre Nora, vol. 1 (Paris: Gallimard, 1997), pp. 199- 223.

15. Circular from Michelin sent to the colonels of the French army dated March 1920, Michelin archives.

16. Fernand Gillet, "Cent ans d'industrie: Histoire anecdotique de la Maison Michelin," 1929, vol. 2, p. 114, typewritten manuscript, Michelin archives.

17. Champeaux, "Les guides illustrés," p. 53.

18. Berger-Levrault, "Nos champs de bataille," p. 1 (unnumbered), a brochure advertising the appearance of the guides, in the file about the guides, Michelin archives; Henry Defert, "Guide pour la visite des champs de bataille," *Revue mensuelle du Touring Club,* July–August 1917, p. 78.

19. Michelin wrote that the readers of its advertisements for the guides no doubt thought that Michelin had something to gain by selling the guides. The company reassured them that the entirety of the profits would go to the Alliance Nationale. "Le Troisième Samedi de Michelin: Les Guides Illustrés Michelin des Champs de Bataille," *L'Illustration,* 3 May 1919, advertising section.

20. Gillet, "Cent ans d'industrie," vol. 2, p. 114.

21. Henry Defert, "Guide pour la visite des champs de bataille," *Revue mensuelle du Touring Club,* July–August 1917, p. 78.

22. Champeaux, "Les guides illustrés," p. 55.

23. André Michelin's presentation, printed in Berger-Levrault, "Nos champs de bataille," pp. 18–20. German soldiers, fearful and mindful of French guerrilla actions of 1870–71, did commit a number of atrocities on the western front. John Horne and Alan Kramer, "German 'Atrocities' and Franco-German Opinion, 1914: The Evidence of German Soldiers' Diaries," *Journal of Modern History* 66 (March 1994): 1–33.

24. Berger-Levrault, "Nos champs de bataille," pp. 4–5, 11–14, 21; "Visite au champ de bataille de l'Ourcq," *Revue mensuelle du Touring Club,* September–October 1917, p. 103. For Ernest Lavisse's many patriotic efforts during the war, see Martha Hanna, *The Mobilization of Intellect: French Scholars and Writers during the Great War* (Cambridge, Mass.: Harvard University Press, 1996).

25. Works by Paul Fussell (*The Great War and Modern Memory* [New York: Oxford University Press, 1975]) and Modris Eksteins (*The Rites of Spring: The Great War and the Birth of the Modern Age* [Boston: Houghton Mifflin, 1989]), by focusing on the "modernism" of an intellectual elite, have tended to understate the place of religion altogether, whereas Jay M. Winter (*Sites of Memory, Sites of Mourning: The Great War in European Cultural History* [Cambridge: Cambridge University Press, 1996]) and Annette Becker (*La guerre et la foi: De la mort à la mémoire [1914–1930]* [Paris: Colin, 1994]) argue that traditional religious faith provided an important means for contemporaries to understand the devastation of the war. George L. Mosse (*Fallen Soldiers: Reshaping the Memory of the World Wars* [New York: Oxford University Press, 1990]) asserts that traditional religious faith was not an important component in remembrance during and after the war because the secular had essentially appropriated traditional sacred expression. Similarly, Antoine Prost has argued that commemoration became a sort of civic religion, at least in France, thus placing

NOTES TO PAGES 97–99

traditional religious language in the service of the French state ("Les monuments aux morts," in *Les Lieux de mémoire*, ed. P. Nora, vol. 1. [Paris: Gallimard, 1997], pp. 199–223). Daniel J. Sherman has recently gone beyond the old modern/traditional debate in order to focus instead on other, more embedded assumptions operating in the larger process of commemorating World War I: *The Construction of Memory in Interwar France* (Chicago: University of Chicago Press, 1999).

26. George Mosse notes the importance of the distinction between pilgrimage and tourism for British travelers to the battlefields as well. Mosse, *Fallen Soldiers*, pp. 153–59. Jay Winter essentially accepts the distinction when writing of pilgrimages in *Sites of Memory, Sites of Mourning*, p. 52. In contrast, David Lloyd has recently emphasized the ambiguity of the distinction while sometimes still making it. David W. Lloyd, *Battlefield Tourism: Pilgrimage and the Commemoration of the Great War in Britain, Australia, and Canada, 1919–1939* (New York: Berg, 1998), especially pp. 6–7.

27. Suzanne Kaufman, "Selling Lourdes: Pilgrimage, Tourism and the Mass-Marketing of the Sacred in Nineteenth-Century France," in *Tourism, Commercial Leisure and National Identities in Nineteenth and Twentieth-Century Europe and North America*, ed. Shelley Baranowski and Ellen Furlough (Ann Arbor: University of Michigan Press, 2001). On Lourdes and pilgrimage to it, see also Ruth Harris, *Lourdes: Body and Spirit in the Secular Age* (New York: Penguin, 1999), and Thomas A. Kselman, *Miracles and Prophecies in Nineteenth-Century France* (New Brunswick, N.J.: Rutgers University Press 1983), especially pp. 118–21.

28. Daniel J. Sherman suggested this idea in a private communication. His work approaches battlefield touring from the perspective of the construction of memory, elucidating some of the psychological links between pilgrimage and tourism: *The Construction of Memory*, especially pp. 7, 13–64. It includes examples of the grief-ridden guilt that postwar survivors often had about their own survival.

29. On tourism during the Franco-Prussian War, Harvey J. Levenstein, *Seductive Journey: American Tourists in France from Jefferson to the Jazz Age* (Chicago: University of Chicago Press, 1998), pp. 140–41.

30. Lloyd, *Battlefield Tourism*, pp. 19–22.

31. "Le Quatrième Samedi Touriste de Michelin: En Eclaireur," *L'Illustration*, 10 May 1919, advertising section.

32. Jean Norton Cru, *Témoins: Essai d'analyse et de critique des souvenirs de combattants édités en français de 1915 à 1928* (Paris: Les Etincelles, 1929). Recent historiography focusing on the experience of ordinary soldiers includes Stéphane Audoin-Rouzeau, *Men at War, 1914–1918: National Sentiment and Trench Journalism in France during the First World War*, trans. Helen McPhail (Providence, R.I.: Berg, 1992);

Eric J. Leed, *No Man's Land: Combat and Identity in World War I* (Cambridge: Cambridge University Press, 1979); Leonard V. Smith, *Between Mutiny and Obedience: The Case of the French Fifth Infantry Division during World War I* (Princeton: Princeton University Press, 1994); and Robert Weldon Whalen, *Bitter Wounds: German Victims of the Great War, 1914–1939* (Ithaca, N.Y.: Cornell University Press, 1984). The new Historial de Péronne, a museum opened in the Somme in 1992, epitomizes the new interest in the experiences of individual soldiers.

33. "Le 12e Samedi de Michelin: La roue Michelin," *L'Illustration,* 5 July 1919, advertising section.

34. "Le 16e Samedi de Michelin: Au Cantonnement!" *L'Illustration,* 9 August 1919, advertising section.

35. "Les Samedis de Michelin," *L'Illustration,* 19 July 1919, 9 August 1919, and 30 August 1919, advertising sections.

36. Jean Bernier described the arrival of a general and his staff on horseback as a "very Napoleonic group" that surveys the sight without dismounting or having any contact with the footsoldiers in the sector: *La percée* (Paris: Albin Michel, 1920), cited in Sherman, *The Construction of Memory,* p. 22.

37. Berger-Levraut, "Nos champs de bataille," p. 1.

38. "Le 3e Samedi Touriste de Michelin: Les Guides Illustrés Michelin des Champs de Bataille," *L'Illustration,* 3 May 1919, advertising section.

39. Michelin, *Rheims and the Battles for Its Possession* (Clermont-Ferrand: Michelin, 1920), is a translation of *Reims et les batailles pour Reims* (1919).

40. Michelin, *Rheims,* pp. 17–18.

41. Michelin, *Amiens avant et pendant la guerre* (Clermont-Ferrand: Michelin, 1920); *Lille avant et pendant la guerre* (1919); *Soissons avant et pendant la guerre* (1919).

42. *La bataille de Verdun (1914–1918)* (Clermont-Ferrand: Michelin, 1919).

43. Gérard Canini, "'L'Illustration' et la bataille de Verdun," in *Verdun 1916* (Verdun: Association Nationale du Souvenir de la Bataille de Verdun/Université de Nancy II, 1976), pp. 179–81.

44. Michelin, *The Battle of Verdun (1914–1918)* (Clermont-Ferrand: Michelin, 1919), pp. 9, 12.

45. Antoine Prost, "Verdun," in *Les lieux de mémoire,* vol. 2, ed. Pierre Nora (Paris: Gallimard, 1997), pp. 1763–64; Norton Cru, *Témoins,* pp. 33–36.

46. Michelin, *Verdun-Argonne-Metz (1914–1918)* (Clermont-Ferrand: Michelin, 1919), especially p. 100; and *Verdun-Argonne (1914–1918)* (1937), especially p. 100. The first guide to Verdun, which does not include the legend, is *La bataille de Verdun (1914–1918)* (1919).

47. Norton Cru, *Témoins,* pp. 33–36; Prost, "Verdun," pp. 1763–64.

48. *The Battle of Verdun* (1919), p. 29.

49. Ibid., p. 12.

50. Michelin, *Les batailles de Picardie* (Clermont-Ferrand: Michelin, 1920), p. 7. For a detailed account of the mutinies in the French Fifth Infantry, see Leonard V. Smith, *Between Mutiny and Obedience: The Case of the French Fifth Infantry Division during World War I* (Princeton: Princeton University Press, 1994).

51. Michelin, *Colmar, Mulhouse, Schlestadt* (Clermont-Ferrand: Michelin, 1920), p. 6; *Strasbourg* (Clermont-Ferrand: Michelin, 1919), pp. 8, 14–60. The guidebooks' interpretation of Alsatian history, although very representative of many French views, both before and after the war, was so far removed from the political and social realities of Alsace that it might well be termed pure propaganda. For recent work on Alsace, see Samuel Huston Goodfellow, *Between the Swastika and the Cross of Lorraine: Fascisms in Interwar Alsace* (DeKalb: Northern Illinois University Press, 1998); Stephen L. Harp, *Learning to Be Loyal: Primary Schooling as Nation Building in Alsace and Lorraine, 1850–1940* (DeKalb: Northern Illinois University Press, 1998); Anthony J. Steinhoff, "Protestants in Strasbourg, 1870–1914: Religion and Society in Late Nineteenth-Century Europe" (Ph.D. diss., University of Chicago, 1996); and Bernard Vogler, *Histoire culturelle de l'Alsace: Du Môyen Age à nos jours, les très riches heures d'une région frontière* (Strasbourg: La Nuée Bleue, 1993).

52. Michelin, *L'Alsace et les combats des Vosges (1914–1918),* vol. 1 (Clermont-Ferrand: Michelin, 1920), p. 6.

53. Ibid.

54. Rudy Koshar, "'What Ought to Be Seen': Tourists' Guidebooks and National Identities in Modern Germany and Europe," *Journal of Contemporary History* 33, no. 3 (July 1998): 323–40. On the propaganda directed at children during World War I, see Stéphane Audoin-Rouzeau, *La guerre des enfants, 1914–1918: Essai d'histoire culturelle* (Paris: Colin, 1993).

55. Jacques Bertillon's presentation in Berger-Levrault, "Nos champs de bataille," pp. 22–23.

56. "Le 54e Samedi de Michelin: Demain?" *L'Illustration,* 15 May 1920, advertising section.

57. André Michelin's presentation, printed in Berger-Levrault, "Nos champs de bataille," p. 18.

58. "Le 5e Samedi Touriste de Michelin: Vers les rives de France," *L'Illustration,* 17 May 1919, advertising section.

59. Michelin, *Clermont-Ferrand, Royat and Surroundings* (Clermont-Ferrand: Michelin, n.d. [dépôt légal stamp on the Bibliothèque Nationale's copy is 1919]); Cham-

peaux, "Les guides illustrés," pp. 167–71. On American soldiers in France, see André Kaspi, *Le temps des Américains: Le concours américain à la France en 1917–1918* (Paris: Publications de la Sorbonne, 1976).

60. Michelin, *The Americans in the Great War*, 3 vols. (Clermont-Ferrand: Michelin, 1919).

61. André Michelin's close collaborator Fernand Gillet claimed that 2 million guidebooks were sold (Gillet, "Cent ans d'industrie," vol. 2, p. 115). Company publications have also claimed that only 1.5 million were sold, a number repeated by Champeaux, "Les guides illustrés," p. 68; and Lloyd, *Battlefield Tourism*, p. 103.

62. Gillet, "Cent ans d'industrie," vol. 2, p. 115.

63. Levenstein, *Seductive Journey*, pp. 225, 340.

64. Michelin, *Verdun-Argonne-Metz* (Clermont-Ferrand: Michelin, 1926), *Verdun-Argonnen* (1929); on the war books boom and renewed interest in the battlefields, see Lloyd, *Battlefield Tourism*, pp. 106–8; on the American Legion annual meeting in Paris in 1927 and its role in spurring other international tourists, see both Lloyd and Levenstein (*Seductive Journey*, pp. 271–75).

65. Prices are from the various guides and from "Le 33e Samedi de Michelin: La colonne de Babel," *L'Illustration*, 6 December 1919, advertising section. After the First World War, Michelin began charging for the red guide.

66. Memorandum of 27 August 1917 in the files for Guides to the Battlefields, Michelin archives.

67. "Ce qu'il faut avoir vu sur les champs de bataille de Verdun: Itinéraire dressé et publié par les soins de ACF, ONT, TCF," n.p., n.d.

68. Office National du Tourisme, Ministère des Travaux Publics, *Guide officiel de la zone des armées: La voie sacrée* (Paris: Paul Mellottée [Librairie des Guides Conty], 1920). The title uses the term *voie sacrée* not as the specific famed path for supplying Verdun but the imagined road that ran along the front from the English Channel to Switzerland, on which postwar tourists were to do their pilgrimage. For this other usage, often referred to by the Latin, *via sacra*, see Lloyd, *Battlefield Tourism*, pp. 27–28, and Modris Eksteins, "Michelin, Pickfords et la grande guerre: Le tourisme sur le front occidental," in *Guerres et culture, 1914–1918* (Paris: A. Colin, 1994), p. 418.

69. The English version was entitled *Guide to the War Regions of France and Belgium* (London, n.d.); there is a copy in the University of Akron archives. The following reference is to the French version at the Bibliothèque Trocadéro; it lacks an actual title page as well as all publication information: Goodrich, *Guide*, p. 73, and several unnumbered pages.

70. Champeaux, "Les guides illustrés," p. 160.

71. Chemin du Nord, *Souvenir de la Grande Guerre: Visite des régions dévastées du nord de la France: Albert, Arras, Lens,* n. p., n.d., available in the tourism section at the Bibliothèque Trocadéro.

72. Chemins de Fer du Nord et de l'Est, *Les champs de bataille de France,* n.p., n.d., available in the tourism section at the Bibliothèque Trocadéro.

73. Lloyd, *Battlefield Tourism,* pp. 98–99, 106. Modris Eksteins notes that the Pickfords seem at certain points to have directly copied the Michelin guides to the battlefields ("Michelin, Pickfords," p. 420).

74. See the accounts of visitors to the battlefields interspersed throughout Lloyd, *Battlefield Tourism.*

75. "Le 33e Samedi de Michelin," *L'Illustration,* 6 December 1919, advertising section; "De belles étrennes pour vos enfants," *L'Illustration,* 13 December 1919, advertising section.

76. Champeaux, "Les guides illustrés," p. 28.

Chapter 4. Saving the French Nation

1. Joseph J. Spengler, *France Faces Depopulation: Postlude Edition, 1936–1976* (Durham, N.C.: Duke University Press, 1979), p. 22. In the period from 1900 to 1930, the French population grew by 6 percent, that of Britain and the Irish Free State 18 percent, Germany 14 percent, Italy 27 percent, and Europe as a whole 29 percent.

2. Colin Dyer, *Population and Society in Twentieth Century France* (New York: Hodder and Stoughton, 1978), p. 5. On the declining birthrate in the late nineteenth and twentieth centuries in Europe, see also John R. Gillis, Louise A. Tilly, and David Levine, eds., *The European Experience of Declining Fertility, 1850–1970: The Quiet Revolution* (Cambridge: Blackwell, 1992); Richard A. Soloway, *Demography and Degeneration: Eugenics and the Declining Birthrate in Twentieth-Century Britain* (Chapel Hill: University of North Carolina Press, 1990); and Michael S. Teitelbaum and Jay M. Winter, *The Fear of Population Decline* (San Diego: Academic Press, 1985).

3. For a particularly articulate description of *dénatalité* as a cultural rather than a demographic issue, see Mary Louise Roberts, *Civilization without Sexes: Reconstructing Gender in Postwar France, 1917–1927* (Chicago: University of Chicago Press, 1994), pp. 97–98.

4. Joshua H. Cole, "'There Are Only Good Mothers': The Ideological Work of Women's Fertility in France before World War I," *French Historical Studies* 19, no. 3 (Spring 1996): 639–72; and Jean Elisabeth Pedersen, "Regulating Abortion and Birth Control: Gender, Medicine, and Republican Politics in France, 1870–1920,"

French Historical Studies 19, no. 3 (Spring 1996): 673–98. See also Cole, "The Power of Large Numbers: Population and Politics in the Nineteenth Century" (Ph.D. diss., University of California at Berkeley, 1991). For the larger context of bourgeois men legislating the fate of working-class French women for the sake of increased natality or "morality," before 1914, see Elinor A. Accampo, Rachel G. Fuchs, and Mary Lynn Stewart, eds., *Gender and the Politics of Social Reform in France, 1870–1914* (Baltimore: Johns Hopkins University Press, 1995); Rachel G. Fuchs, *Poor and Pregnant in Paris: Strategies for Survival in the Nineteenth Century* (New Brunswick, N.J.: Rutgers University Press, 1992); Karen Offen, "Depopulation, Nationalism, and Feminism in Fin-de-Siècle France," *American Historical Review* 89 (June 1984): 648–76; Sylvia Schafer, *Children in Moral Danger and the Problem of Government in Third Republic France* (Princeton: Princeton University Press, 1997); and Alisa Klaus, *Every Child a Lion: The Origins of Maternal and Infant Health Policy in the United States and France, 1890–1920* (Ithaca, N.Y.: Cornell University Press, 1993).

5. Cole, "'There Are Only Good Mothers,'" pp. 663–64; Alain Becchia, "Les milieux parlementaires et la dépopulation de 1900 à 1914," *Communications* 44 (1986): 202–5. On the founding of the Alliance Nationale, see Robert Talmy, *Histoire du mouvement familiale en France (1896–1930)*, vol. 1 (Paris: UNCAF, 1962), pp. 66–74.

6. Mary Lynn [McDougall] Stewart, *Women, Work, and the French State: Labour Protection and Social Patriarchy, 1879–1919* (Kingston, Ontario: McGill-Queen's University Press, 1989), pp. 169–70; Susan Pedersen, *Family, Dependence, and the Origins of the Welfare State: Britain and France, 1914–1945* (Cambridge: Cambridge University Press, 1993), pp. 73–74.

7. Richard Tomlinson, "The 'Disappearance' of France, 1896–1940: French Politics and the Birthrate," *Historical Journal* 28 (1985): 407–8; Douglas Porch, *The March to the Marne, 1871–1914* (Cambridge: Cambridge University Press, 1981), p. 194. On the increasingly virulent nationalist rhetoric in the years before the war, see Eugen Weber, *The Nationalist Revival in France, 1905–1914* (Berkeley: University of California Press, 1959).

8. Richard Tomlinson, "The Politics of *Dénatalité*" (Ph.D. diss., Christ's College, Cambridge, 1984), p. 111.

9. American rubber companies provided an extensive social welfare for their own employees, but they did not feature those initiatives in advertising or other national public forums. This assessment results from my current research and Daniel Nelson, *American Rubber Workers and Organized Labor, 1900–1941* (Princeton: Princeton University Press, 1988), pp. 20–21, 56–63, 113.

10. Offen, "Depopulation, Nationalism, and Feminism," pp. 648–76.

11. Françoise Thébaud, "Le mouvement nataliste dans la France de l'entre-deux-guerres: L'Alliance nationale pour l'accroissement de la population française," *Revue d'histoire moderne et contemporaine* 32 (April–June 1985): 278.

12. "Don: M. André Michelin," *Bulletin de l'Alliance Nationale,* October 1913, p. 219. This is the first public mention of a gift from Michelin in the *Bulletin.* It is entirely possible that André Michelin offered earlier financial support that was not reported in the *Bulletin.* In 1937, when Boverat managed at last to get himself named president of the Alliance Nationale, the *Revue* (the *Bulletin* was renamed in 1922) noted that Michelin, who had been impressed by Boverat's pamphlets, had earlier offered the organization a check for 2,000 francs. "Fernand Boverat: Président de l'Alliance Nationale," *Revue de l'Alliance Nationale* (December 1937): 359. For a detailed description of Boverat and his politics, see Cheryl A. Koos, "Engendering Reaction: The Politics of Pronatalism and the Family in France, 1914–1944" (Ph.D. diss., University of Southern California, 1996), chapter 2.

13. Thébaud, "Le mouvement nataliste," p. 278; "Les élections législatives," *Bulletin de l'Alliance Nationale,* April 1914, pp. 353–55.

14. Tomlinson, "The 'Disappearance' of France," p. 408; *Bulletin de l'Alliance Nationale,* May 1914, p. 388.

15. "Notre Comité d'Assistance aux Familles Nombreuses," *Bulletin de l'Alliance Nationale,* January 1915, pp. 568–69.

16. Marie-Monique Huss, "Pronatalism and the Popular Ideology of the Child in Wartime France: The Evidence of the Picture Postcard," in *The Upheaval of War: Family, Work, and Welfare in Europe, 1914–1918,* ed. Richard Wall and Jay Winter (Cambridge: Cambridge University Press, 1989), pp. 329–67; idem, "Pronatalism in the Inter-war Period in France," *Journal of Contemporary History* 25 (1990): 39–68.

17. Tomlinson, "The 'Disappearance' of France," p. 409.

18. Ibid., p. 405.

19. Roberts, *Civilization without Sexes,* pp. 93–119, 160–61; Robert Nye, *Crime, Madness, and Politics in Modern France: The Medical Concept of Decline* (Princeton: Princeton University Press, 1984), pp. 139–40; and Jean Elisabeth Pedersen, "Regulating Abortion and Birth Control: Gender, Medicine, and Republican Politics in France, 1870–1920," *French Historical Studies* 19, no. 3 (Spring 1996): 673–98. For the text of the law of 1920, see Susan Groag Bell and Karen M. Offen, eds., *Women, the Family, and Freedom: The Debate in Documents,* vol. 2 (Stanford: Stanford University Press, 1983), pp. 309–10.

20. "Création d'un Conseil Supérieur de la Natalité," *Bulletin de l'Alliance Nationale,* February 1920, pp. 219–22; Andrés Horacio Reggiani, "Birthing the French Welfare State: Political Crises, Population, and Public Health, 1914–1960" (Ph.D. diss., State University of New York, Stony Brook, 1998), p. 127.

21. Tomlinson, "The 'Disappearance' of France," p. 412. The membership number fell to about twenty thousand by 1935. Tomlinson attributes the drop in the 1930s to the Depression. For more on the Alliance Nationale in the interwar period, see Cheryl A. Koos, "Gender, Anti-individualism, and Nationalism: The Alliance Nationale and the Pronatalist Backlash against the *Femme moderne*, 1933–1940," *French Historical Studies* 19, no. 3 (Spring 1996): 699–723; Koos, "Engendering Reaction"; Thébaud, "Le mouvement nataliste"; Richard Tomlinson, "The 'Disappearance' of France"; Huss, "Pronatalism in the Inter-war Period," pp. 39–68.

22. Berger-Levrault, "Nos champs de bataille" (1917), Michelin archives, reprinted in Antoine Champeaux, "Les guides illustrés Michelin des champs de bataille, 1914–1918" (master's thesis, Université de Paris IV, 1984), pp. 173–82. Bertillon's speech also appeared in "Guides Michelin," *Bulletin de l'Alliance Nationale*, October 1917, pp. 382–86. On postcards and "Un bon coup de baïonette" in particular, see Huss, "Pronatalism and the Popular Ideology," pp. 329–67, especially p. 342.

23. "L'Ame d'un peuple en folie," "Le Samedi de Michelin," *L'Illustration*, 21 February 1920.

24. "Demain? Le 54e samedi de Michelin," *L'Illustration*, 15 May 1920, advertising section.

25. Ibid.

26. Cole, "'There Are Only Good Mothers,'"; Cole, "The Power of Large Numbers"; Hervé Le Bras, *Marianne et les lapins: L'obsession démographique* (Paris: Olivier Orban, 1991).

27. "Demain? Le 54e samedi de Michelin," *L'Illustration*, 15 May 1920, advertising section.

28. Ibid.

29. For Bibendum with the *jumelés*, or dual tires, see "Le Lundi de Michelin," *Le Journal*, 21 August 1911, p. 5; "Le Lundi de Michelin," *Le Journal*, 11 September 1911, p. 5; "Le Lundi de Michelin," *Le Journal*, 13 November 1911, p. 5; "Le Lundi de Michelin," *Le Journal*, 11 December 1911, p. 5; "Le Lundi de Michelin," *Le Journal*, 29 July 1912, p. 5.

30. "Souvenir du Salon de l'Aviation, Noël 1919," Michelin archives; there is a photograph of the Michelin booth at the aviation salon in René Bletterie, *Michelin: Clermont-Ferrand, Capitale du pneu 1900/1920* (Paris: Civry, 1981), p. 133.

31. In the announcement of the winners of the Prix Michelin de la Natalité, Paul Haury "a eu déjà trois enfants dont deux sont vivants," whereas the teacher [*instituteur*] Marcel Launay "a déjà eu trois enfants dont deux sont vivants." In addition to the unusual usage of "a eu," such lists in the *Revue* are reminders of the high rate of infant mortality, a subject that received comparatively little of the Alliance's

attention. "Prix Michelin de la Natalité," *Revue de l'Alliance Nationale,* May 1923, p. 145.

32. Treasurer's report by Jules Jacob, *Bulletin de l'Alliance Nationale,* March 1920, p. 298.

33. "La campagne électorale de l'Alliance Nationale," *Revue de l'Alliance Nationale,* April 1924, pp. 100–104; "Messieurs André and Edouard Michelin font un nouveau don," *Revue de l'Alliance Nationale,* December 1924, p. 355; "Un nouveau don de MM. Michelin," *Revue de l'Alliance Nationale,* May 1926, p. 132; "Un nouveau don de MM. Michelin," *Revue de l'Alliance Nationale,* April 1928, p. 105.

34. The Etienne Lamy prizes awarded 10,000 francs to "large families of French and Catholic peasants chosen among the poorest, most prolific, the most religious, and of the purest morality." The Cognacq-Jay prizes were ninety (one per department) 25,000 franc awards given to a couple with at least nine children as well as 1,000 francs to couples with five legitimate children: Reggiani, "Birthing the French Welfare State," p. 133, citing *La Famille* 6 (April 1926); see also Pedersen, *Family, Dependence, and the Origins of the Welfare State,* p. 363.

35. "Prix Michelin de la natalité," *Revue de l'Alliance Nationale,* June 1922, pp. 161–64.

36. Roberts, *Civilization without Sexes,* p. 47.

37. "Programme du concours," *Revue de l'Alliance Nationale,* June 1922, p. 165; "Programme du prix Michelin de la natalité," *Revue de l'Alliance Nationale,* June 1922, pp. 163–64.

38. "Programme du concours," *Revue de l'Alliance Nationale,* June 1922, p. 165.

39. "Prix Michelin de la Natalité," *Revue de l'Alliance Nationale,* January 1923, p. 31; "Prix Michelin de la Natalité," *Revue de l'Alliance Nationale,* May 1923, p. 145. For more on Haury, see Koos, "Engendering Reaction," chapter 3.

40. Paul Haury, "La vie ou la mort de la France: Prix Michelin de la Natalité, Concours doté de 120,000 francs de prix organisé par l'Alliance Nationale pour l'accroissement de la population française, brochure classée première (prix de 50,000 francs)," pp. 1–5. On the importance of doctors in the parliaments of the Third Republic and the medical rhetoric, see Jack D. Ellis, *The Physician-Legislators of France: Medicine and Politics in the Early Third Republic, 1870–1914* (Cambridge: Cambridge University Press, 1990); Roberts, *Civilization without Sexes,* pp. 94–119; and Pedersen, "Regulating Abortion and Birth Control," pp. 673–98.

41. Haury, "La vie ou la mort," pp. 9–11.

42. "La natalité et la sécurité du pays," *Revue de l'Alliance Nationale,* November 1924, p. 331.

43. Haury, "La vie ou la mort," p. 13.

44. Ibid., p. 17.

45. Ibid., pp. 18–22, 29. On the *garçonne*, see Roberts, *Civilization without Sexes*, pp. 19–88.

46. Haury, "La vie ou la mort," p. 32.

47. "Allocations de famille," *Bulletin de l'Alliance Nationale*, October 1916, p. 144; "Michelin pendant la guerre, 1914–1917" (Clermont-Ferrand: Michelin, 1917), pp. 34–36. In 1914, the company began to pay 30 francs for each child born and 90 francs if the mother had worked for Michelin at least nine months.

48. Henri Hatzfeld, *Du paupérisme à la sécurité sociale: Essai sur les origines de la sécurité sociale en France, 1850–1940* (Paris: Armand Colin, 1971), p. 174; Pedersen, *Family, Dependence, and the Origins of the Welfare State*, pp. 75–76.

49. Pedersen, *Family, Dependence, and the Origins of the Welfare State*, pp. 73–77; Hatzfeld, *Du paupérisme à la sécurité sociale*, p. 174.

50. Pedersen, *Family, Dependence, and the Origins of the Welfare State*, pp. 227–28; André Gueslin, "Le système social Michelin (1889–1940)," in *Michelin, les hommes du pneu: Les ouvriers Michelin, à Clermont-Ferrand, de 1889 à 1940*, ed. André Gueslin (Paris: Editions de l'Atelier/Editions Ouvrières, 1993), p. 109.

51. Pedersen, *Family, Dependence, and the Origins of the Welfare State*, pp. 116–18.

52. "Les allocations de famille," *Bulletin de l'Alliance Nationale*, October 1916, p. 145; Gueslin, "Les système social Michelin," p. 110. In the case of death of the head of the family, assumed to be the father, a family with two children received 25 francs monthly and one with three children 70 francs monthly; those with four and five children, 88 and 100 francs respectively. If the father died on the field of battle, Michelin granted 400 francs for one child, 600 for two, and 800 for three. "Allocations familiales," *Bulletin de l'Alliance Nationale*, October 1916, pp. 145–46.

53. "Allocations et rentes pour les familles nombreuses" (Clermont-Ferrand: Michelin, 1923), p. 3.

54. "Une expérience de natalité" (Clermont-Ferrand: Michelin, 1926), pp. 3, 11.

55. "Assemblée Générale," *Revue de l'Alliance Nationale*, January 1927, p. 12.

56. "Une expérience de natalité" (1926), pp. 8–9, 13–15; "Les industriels et la natalité," *Revue de l'Alliance Nationale*, February 1926, pp. 33–42.

57. G. Dequidt and G. Forestier, "Une expérience de natalité: L'efficacité des allocations familiales sur la natalité est-elle démontrée?" *Le Mouvement sanitaire* (31 July 1926), pp. 485–503, Michelin archives.

58. Michelin, "Une expérience de natalité," *Prospérité* 1, no. 4 (January–March 1929).

59. "Une expérience de natalité" (1926), p. 12.

60. "Les industriels et la natalité," *Revue de l'Alliance Nationale*, February 1926, p. 42.

61. Ibid., p. 41. There were in March 1925, 13,890 French workers and 1,309

"foreign" ones, while there were 1,405 salaried employees. See also Ralph Schor, *L'opinion française et les étrangers en France, 1919–1939* (Paris: Publications de la Sorbonne, 1985).

62. "Travaux du Conseil Supérieur de la Natalité," *Bulletin de l'Alliance Nationale*, September 1920, p. 448; "A la merci des étrangers," *Revue de l'Alliance Nationale*, December 1925, p. 362.

63. "Travaux du Conseil Supérieur de la Natalité," *Bulletin de l'Alliance Nationale*, September 1920, p. 448; see also the practices of *caisses* in the metallurgical industries in Paul V. Dutton, "The *Salaire Vital:* Family Allowances and the French Welfare State" (Ph.D. diss., University of California at San Diego, 1997).

64. "Une expérience de natalité" (1926), pp. 5–7.

65. See Susan Pedersen's extended description of the Consortium Textile de Roubaix-Tourcoing, *Family, Dependence, and the Origins of the Welfare State*, pp. 239–61, especially p. 246.

66. "Allocations et rentes," p. 7.

67. "Conférence aux chefs de service," 12 July 1924, in "Notes d'Edouard Michelin," Michelin archives.

68. Pedersen, *Family, Dependence, and the Origins of the Welfare State*, pp. 236–61, especially p. 244; on Michelin's wage rates, see Gueslin, "Le système social Michelin," especially pp. 98–99.

69. Pedersen, *Family, Dependence, and the Origins of the Welfare State*, pp. 232–33.

70. Hatzfeld, *Du paupérisme à la sécurité sociale*, p. 175.

71. The Alliance Nationale sent "Une expérience de natalité" as a supplement to the *Revue* to its thirty-five thousand members and André Michelin's "Les industriels et la natalité" (*Revue de l'Alliance Nationale*, February 1926, pp. 33–45) to all thirty thousand doctors practicing in France. "Assemblée Générale," *Revue de l'Alliance Nationale*, January 1927, p. 12.

72. There were only 6 *caisses* in France in 1920, but 230 by 1930. The number of firms participating grew from 230 to 32,000, and the number of employees affected grew from 50,000 to 1,880,000. Dominique Ceccaldi, *Histoire des prestations familiales en France* (Paris: UNCAF, 1957), p. 21.

73. "Avantages acquis aux familles nombreuses en 1923–1924," *Revue de l'Alliance Nationale*, May 1924, pp. 156–57; "Une expérience de natalité" (1926), p. 6.

74. Michelin, "On peut faire reculer la tuberculose: Compte-rendu d'une expérience," *Prospérité* (April–June 1937): 5.

75. Pedersen, *Family, Dependence, and the Origins of the Welfare State*, p. 378. This section owes much to Pedersen's general analysis of the transformation of employers' *caisses* into a veritable family allowance system in France.

76. Ibid., p. 370; "Mme. André Michelin et M. Jean Michelin," *Revue de l'Alliance Nationale*, June 1931, p. 522; "Appel à la nation," *Revue de l'Alliance Nationale*, July 1934, pp. 194–95.

Chapter 5. Advocating Aeronautics

1. Jean Arren, *Sa majesté la publicité* (Tours: Mame, 1914), p. 300.

2. Fernand Gillet, "Michelin et l'aviation," typewritten manuscript, p. 1, Michelin archives. Gillet was a close collaborator of André Michelin in the Parisian tourist office before the First World War; he worked with Michelin on advertising after the war.

3. Daniel Bordet, Frédérique Decoudin, and Jacques Dreux, *Pneu Continental: Le temps des pionniers, 1890–1920* (Paris: Somogy, 1996), pp. 22, 64.

4. Robert Wohl, *A Passion for Wings: Aviation and the Western Imagination, 1908–1918* (New Haven, Conn.: Yale University Press, 1996), pp. 29, 41–42, 46–48; Antoine Champeaux, "Michelin et l'aviation, 1896–1936" (Mémoire de diplôme d'études approfondies, Universités de Montpellier III and Paris I, 1988), p. 29; Emmanuel Chadeau, *De Blériot à Dassault: Histoire de l'industrie aéronautique en France, 1900–1950* (Paris: Fayard, 1987), pp. 17, 27, 33. Chadeau notes that only 260,000 francs would actually be distributed in 1908, with Farman receiving 128,000 francs, Blériot 27,500 francs, and Delagrange 25,000; while the prizes may have stimulated the development of aviation, Chadeau points out that they did not cover the costs of maintaining the airplanes, workshops, and assistants.

5. Gillet, "Michelin et l'aviation," p. 4.

6. According to the Michelin company, after seeing a film of Santos-Dumont's flight, André Michelin attempted to convince his brother to undertake airplane production, but Edouard refused, suggesting instead the sponsorship of a prize for the aviator capable of passing over the Arc de Triomphe and then landing 400 kilometers away at the top of the Puy-de-Dôme, a peak just outside Michelin's hometown of Clermont-Ferrand, at an altitude of 1,465 meters. The story may well be true, but the primary sources date from more than fifty years after the fact. *Bulletin Intérieur Michelin*, 10 March 1961, pp. 2–3; "La rubrique d'un siècle," *Michelin Magazine*, August–September 1989, p. 3.

7. "Prizes are wanted to encourage aviation," *New York Herald*, 26 March 1908. This is one of the hundreds of articles that Michelin gathered and marked "Publicité" for its own files, Michelin archives.

8. The actual rules issued by the Aéro-Club de France can be found in the Michelin archives and have been reproduced in Champeaux, "Michelin et l'aviation," pp. 34, 61–62.

9. "Le Lundi de Michelin," *Le Journal*, 20 July 1908, p. 5; Gillet, "Michelin et l'aviation," pp. 8–9; "Le Lundi de Michelin," *Le Journal*, 21 December 1908, p. 5; "Le Lundi de Michelin," *Le Journal*, 28 December 1909, p. 5.

10. Champeaux, "Michelin et l'aviation," pp. 72–74; Wohl, *A Passion for Wings*, pp. 33–46.

11. Gillet, "Michelin et l'aviation," pp. 12–13.

12. This point is based on the number of articles that were cut out and pasted into Michelin's volumes tracking press coverage. Publicité-Aviation, Michelin archives.

13. Champeaux, "Michelin et l'aviation," pp. 82–85.

14. "Le Lundi de Michelin," *Le Journal*, 31 March 1913, p. 5.

15. Wohl, *A Passion for Wings*, p. 58; Emmanuel Chadeau, *Le rêve et la puissance: L'avion et son siècle* (Paris: Fayard, 1996), p. 52. Chadeau notes that *Le Matin* doubled the Ruinart offer of 12,500, making the total 25,000 francs (Chadeau, *De Blériot à Dassault*, p. 32).

16. Wohl, *A Passion for Wings*, pp. 48–66, especially p. 63. On the ways that Gallic heroes were used to construct a certain French national identity since the nineteenth century, see Christian Amalvi, "Vercingétorix dans l'enseignement primaire," in *Nos ancêtres les Gaulois* [Actes du colloque international de Clermont-Ferrand] , ed. Paul Viallaneix and Jean Ehrard (Clermont-Ferrand: Université de Clermont-Ferrand II, 1982), pp. 349–55; Amalvi, "Vercingétorix ou les métamorphoses idéologiques et culturelles de nos origines nationales," in *De l'art et de la manière d'accommoder les héros de l'histoire de la France*, ed. C. Amalvi (Paris: Albin Michel, 1988), pp. 53–87; André Simon, *Vercingétorix et l'idéologie française* (Paris: Imago, 1989); and Charlotte Tacke, *Denkmal im sozialen Raum: Nationale Symbole in Deutschland und Frankreich im neunzehnten Jahrhundert* (Göttingen: Vandenhoeck und Ruprecht, 1995).

17. Antoine Champeaux found a handbill entitled "Tissus caoutchoutés Michelin," and a photograph of what appear to be Michelin dirigibles at the first aeronautical salon in 1909. Champeaux reproduced the handbill and photograph in "Michelin et l'aviation," pp. 24–25. The handbill notes that Latham's plane was outfitted with Michelin fabric and then provides a detailed price list for different weights of the fabric. Fernand Gillet noted that Michelin experimented with rubberized cloth for dirigibles but found them wanting and abandoned the experiments. Gillet, "Cent ans d'industrie: Histoire anecdotique de Michelin," vol. 2, pp. 58–60, Michelin archives.

18. Bordet et al., *Pneu Continental*, p. 66.

19. Gillet, "Michelin et l'aviation," pp. 10–12; the letter was published in *Le Journal*, 18 October 1909; "Le Lundi de Michelin," *Le Journal*, 14 February 1910, p. 5.

20. "Le Lundi de Michelin," *Le Journal*, 14 February 1910, p. 5. Raymond Poide-

vin, *Les relations économiques et financières entre la France et l'Allemagne de 1898 à 1914* (Paris: Armand Colin, 1969), p. 767. In what is apparently a misprint, Poidevin places the *République* crash in 1911.

21. *L'Aurore,* 8 March 1908; Gillet, "Michelin et l'aviation," pp. 35–36; "Le Lundi de Michelin," *Le Journal,* 16 March 1908, p. 5. According to Gillet ("Michelin et l'aviation," p. 6), no one took Michelin up on the bet.

22. André Michelin, "Les paroles qu'il faut dire," *L'Aéro,* 30 March 1911, p. 1; Gillet, "Michelin et l'aviation," pp. 17–18.

23. Gillet, "Michelin et l'aviation," p. 13.

24. Peter Fritzsche, *A Nation of Fliers: German Aviation and the Popular Imagination* (Cambridge, Mass.: Harvard University Press, 1992), pp. 9–58. On the extent to which aviation exemplified modernity, see also Chadeau, *Le rêve et la puissance;* Laurence Goldstein, *The Flying Machine and Modern Literature* (Bloomington: Indiana University Press, 1986); and especially Wohl, *A Passion for Wings.*

25. Chadeau, *Le rêve et la puissance,* p. 71.

26. Wohl, *A Passion for Wings,* pp. 69–94; Fritzsche, *A Nation of Fliers,* pp. 38–41.

27. Letter from Michelin brothers to the president of the Aéro-Club de France, 22 August 1911, in Gillet, "Michelin et l'aviation," pp. 20–21.

28. Ibid.; and the Règlement issued by the Aéro-Club de France, reproduced in Champeaux, "Michelin et l'aviation," pp. 108–9. Champeaux, citing Pierre Lorain, claims that the rectangular target was a clear reference to the dirigible hangar, potentially enabling French pilots quickly to attack the hangars before the dirigibles would be ready for action: "Michelin et l'aviation," p. 110.

29. Henri Desgrange in *L'Auto,* 23 August 1911; "Les Bombardeurs," *L'Aurore,* 24 August 1911, p. 1; Georges Prade, "La guerre dans les airs," *Le Journal,* 20 November 1911, p. 7. Michelin gathered such articles into three large volumes labeled "Publicité-Aérocible," Michelin archives.

30. Six of the Michelin's original twelve postcards have been reproduced in Pierre-Gabriel Gonzalez, *Bibendum: Publicité et objets Michelin* (Paris: Editions du Collectionneur, 1994), p. 25; "Le Lundi de Michelin," *Le Journal,* 15 January 1912, p. 5; "Le Lundi de Michelin," *Le Journal,* 4 March 1912, p. 5; "Le Lundi de Michelin," *Le Journal,* 11 March 1912, p. 5.

31. "Le Lundi de Michelin," *Le Journal,* 28 August 1911, p. 5.

32. Michelin, leading article, *Le Matin,* 7 December 1911, p. 1.

33. The numbers are from the article "Aéronautique: Les avions départementaux," *L'Action,* 27 February 1912, p. 5.

34. Michelin, "Notre avenir est dans l'air" (n.p., 1912), Michelin archives.

35. Although Robert Wohl clearly did not know about the pamphlet by Michelin

and attributes the title to Picasso, his account of the cultural impact of aviation remains the best single attempt to explain just how important the idea of aviation was in early-twentieth-century Europe. Wohl, *A Passion for Wings*, especially p. 272.

36. Michelin speech at the awarding of the Aérocible prize, printed in *L'Aérophile*, 15 January 1914, pp. 45–47.

37. Gillet, "Michelin et l'aviation," p. 25. Noting that several sources refer to such a donation, Chadeau did not find reference to it in military files, *De Blériot à Dassault*, p. 78.

38. Michelin, "Notre sécurité est dans l'air" (n.p., 1919), pp. 1, 10.

39. Emmanuel Chadeau estimates that France had 450 airplanes in service by the summer of 1914, in Chadeau, *De Blériot à Dassault*, pp. 38–40, but he does not cite the sources for his number. John Morrow claims that France had 141 planes, but he does not list a source either (*The Great War in the Air: Military Aviation from 1909 to 1921* [Washington, D.C.: Smithsonian, 1993], p. 35). On the role of airplanes in the war, see Chadeau, *Le rêve et la puissance*, pp. 86–87.

40. Rapport d'Aubigny, 5 July 1915, Archives Nationales C 7615, Dossier 2677, cited in Annie Moulin-Bourret, *Guerre et industrie, Clermont-Ferrand, 1912–1922: La victoire du pneu*, vol. 1 (Clermont-Ferrand: Publications de l'Institut d'Etudes du Massif Central, 1997), p. 204.

41. Ibid., pp. 276–95, especially p. 280; Champeaux, "Michelin et l'aviation," pp. 169–288.

42. Moulin-Bourret, *Guerre et industrie*, vol. 1, pp. 282–95; Champeaux, "Michelin et l'aviation," pp. 293–97.

43. Morrow, *The Great War in the Air*, pp. 212–14; Chadeau, *Le rêve et la puissance*, pp. 292–94; Moulin-Bourret, *Guerre et industrie*, vol. 1, pp. 297–99.

44. Morrow, *The Great War in the Air*, pp. 102, 376; Fritzsche, *A Nation of Fliers*, pp. 69–70.

45. Moulin-Bourret, *Guerre et industrie*, vol. 1, p. 307.

46. Michelin, "Notre sécurité est dans l'air," p. 27.

47. Form letter dated January 1920 from Michelin to booksellers, reproduced in Champeaux, "Michelin et l'aviation," p. 394.

48. Michelin, "Notre sécurité est dans l'air," p. 6.

49. Ibid., pp. 3, 18–19, 27.

50. Ibid., pp. 28–32.

51. Ibid., pp. 33, 36.

52. Ibid., p. 37.

53. Ibid., pp. 37, 39–40.

54. "Pourquoi il nous faut une flotte aérienne," *Journal des Débats*, 20 December 1919, p. 5.

55. Ibid.

56. The brochure carefully noted that of 1,884 Breguets produced, 147 were for the American army. Of the 342,000 bombs, 27,700 had been for the Americans. Michelin, "Notre sécurité est dans l'air," p. 27.

57. Champeaux, "Michelin et l'aviation," pp. 86–102.

58. "M. André Michelin," *L'Aérophile*, 1 March 1920, p. 75.

59. André Michelin was speaking in his capacity as president of the Aéro-Club de France at the Salon de l'Aéronautique during a banquet presided over by Barthou on 24 November 1921. Gillet, "Michelin et l'aviation," p. 35; Champeaux, "Michelin et l'aviation," p. 452.

60. "Pour la puissance de l'Aviation française," *L'Aérophile*, 1 October 1920, pp. 289–90; Moulin-Bourret, *Guerre et industrie*, vol. 2, p. 711.

61. Gillet claims that Michelin realized the limits of the Aéro-Club for his agenda. "Michelin et l'aviation," p. 36; Champeaux, "Michelin et l'aviation," pp. 455–56, 465.

62. Champeaux, "Michelin et l'aviation," pp. 466–83. Although Champeaux does not note the social composition of the audience, his list of the group's activities makes the profile of the intended audience obvious.

63. "Le Danger allemand: Aviation et guerre chimique: Conférence faite les 9 et 14 Décembre 1922 par M. André Michelin," (n.p., n.d.), p. 9, Michelin archives. This was the pamphlet version of his speech.

64. Ibid., p. 25. Peter Fritzsche has shown that initially gliding had not connoted resistance to the terms of the Treaty of Versailles but that it increasingly came to embody German nationalism (Fritzsche, *A Nation of Fliers*, pp. 103–31). On the implementation of the treaty, which itself did not ban civilian aircraft but did place restrictions on horsepower, speed, altitude, and payload, see Fritzsche, pp. 106–7; Morrow, *The Great War in the Air*, pp. 350–56; Chadeau, *Le rêve et la puissance*, pp. 152–55.

65. "Le Danger aérien allemand: Discours prononcés à la Sorbonne le 31 octobre 1923," (Paris: *L'Echo de Paris*, 1923), p. 47. This is the published version of his speech.

66. Champeaux, "Michelin et l'aviation," pp. 527–36; on Luft Hansa, see Fritzsche, *A Nation of Fliers*, pp. 146–47, 162–78.

67. Léon Barthou, "Le tourisme aérien," *Revue mensuelle du Touring Club*, February 1911, pp. 77–78.

68. Chadeau, *De Blériot à Dassault*, p. 36.

69. Continental, *Guide routier et aérien: France-Corse-Algérie-Tunisie* (n.p., 1912).

70. Lt. Thoret, "Le Tourisme aérien au Mont-Blanc," *La Revue du Touring Club,* July 1928, pp. 141–43.

71. "Le Voyage aérien du Touring Club de France à Strasbourg," *Revue du Touring Club,* November 1929, pp. 265–66.

72. Michelin, *Guide aérien: France, Afrique du Nord, A.O.F.* (Clermont-Ferrand: Michelin, 1930), title page, p. 2.

73. Michelin, *Guide aérien France, 1935–36* (n.p., n.d.; available at the Bibliothèque Nationale).

74. Champeaux, "Michelin et l'aviation," pp. 561–68; Moulin-Bourret, *Guerre et industrie,* vol. 2, p. 712.

Chapter 6. Advocating Americanization?

1. Georges Duhamel, *America the Menace: Scenes from the Life of the Future,* trans. Charles Miner Thompson (Boston: Houghton Mifflin, 1931).

2. Jean-Louis Cohen, *Scenes of the World to Come: European Architecture and the American Challenge, 1893–1960* (Paris: Flammarion, 1995), pp. 39–83; J.-L. Cohen and H. Damisch, *Américanisme et modernité: L'idéal américain dans l'architecture* (Paris: EHESS, 1993); and Isabelle Jeanne Gournay, "France Discovers America, 1917–1939: French Writings on American Architecture" (Ph.D. diss., Yale University, 1989).

3. Richard Abel, *French Cinema: The First Wave, 1915–1929* (Princeton: Princeton University Press, 1984); Victoria de Grazia, "Mass Culture and Sovereignty: The American Challenge to European Cinemas, 1920–1960," *Journal of Modern History* 61 (March 1989): 53–87.

4. Jean-Pierre Bardou et al., *The Automobile Revolution: The Impact of an Industry,* trans. James M. Laux (Chapel Hill: University of North Carolina Press, 1982), p. 91; Michelin, "Des faits et des chiffres sur l'industrie automobile française," *Prospérité* 4, no. 10 (July–September 1931): 3, 22. The actual numbers were 4,034,012 in the United States and 298,000 in France.

5. On Citroën's rationalization of production and publicity, see Sylvie Schweitzer, *André Citroën, 1878–1935: Le risque et le défi* (Paris: Fayard, 1992); idem, *Des engrenages à la chaîne: Les usines Citroën, 1915–1935* (Lyon: Presses Universitaires de Lyon, 1982); Jean-Louis Loubet, "La société André Citroën (1924–1968): Etude historique" (Thèse de 3e cycle, Université de Paris X, Nanterre, 1979); Jacques Wolgensinger, *André Citroën* (Paris: Arthaud, 1996), especially photograph with Ford in the front matter. Louis Renault made a well-publicized trip to Ford's works before com-

pletion of his expanded plant on the Ile Seguin: Patrick Fridenson, *Histoire des usines Renault: Naissance de la grande entreprise, 1898–1939* (Paris: Seuil, 1972), pp. 191–92.

6. Lionel Dumond, "L'industrie française du caoutchouc, 1828–1938: Analyse d'un secteur de production" (Doctorat Nouveau Régime, Université de Paris VII, 1996), especially pp. 43, 559, 564, 659; idem, "L'arrière plan technique et commercial," in Michelin, *Les hommes du pneu: Les ouvriers Michelin, à Clermont-Ferrand, de 1889 à 1940*, ed. André Gueslin (Paris: Editions Ouvrières, 1993), especially pp. 35–72; Louis Castellan, *L'industrie caoutchoutière* (Thèse de droit de l'Université de Paris; Thiers: A. Fayé, 1915), p. 152; Mansel G. Blackford and K. Austin Kerr, *BFGoodrich: Tradition and Transformation, 1870–1995* (Columbus: Ohio State University Press, 1997), pp. 60–61.

7. Michael J. French, "The Emergence of a US Multinational Enterprise: The Goodyear Tire and Rubber Company," *Economic History Review* 40, no. 1 (1987): 64–79, especially p. 69; idem, *The U.S. Tire Industry: A History* (Boston: Twayne, 1990), chapters 3–4; Dan Nelson, *American Rubber Workers and Organized Labor, 1900–1941* (Princeton: Princeton University Press, 1988), pp. 79–113; Dumond, "L'arrière plan," pp. 53–54.

8. Dumond, "L'arrière plan," pp. 38–45; Dumond, "L'industrie française," especially pp. 595–97.

9. Fridenson, *Histoire des usines Renault,* p. 159.

10. Dumond, "L'industrie française," pp. 561, 597–98, 666; Dumond, "L'arrière plan," pp. 58, 63, 61–62; Pierre Couderc, *Dunlop-Montluçon: 75 ans d'histoire partagée* (Premilhat: n.p., 1996), pp. 39–43.

11. Michelin, "Des faits et des chiffres sur l'industrie automobile française," *Prospérité* 4, no. 10 (July–September 1931): 10–14; idem, "Des faits et des chiffres," *Prospérité* 6, no. 14 (July–September 1933): 24–27; Fridenson, *Histoire des usines Renault,* pp. 158–59; Dumond, "L'arrière plan," pp. 70–72; idem, "L'industrie française," pp. 659–61.

12. Michelin, "La Micheline: La Micheline entre dans l'indicateur des chemins de fer" (Clermont-Ferrand: Michelin, 1932); Dumond, "L'arrière plan," pp. 66–69; Alain Jemain, *Michelin: Un siècle de secrets* (Paris: Calmann-Lévy, 1982), pp. 114–16.

13. Dumond, "L'industrie française," p. 669.

14. Moutet, Aimée, "Les origines du système Taylor, le point de vue patronale (1907–1914)," *Le Mouvement Social* 93 (1975): 15–49, especially pp. 15–16; Charles S. Maier, "Between Taylorism and Technocracy: European Ideologies and the Vision of Industrial Productivity in the 1920s," *Journal of Contemporary History* 5 (1970): 27–62; Maurice de Montmollin and Olivier Pastré, eds., *Le taylorisme* [Actes du colloque international sur le taylorisme organisé par l'Université de Paris XIII, 2–4

May 1983] (Paris: La Découverte, 1984). Taylor, of course, had no monopoly on the notion of efficiencies that might result from the approach of "science" to labor: William H. Schneider, "The Scientific Study of Labor in Interwar France," *French Historical Studies* 17, no. 2 (Fall 1991): 410–46.

15. Ellen Furlough, "Selling the American Way in Interwar France," *Journal of Social History* 26 (1993): 491–520; Robert L. Frost, "Machine Liberation: Inventing Housewives and Home Appliances in Interwar France," *French Historical Studies* 18, no. 1 (Spring 1993): 109–30; and Mary Nolan, *Visions of Modernity: American Business and the Modernization of Germany* (New York: Oxford University Press, 1994), chapter 10.

16. On Fayol, see Paul Devinat, *Scientific Management in Europe* (Geneva: International Labor Office, 1927), 31–33; L. Urwick and E. F. L. Brech, *The Making of Scientific Management: Thirteen Pioneers*, vol. 1 (London: Pitman, 1956), pp. 39–47; Moutet, *Les logiques de l'entreprise*, pp. 18, 38, 40–42, and passim; and Marjorie A. Beale, *The Modernist Enterprise: French Elites and the Threat of Modernity, 1900–1940* (Stanford: Stanford University Press, 1999), chapter 3.

17. Daniel Nelson, *Frederick W. Taylor and the Rise of Scientific Management* (Madison: University of Wisconsin Press, 1980); and Judith A. Merkle, *Management and Ideology: The Legacy of the International Scientific Management Movement* (Berkeley: University of California Press, 1980), chapters 1–2.

18. Patrick Fridenson, "Un tournant Taylorien de la société française, 1900–1914," *Annales ESC* 42 (1987): 1031–60; Aimée Moutet, "Les origines du système de Taylor en France: Le point de vue patronal (1907–1914)," *Le Mouvement Social* 93 (1975): 15–49; Merkle, *Management and Ideology*, chapter 5; Maier, "Between Taylorism and Technocracy"; Daniel Nelson, ed., *A Mental Revolution: Scientific Management since Taylor* (Columbus: Ohio State University Press, 1992), introduction; Gary Cross, *A Quest for Time: The Reduction of Work in Britain and France, 1840–1940* (Berkeley: University of California Press, 1989), pp. 106–10.

19. Aimée Moutet, "La première guerre mondiale et le taylorisme," in *Le taylorisme*, pp. 67–81; Fridenson, "Un tournant Taylorien"; George C. Humphreys, *Taylorism in France, 1904–1920: The Impact of Scientific Management on Factory Relations and Society* (New York: Garland, 1986), chapter 4; Richard F. Kuisel, *Capitalism and the State in Modern France: Renovation and Economic Management in the Twentieth Century* (New York: Cambridge University Press, 1981), chapter 2.

20. Report to Mr. Marcel Michelin from H. K. Hathaway, 20 September 1912. F. W. Taylor Collection, Stevens Institute of Technology; Moutet, "Les origines," pp. 37–38; Aimée Moutet, *La rationalisation industrielle dans l'économie française au vingtième siècle: Etude sur les rapports entre changements d'organisation technique et pro-*

blèmes sociaux (1900–1939) (Doctorat d'état, Université de Paris X-Nanterre, 1992), especially pp. 60–63; and André Gueslin, "Le système social Michelin (1889–1940)," in *Michelin, les hommes du pneu*, ed. André Gueslin, pp. 88–89.

21. Annie Moulin-Bourret, *Guerre et industrie: Clermont-Ferrand, 1912–1922: La victoire du pneu*, vol. 1 (Clermont-Ferrand: Publications de l'Institut d'Etudes du Massif Central, 1997), p. 278; Gueslin, "Le système social Michelin," p. 89.

22. Aimée Moutet, *Les logiques de l'entreprise: La rationalisation dans l'industrie française de l'entre-deux-guerres* (Paris: Ecole des Hautes Etudes en Sciences Sociales, 1997), especially p. 447; idem, *La rationalisation industrielle*, especially p. 63.

23. This assessment is based on the comprehensive work on the rationalization of French industry done by Aimée Moutet: *Les logiques de l'entreprise*, and *La rationalisation industrielle*.

24. Moutet, *Les logiques de l'entreprise*, p. 36; Michelin, "Un grand pas en avant dans l'art du chronométrage," *Prospérité* 4, no. 9 (April–June 1931): 1.

25. Moutet, *La rationalisation industrielle*, pp. 390–93; idem, *Les logiques de l'entreprise*, pp. 36–38.

26. Michelin, "Un grand pas en avant."

27. Michelin, "Comment nous avons taylorisé notre atelier de mécanique d'entretien" (Clermont-Ferrand: Michelin, 1917), p. 4. Aimée Moutet has made the point that the rationalization of production in accordance with Taylor's ideas was most practical in instances where long production runs were not possible: *Les logiques de l'entreprise*, pp. 102–9.

28. Michelin, "Pourquoi et comment chronométrer," *Prospérité* 1, no. 1 (April–June 1928), especially p. 29.

29. Michelin, "Sur le tas, ou Conseils pour débuter dans la méthode Taylor," *Prospérité* 2, supplement (December 1929): 24.

30. "Notes d'Edouard Michelin: Recueillies de 1919 à 1939," no. 18, pp. 5–6, Michelin archives.

31. Moutet, *Les logiques de l'entreprise*, pp. 92–101.

32. Gilbreth began as a builder who greatly increased productivity by using time-motion study and more standardized building materials as worksites. His partner and wife, Lillian Gilbreth, argued for the adoption of Taylor's methods in household management. Their children idealized the couple's home life in the best-selling *Cheaper by the Dozen*. Nelson, *Taylor and Scientific Management*, pp. 131–36, 229–41; and L. Urwick and E. F. L. Crech, *The Making of Scientific Management*, vol. 1 (London: Pitman, 1951), chapter 12.

33. Michelin, "Deux exemples d'application de la méthode Taylor chez Michelin" (Clermont-Ferrand: Michelin, 1925), pp. 2, 16.

34. Moutet, *Les logiques de l'entreprise,* p. 140.

35. Michelin, "Sa majesté le client," *Prospérité* 1, supplement to no. 4 (March 1929): 8–9, 16.

36. Gueslin, "Le système social Michelin," pp. 97–100.

37. The size of Michelin's workforce is from Dumond, "L'arrière plan," pp. 54–58; Edouard Michelin also noted the difficulties of recruitment in 1924 in his "Conférence aux chefs de service," 11 July 1924, in "Notes d'Edouard Michelin," no. 5, Michelin archives.

38. Gueslin, "Le système social Michelin," pp. 92–93.

39. Edouard Michelin, "Conférence: L'esprit d'observation," July 1929, in "Notes d'Edouard Michelin," no. 18, Michelin archives.

40. Gueslin, "Le système social Michelin," pp. 93–95.

41. Moutet notes, however, that Renault took the process less seriously and soon abandoned it: Moutet, *Les logiques,* pp. 328–29. On the Chemins de Fer de l'Etat, whose system was somewhat less extensive than Michelin's, see Moutet, *La rationalisation industrielle,* p. 719. Gueslin has pointed out that there were also nineteenth-century French precedents: Gueslin, "Le système social Michelin," p. 108.

42. Michelin, "Suggestions: Comment nous avons amené notre personnel à collaborer avec nous à la recherche des progrès et des économies," *Prospérité* 6, no. 15 (October–December 1933): 1–3; Edouard Michelin's frustration with the number of suggestions in the Clermont-Ferrand factory is evident in his praise of the comparatively higher number of suggestions in the company's Lyon warehouse, which he viewed as a model for the factory. "Notes d'Edouard Michelin," no. 15, 31 July 1928, Michelin archives.

43. Moutet, *Les logiques de l'entreprise,* p. 330.

44. "Notes d'Edouard Michelin," no. 35, 18 February 1936, pp. 2–3, Michelin archives.

45. "Suggestions," p. 32.

46. For the full argument that Michelin was in essence a paternalist firm, see Gueslin, "Le système social Michelin," pp. 73–154; and for context, Gueslin, "Le paternalisme revisité en Europe occidentale (seconde moitié du XIXe, début XXe siècle)," *Genèses* 7 (March 1992): 210–11. On the "esprit Michelin," see both Gueslin and Moutet, *La rationalisation industrielle,* pp. 1230–33.

47. "Cela vaut-il la peine de s'occuper de la méthode Taylor?" (Clermont-Ferrand: Michelin, 1927). The later reedition was "Aux dépens du gaspillage ou cela vaut-il la peine de s'occuper de la méthode Taylor," *Prospérité* 1, supplement to no. 3 (January 1929): 15.

48. Moutet, *Les logiques de l'entreprise,* pp. 62–63, 267–74.

49. Michelin, "Prospérité ou Sam et François" (Clermont-Ferrand: Michelin, 1927), pp. 7, 10–11.

50. Moutet, *Les logiques de l'entreprise,* pp. 222, 305, 446–47; Moutet, *La rationalisation industrielle,* pp. 975–76.

51. Between 1920 and 1924, Michelin undertook a virulent postcard and poster campaign to convince the Parisian Société des Transports that it needed to equip its buses with pneumatic tires. By portraying pigs riding on pneumatic tires while Parisians rode on the bumpier solid rubber tires of Michelin's competitors, Michelin succeeded in creating enough public attention that the Société yielded to Michelin's demand. Jemain, *Un siècle des secrets,* pp. 111–12.

52. "Notes d'Edouard Michelin," Michelin archives.

53. Michelin, "Des faits et des chiffres sur l'industrie automobile française," *Prospérité* 4, no. 10 (July–September 1931): 14; idem, "Des faits et des chiffres," *Prospérité* (July–September 1932): 58. Michelin encouraged buyers of new Ford cars immediately to replace American "straight side" tires with Michelin tires and wheels: "Le Samedi de Michelin," *L'Illustration,* 15 October 1921, advertising section.

54. "Prospérité ou Sam et François" (Clermont-Ferrand: Michelin, 1927), pp. 1–2.

55. Michelin, "Cheval et auto: Ce que disent les agriculteurs qui s'en servent" (Clermont-Ferrand: Michelin, 1925), pp. 2–3.

56. Michelin, "L'automobile, source de la richesse: L'exemple de l'Aveyron," *Prospérité* 7 (April–June 1934): 2–3, 5, 11, 20.

57. Michelin, "Ce que l'auto coûte réellement: Les bénéfices qu'elle propose" (Clermont-Ferrand: Michelin, n.d., but obviously from the early 1920s), pp. 3–4, 8–9, 11, back cover.

58. Michelin, "Les voyageurs de commerce! L'automobile augmentera vos bénéfices et votre confort: Les résultats de l'expérience Michelin" (Clermont-Ferrand: Michelin, n.d. [mid-1920s]).

59. Michelin, "L'auto contre la crise," *Prospérité* 8, no. 19 (January–March 1935).

60. Michelin, "On veut tuer l'automobile" (Clermont-Ferrand: Michelin, n.d., but obviously from 1923), p. 6.

61. Fernand Gillet, "Cent ans d'industrie: Histoire anecdotique de la Maison Michelin," vol. 2, p. 141, Michelin archives.

62. Michelin, "L'avenir de l'industrie automobile française" (Clermont-Ferrand: Michelin, 1927), pp. 13–15, 17.

63. Michelin, "Des faits et des chiffres sur l'automobile en France," *Prospérité,* 6, no. 14 (July–September 1933): 1.

64. Michelin, "Des faits et des chiffres sur l'automobile en France," *Prospérité* 5, no. 13 (July–September 1932): 1.

65. "Des faits et des chiffres sur l'industrie automobile française," *Prospérité* (July–September 1931, 1932, 1933).

66. Poster for the "Enquête Nationale de l'Automobile Populaire," 1923, poster collection, Michelin archives.

67. Gillet, "Cent ans," vol. 2, p. 127.

68. Louis Baudry de Saunier, "Causerie sur le Salon de 1923," *L'Illustration* 81 (6 October 1923), pp. 306–8. Interestingly, Baudry de Saunier, while noting that auto makers needed to standardize parts and offer credit, placed much of the burden on potential buyers with a class consciousness taken for granted in Michelin's own publications: "The client with limited means needs to practice that old French virtue of saving. It is franc by franc that he will realize his dream. I know that in our time . . . it is a lot more difficult to slip a five-franc bill into a piggy bank. The cinema might [suffer] perhaps, but public health would benefit. One must realize that the most bitter social struggles will be pacified when, by the automobile, tourism is placed within reach of everyone." In short, if only the working class saved its money instead of spending it foolishly on the movies, they could afford a car, and all would benefit from social peace (p. 308).

69. Both James Laux and Patrick Fridenson have relied on the "Faits et des chiffres" for many of their own statistics about automobile production and consumption in prewar and interwar France: James M. Laux, *In First Gear: The French Automobile Industry to 1914* (Montreal: McGill-Queen's University Press, 1976); Fridenson, *Histoire des usines Renault.*

70. Michelin, "L'auto d'occasion" (Clermont-Ferrand: Michelin, 1927).

71. On the development of small, popular vehicles in France and particularly chez Citroën, see Patrick Fridenson, "Genèse de l'innovation: la 2 CV Citroën," *Revue française de gestion* 70 (September–October 1988): 35–44; idem, "La question de la voiture populaire en France de 1930 à 1950," *Culture technique* 19 (1989): 205–10; Jean-Louis Loubet, "La société André Citroen (1924–1968)" (Thèse de 3e cycle, Université de Paris X-Nanterre, 1979), pp. 291–97; Moutet, *La rationalisation industrielle,* pp. 1593–95. On the Citroën takeover, see as well Sylvie Schweitzer, *André Citroën, 1878–1935: Le risque et le défi* (Paris: Fayard, 1992), chapter 8; Jacques Wolgensinger, *André Citroën* (Paris: Arthaud, 1996), chapter 21.

72. This perspective has been influenced by Ellen Furlough, who has very clearly articulated the ways that contemporary French observers could pick and choose aspects of "Americanization" they found useful, discarding those they deemed threatening: Furlough, "Selling the American Way in Interwar France." Nor were French retailers and organizers of the Salons des Arts Ménagers the only ones using such selectivity; advertisers, owners of film production companies, and architects, not to

mention industrialists, also used "America" as a veritable template of ideas that could be adopted, adapted, condemned, or ignored, depending on the social agenda of the chooser: Marie-Emmanuelle Chessel, *La publicité: Naissance d'une profession, 1900–1940* (Paris: CNRS, 1998). Victoria de Grazia has implied that consumer society did indeed erase class differences in the United States, at least as compared with Europe; whether one accepts that assessment or not, it is in any case true that European commentators on the United States certainly believed that mass consumerism brought changes in social distinctions: de Grazia, "Changing Consumption Regimes in Europe, 1930–1970: Comparative Perspectives on the Distribution Problem," in *Getting and Spending: European and American Consumer Societies in the Twentieth Century,* ed. Susan Strasser, Charles McGovern, and Matthias Judt (Washington, D.C.: German Historical Institute; Cambridge: Cambridge University Press, 1998), pp. 59–83.

73. Duhamel, *America the Menace,* pp. 67–68.

74. See Mary Louise Roberts, *Civilization without Sexes: Reconstructing Gender in Postwar France, 1917–1927* (Chicago: University of Chicago Press, 1994). For the idea that the nostalgic portrayals of "traditional" France were themselves intellectual creations in the interwar years, see both Roberts and Beale, *The Modernist Experience.*

75. "Cheval et auto," p. 6.

76. See, for example, the Citroën ad that appeared in *Femina* in January 1921 (and is reproduced in Roberts, *Civilization without Sexes,* figure 11). For U.S. auto makers and tire makers, see *The Saturday Evening Post* during the 1920s and 1930s, although I have based this assessment on virtually all of the newspaper and magazine advertising of BF Goodrich, Firestone, General, and Goodyear during the interwar years.

77. Michelin, "Oeuvres sociales de Michelin et Cie" (Clermont-Ferrand: Michelin, 1927); idem, "Une expérience d'éducation physique," *Prospérité* 5, no. 12 (April–June 1932). On Michelin's *oeuvres sociales* in comparison with those of other French firms and an interpretation of the Michelin system as social control, see Gueslin, "Le système social Michelin," especially pp. 98–110; and idem, "Le paternalisme"; Moutet, *La rationalisation industrielle,* pp. 541–50; as well as Christian Lamy and Jean-Pierre Fornaro, *Michelin-Ville: Le logement ouvrier de l'entreprise Michelin, 1911–1987* (Nonette, France: Créer, 1990).

78. "Oeuvres sociales" (Clermont-Ferrand: Michelin, 1927), p. 9.

79. "Une dépense qui paie: Un service médical," *Prospérité* 1, no. 3 (October–December 1928): 2.

80. "La jambe de Ben Kacem ou Comment nous appliquons la loi sur l'accident du travail," *Prospérité* 2, supplement to no. 5 (May 1929): 1–2, 4, 6.

81. Stephen Meyer, *The Five Dollar Day: Labor Management and Social Control in the Ford Motor Company, 1908–1921* (Albany: State University of New York Press, 1981).

82. Nelson, *American Rubber Workers,* pp. 20–21, 56–61; Maurice O'Reilly, *The Goodyear Story* (Elmsford, N.Y.: n.p., 1983), pp. 30–31. It is noteworthy that American tire makers, while developing a less comprehensive social network than Michelin, nevertheless also instituted company welfare, ignoring Taylor's admonitions about its irrelevance for increasing production.

83. J. Lavaud, "Michelin ou la féodalité industrielle," *Europe,* 28, no. 111 (1932): 473–85, especially p. 473. On *Europe* and its direction, see Pierre Albert, "La presse française de 1871 à 1940," in *Histoire générale de la presse française,* vol. 3, *1871–1940,* ed. Claude Bellanger (Paris: Presses Universitaires de France, 1972), p. 594.

Chapter 7. Defining France

1. "Le tourisme: Important facteur économique," *Revue mensuelle du Touring Club,* July 1912, p. 305. For an excellent and nuanced analysis of the Touring Club's approach to French regions in this earlier period, see Patrick Young, "*La vieille France* as Object of Bourgeois Desire: The Touring Club de France and the French Regions, 1890–1918," in *Histories of Leisure,* ed. Rudy Koshar (New York: Berg, forthcoming).

2. Louis Baudry de Saunier citing Forest's book in "La grande industrie d'avenir," *Revue mensuelle du Touring Club,* October 1915, pp. 91–92; Abel Ballif, "Pour 'notre Revanche,'" *Revue mensuelle,* May–June 1916, p. 85; idem, "Pour nos montagnes," *Revue mensuelle,* September–October 1917, pp. 98–100.

3. Gaston Mortier, *Le tourisme et l'économie nationale: Un passé encourageant . . . Vers un meilleur avenir* (Grenoble: B. Arthaud, 1941), p. 209.

4. René Duchet, *Le tourisme à travers les âges: Sa place dans la vie moderne* (Paris: Vigot Frères, 1949), pp. 172–73, citing Peyromaure-Debord, *Le tourisme, le thermalisme, et le climatisme* (Paris: Imprimerie Nationale, 1935), pp. 45–46.

5. Léon Auscher, "Le tourisme aux Etats-Unis," *Revue mensuelle du Touring Club,* January 1905, pp. 82–84, and February 1905, pp. 78–80.

6. Auscher, who was by the late 1920s not only vice president of the Touring Club but also a member of the board (*conseil d'administration*) of the Office National du Tourisme, originally gave this as a report to the Conseil National Economique. The title page reads "Ministère des Travaux Publics, Office National du Tourisme, 'L'importance économique du tourisme: Rapport présenté au Conseil National Economique (Session du juillet 1927) par M. Léon Auscher'" (Paris: Imprimerie Nationale, 1928), p. 5.

7. Henry Defert, "Donnez vos prix!" *Revue du Touring Club,* March 1922, pp. 91–92; Karl S. Cate, "L'opinion d'un américain sur le prix de la vie touristique en France," *Revue du Touring Club,* May 1922, pp. 171–72. On American perceptions of French price-gouging and its effect on tourism after the war, see also Harvey Levenstein, *Seductive Journey: American Tourists in France from Jefferson to the Jazz Age* (Chicago: University of Chicago Press, 1998), p. 225.

8. Henry Defert, "Les chambres de commerce et le crédit hôtelier," *Revue mensuelle du Touring Club,* January–February, 1918, p. 27; Louis Baudry de Saunier, "Le crédit hôtelier," *Revue du Touring Club,* February 1924, pp. 51–53; Léon Delamarche and Léon Auscher, "Le crédit national hôtelier," *Revue du Touring Club,* January 1926, pp. 3–4.

9. Catherine Bertho Lavenir, *La roue et le style: Comment nous sommes devenus touristes* (Paris: Editions Odile Jacob, 1999), pp. 250–51.

10. "Nouvelle signalisation des routes," *Revue du Touring Club,* May–June 1920, p. 113; Georges Ribeill, "Du pneumatique à la logistique routière: André Michelin, promoteur de la 'révolution automobile,'" *Culture technique* 19 (1989): 200; Marina Duhamel, *Un demi-siècle de signalisation routière: Naissance et évolution du panneau de signalisation routière en France, 1894–1946* (Paris: Presses des Ponts et Chaussées, 1994), pp. 40–44.

11. Fernand Gillet, "Cent ans d'industrie: Histoire anecdotique de la Maison Michelin," vol. 2, typewritten manuscript, pp. 122–23, Michelin archives.

12. "Nouvelle signalisation," *Revue du Touring Club,* May–June 1920, p. 113; Duhamel, *Un demi-siècle de signalisation routière,* pp. 40–44; Ribeill, "Du pneumatique à la logistique routière," p. 200.

13. "La signalisation des routes," *Revue du Touring Club,* October–December 1920, p. 205; "La signalisation des routes," *Revue du Touring Club,* August–September 1921, p. 266; Duhamel, *Un demi-siècle de signalisation routière,* pp. 40–44, 61, 93, 96; Ribeill, "Du pneumatique à la logistique routière," p. 200.

14. Quotation and statistics from Duhamel, *Un demi-siècle de signalisation routière,* p. 92.

15. *Revue du Touring Club,* May 1922, p. 204; *Revue du Touring Club,* April 1927, p. 48; *Revue du Touring Club,* May 1932, p. 160.

16. Pierre Couderc, *Dunlop-Montluçon: 75 ans d'histoire partagée* (Premilhat: n.p., 1996), pp. 29–30.

17. Baudry de Saunier, "La signalisation des routes," *Revue du Touring Club,* October 1927, p. 190; copy of the letter from L. Auscher of the Touring Club to Gaston Gérard, the sous-secrétaire d'état aux Travaux Publics, appearing in the *Revue du Touring Club,* May 1932, p. 160.

18. Duhamel, *Un demi-siècle de signalisation routière*, pp. 45–49; Ribeill, "Du pneumatique à la logistique routière," pp. 200–201. Paris had some street signs made of the lava as early as 1835, and the department of Puy-de-Dôme had similar road signs in the early twentieth century. Louis Baudry de Saunier, "La signalisation routière: La borne d'angle" (Paris: Flammarion, 1931), pp. 4–9; Gillet, "Cent ans d'industrie," vol. 2, pp. 121–22.

19. "La tournée d'étude des bornes d'angle Michelin," *Revue du Touring Club*, December 1927, p. 239.

20. Ibid., p. 219; Duhamel, *Un demi-siècle de signalisation routière*, pp. 51, 77; Ribeill, "Du pneumatique à la logistique routière," p. 201. Louis Baudry de Saunier, after criticizing the effects of weather as well as vandals who shot existing signs literally destroying them, strongly advocated the adoption of Michelin *bornes* for their durability. Baudry de Saunier, "La signalisation des routes," *Revue du Touring Club*, October 1927, pp. 189–90.

21. Baudry de Saunier, "La signalisation routière," pp. 2–4.

22. Harvey Levenstein, *Seductive Journey: American Tourists in France from Jefferson to the Jazz Age* (Chicago: University of Chicago Press, 1998), part 3.

23. Stephen Mennell, *All Manner of Food: Eating and Taste in England and France from the Middle Ages to the Present* (Oxford: Blackwell, 1985), chapters 5–10; Pascal Ory, "La Gastronomie," in *Les lieux de mémoire*, ed. Pierre Nora, vol. 3 (Paris: Gallimard, 1997), pp. 3743–69; and Priscilla Parkhurst Ferguson, "A Cultural Field in the Making: Gastronomy in 19th-Century France," *American Journal of Sociology* 104, no. 3 (November 1998): 597–641.

24. Anne-Marie Thiesse, *Ecrire la France: Le mouvement littéraire régionaliste de langue française entre la Belle Epoque et la Libération* (Paris: Presses Universitaires de France, 1991), especially pp. 213–14, 243–46; Jean-Yves Guiomar, "Le 'Tableau de la géographie de la France' de Vidal de la Blache," in *Les lieux de mémoire*, ed. Pierre Nora, vol. 1, (Paris: Gallimard, 1997), pp. 1073–98.

25. Weber considers the linguistic assimilation of the provinces as part of a larger process of modernization. Eugen Weber, *Peasants into Frenchmen: The Modernization of Rural France, 1870–1914* (Stanford: Stanford University Press, 1976). Other scholars, such as Charles Tilly and Maurice Agulhon, critique Weber's use of modernization theory in their consideration of the process of acculturation into national economic and political norms which they claim to have occurred before 1870. Charles Tilly, "Did the Cake of Custom Break?" in *Consciousness and Class Experience in Nineteenth-Century Europe*, ed. John M. Merriman (New York: Holmes and Meier, 1979), pp. 17–44; Maurice Agulhon, *La République au village* (Paris: Seuil, 1979). In contrast, Mendras and Wright have placed the integration of the provinces as having

occurred after the Second World War. Henri Mendras, *La fin des paysans: Innovation et changement dans l'agriculture française* (Paris: Sédéis, 1967); Gordon Wright, *Rural Revolution in France: The Peasantry in the Twentieth Century* (Stanford: Stanford University Press, 1964). Recent, nuanced investigations that point out how regionalism became forged within the nationalist discourse rather than displaced by it include Susan Carol Rogers, *Shaping Modern Times in Rural France: The Transformation and Reproduction of an Aveyronnais Community* (Princeton: Princeton University Press, 1991); Shanny Peer, *France on Display: Peasants, Provincials, and Folklore in the 1937 Paris World's Fair* (Albany: State University of New York Press, 1998).

26. G. Bruno [Augustine Fouillée], *Le tour de la France par deux enfants: Livre de lecture courante pour le cours moyen* (Paris, 1877); Jacques and Mona Ozouf, "Le tour de la France par deux enfants: Le petit livre rouge de la République," in *Les lieux de mémoire*, ed. Pierre Nora, vol. 1 (Paris: Gallimard, 1997), pp. 277–301.

27. "Le tourisme gastronomique," *Revue mensuelle du Touring Club*, April 1912, pp. 146–50. For a fascinating description of how provincial dishes, at best the preserve of wealthy notables on feast days, were remade to suit tourists' palates, see Bertho Lavenir, *La roue et le stylo*, pp. 233–39.

28. Touring-Club de France, *Annuaire Générale: France-Nord* (Paris: n.p., 1914), especially pp. 408–9.

29. Michelin, *Guide Michelin France* (1914), especially p. 53.

30. A. Liégard, "Le tourisme et la cuisine," *Revue mensuelle du Touring Club*, July 1919, p. 168.

31. Baudry de Saunier, "L'Inventaire de nos richesses gastronomiques," *Revue du Touring Club*, February 1920, p. 39.

32. Baudry de Saunier, "L'Inventaire de nos richesses gastronomiques," *Revue du Touring Club*, May–June 1920, p. 144.

33. P. Kauffmann, "Nos richesses gastronomiques: Un peu de cuisine alsacienne," *Revue du Touring Club*, March 1921, p. 75.

34. Curnonsky (Maurice-Edouard Sailland), *Souvenirs littéraires et gastronomiques* (Paris: Albin Michel, 1958); Mennell, *All Manners of Food*, pp. 276–77; René Chauvelot, *La vie de Curnonsky: Prince des gastronomes; L'adolescence angevine* (Paris: Clés du Monde, 1983).

35. Curnonsky and Marcel Rouff, *La France gastronomique: Guide des merveilles culinaires et des bonnes auberges françaises: L'Alsace* (Paris: F. Rouff, 1921), pp. 10, 13–15.

36. Ibid., pp. 33–34, 81–82. On the importance of German entrepreneurs in the French champagne industry, see Kolleen M. Guy, "'Oiling the Wheels of Social Life': Myths and Marketing in Champagne during the Belle Epoque," *French Histori-*

cal Studies 22, no. 2 (Spring 1999): 211–39. On the TCF's gendering of regions, see Young, "*La vieille France* as Object."

37. Georges Duhamel, *America the Menace: Scenes from the Life of the Future*, trans. Charles Miner Thompson (Boston: Houghton Mifflin, 1931), pp. 93–112.

38. Curnonsky and Marcel Rouff, *Paris, the Environs of Paris, Normandy*, vol. 1, of "The Epicure's Guide to France" (New York: Harper, 1926), p. xiii. Just as French cuisine could embody French civilization, Chinese cuisine merited respect as the epitome of a great Eastern culture: on the parallel between Chinese and French cooking, familiar in France at the time, see Ferguson, "A Cultural Field," especially p. 632.

39. Curnonsky and Rouff, *L'Alsace*, p. 88.

40. J.-A.-P. Cousin, *Voyages gastronomiques au pays de France: Le Lyonnais* (Paris: Flammarion, 1927); *Voyages gastronomiques au pays de France: Paris et la région parisienne. Avec sept itinéraires gastronomiques de Paris à Nice* (Paris: Flammarion, 1925); *Voyages gastronomiques au pays de France: Le Sud-Ouest de Nîmes à Bordeaux* (Paris: Flammarion, 1928).

41. Edouard Dulac, *Le tour de France gastronomique: Guide du touriste gourmand à Paris et dans les vieilles provinces françaises* (Paris: Editions de France, 1926).

42. Austin de Croze, *Les plats régionaux de France: 1400 succulentes recettes traditionnelles de toutes les provinces françaises* (Paris: Montaigne, 1928); Curnonsky and Austin de Croze, *Le trésor gastronomique de France: Répertoire complet des spécialités gourmandes des trente-deux provinces françaises* (Paris: Delagrave, 1933); and the overview in Julia Csergo, "The Emergence of Regional Cuisines," in *Food: A Culinary History from Antiquity to the Present*, ed. Jean-Louis Flandrin, Massimo Montanari, and Albert Sonnenfeld (New York: Columbia University Press, 1999), pp. 500–515.

43. See Pierre Béarn, *Paris Gourmand: Ce que doit savoir un gourmand pour devenir gastronome*, 6th ed. (Paris: Gallimard, 1929); R. Bodet, *Toques blanches et habits noirs: L'hôtellerie et la restauration autrefois et aujourd'hui, recettes nouvelles, le service de table, les vins, gastronomie, tourisme* (Paris: Dorbon, 1939), as well as the periodicals *Le bon vin de France: Gastronomie, hostellerie et tourisme* and *L'art culinaire*.

44. Peer, *France on Display*.

45. Jean Fourastié with Jacqueline Fourastié, *Pouvoir d'achat, prix et salaires* (Paris: Gallimard, 1977), p. 65.

46. Alain Jemain, *Michelin: Un siècle de secrets* (Paris: Calmann-Lévy, 1982), p. 107; Ribeill, "Du pneumatique à la logistique routière," p. 202.

47. Ribeill, "Du pneumatique à la logistique routière," p. 202; Herbert Lottman, *Michelin: 100 ans d'aventure* (Paris: Flammarion, 1998), p. 177; *Guide Michelin France* (1922), p. 772.

48. "Les guides et les cartes Michelin," *Bulletin Intérieur Michelin,* 20 June 1946, pp. 1–2.

49. Ibid. On the publication runs of popular novels, Mary Louise Roberts, *Civilization without Sexes: Reconstructing Gender in Postwar France, 1917–1927* (Chicago: University of Chicago Press, 1994), p. 47.

50. *Guide Michelin France* (1923), p. 9.

51. *Guide Michelin France* (1927), pp. 8–9.

52. Ibid., p. 9; *Guide Michelin France* (1929), p. 9.

53. *Guide Michelin France* (1932), p. 1.

54. *Guide Michelin France* (1939), p. 11.

55. Ibid., pp. 29–35.

56. For one of the twelve, nothing at all was listed: *Guide Michelin France* (1939), pp. 712–13.

57. Ibid., pp. 994–95.

58. *Guide du pneu Michelin: Belgique, Luxembourg, Hollande* (1936–37), p. 8.

59. *Guide Michelin: Suisse, Lacs Italiens, Dolomites, Tyrol* (1931–32).

60. *Guide Michelin: Maroc, Algérie, Tunisie* (1930), p. 25.

61. Erica Peters, "National Preferences and Colonial Cuisine: Seeking the Familiar in French Vietnam," paper presented at the Western Society for French History, Monterey, Calif., 2 November 1999.

62. *Guide Michelin: Maroc, Algérie, Tunisie* (1930), p. 168.

63. R. Bodet, *Toques blanches et habits noirs: L'hôtellerie et la restauration d'autrefois et aujourd'hui, recettes nouvelles, le service de table, les vins, gastronomie, tourisme* (Paris: Dorbon, 1939), pp. 79–87.

64. Pierre Béarn, *Paris gourmand: Ce que doit savoir un gourmand pour devenir gastronome,* 6th ed. (Paris: Gallimard, 1929), pp. 135–37.

65. Jean-François Mesplède, *Trois étoiles au Michelin: Une histoire de la haute gastronomie française* (Paris: Gründ, 1998), pp. 19–21; Bodet, *Toques blanches,* p. 135.

66. Karl Baedeker, *Les bords du Rhin depuis Bâle jusqu'à la frontière de Hollande: Manuel du voyageur* (Coblenz: Baedeker, 1862), p. v; Baedeker, *Paris et ses environs: Manuel du voyageur* (Paris: Paul Ollendorff, 1900), p. v.

67. Marcel Monmarché, *Normandie* (Paris: Hachette, 1926), front matter; idem, *Provence* (Paris: Hachette, 1925).

68. Touring Club de France, *Annuaire générale* (Paris: n.p., 1921).

69. The *concours* solicited individual restaurants and inns to commit to a set price for a fine meal including wine to be served during the summer tourist season. The establishments would then be ranked by judges: the first place entry received 5,000 francs, the second 3,000 francs, the third 1,000 francs. "Concours de la bonne cuisine," *Revue du Touring Club,* May 1930, p. 145; "Concours de la bonne cuisine,"

Revue du Touring Club, July 1930, p. 212; "Résultats du concours de la bonne cuisine," *Revue du Touring Club,* January 1931, p. 25.

70. Michelin's reputation for secrecy, particularly in the realm of the red guides,
is of course legion. For recent popular portrayals of the mystery of Michelin rankings, see Mesplède, *Les trois étoiles;* and Lottman, *Michelin.*

71. Guy, "'Oiling the Wheels of Social Life,'" pp. 211–39.

72. James Buzard, *The Beaten Track: European Tourism, Literature, and the Ways
to Culture, 1800–1918* (New York: Oxford University Press, 1993), pp. 49–77. On
Cook, see also E. Swinglehurst, *Cook's Tours: The Story of Popular Travel* (Poole,
Dorset: Blandford Press, 1982).

73. Michelin, "Des faits et des chiffres," *Prospérité* 4 (July–September 1931): 15.

74. *Guide Michelin France* (1925), p. 411; and Georges Ribeill, "Du pneumatique
à la logistique routière," p. 196.

75. Henry Defert, "Guides à proscrire," *Revue mensuelle du Touring Club,* August
1914–April 1915, p. 28; "A propos des guides," *Revue mensuelle du Touring Club,*
July–August 1916, p. 110; "Guides plagiaires," *Revue mensuelle du Touring Club,*
November–December 1917, p. 128. German-language Baedeker guides to Paris and
other areas in France did appear in the interwar years, but the only new edition to
France in French were guidebooks to Paris in 1924 and 1931 (Alex W. Hinrichsen,
Baedeker-Katalog: Verzeichnis aller Baedeker-Reiseführer von 1832–1987 [Holz-
minden: Ursula Hinrichsen, 1988], pp. 73–74). Edward Mendelson claims that the
other guidebooks to France were still reprinted in the interwar years, but that there
were no new editions ("Baedeker's Universe," *Yale Review* 74, no. 3 [April 1985]:
386–403).

76. Henry Defert, "Un nouveau guide français," *Revue mensuelle du Touring Club,*
February 1916, p. 67. This gave the TCF some leverage, and it happily announced
that only hotels with French or Allied personnel could appear in either the TCF's
own *annuaire* or, more significantly, in the blue guides: Henry Defert, "L'hôtellerie
française aux Français," *Revue mensuelle du Touring Club,* May 1919, p. 98.

77. Marc Francon, "Le guide vert Michelin: L'invention du tourisme culturel populaire" (Thèse de doctorat du Nouveau Régime, Université de Paris VII, 1998),
p. 95. It is noteworthy that the Michelin guides *Morocco, Algérie, Tunisie* and to other
countries included substantial sections charting itineraries as well as the information
about hotels, restaurants, and *stockistes* to which the Michelin red guides to France
were dedicated, making the guides to areas outside metropolitan France a mix of the
two forms of guides, minus substantive descriptions of tourist sights.

78. This was the case, for example, in *Guide Régional: Auvergne* (Clermont-
Ferrand: Michelin, 1937); *Guide Régional: Les châteaux de la Loire* (1928); and *Guide
Régional: Pyrénées, Côte d'Argent* (1930).

79. *Guide Régional Michelin: Bretagne* (Clermont-Ferrand: Michelin, 1926), pp. i–ii.

80. Ibid., p. xvii.

81. *Guide Régional Michelin: Auvergne* (1937–38), pp. xix–xxiii.

82. Rudy Koshar, "'What Ought to Be Seen': Tourists' Guidebooks and National Identities in Modern Germany and Europe," *Journal of Contemporary History* 33, no. 3. (1998): 331.

83. Baedeker, *Paris et ses environs: Manuel du voyageur* (Paris: Paul Ollendorff, 1900).

84. Marcel Monmarché, ed., *Vosges, Champagne (Sud), Lorraine, Alsace* (Paris: Hachette, 1928), pp. 403–8.

85. *Guide Régional Michelin: Les châteaux de la Loire,* (1928–29), pp. iii, xxxix.

86. On the nearly telegraphic style of interwar Baedeker guides, see Koshar, "Guidebooks and National Identities," p. 331.

87. *Guide Régional Michelin: Normandie* (1933–34), p. 576.

88. Paul Fussell, *Abroad: British Literary Traveling between the Wars* (New York: Oxford University Press, 1980); Levenstein, *Seductive Journey,* introduction; see Levenstein's notes for accounts of both contemporary distinctions between travelers and tourists as well as for historiography that has accepted such categories. The quite critical assumptions about post–World War II tourists are part and parcel of appraisals of the Guides Bleus: see, for example, Roland Barthes, "Le Guide Bleu," *Mythologies* (Paris: Seuil, 1957), pp. 136–39; Jules Gritti, "Les contenus culturels du Guide Bleu: Monuments et sites à voir," *Communications,* 10 (1976): 51–64.

89. See, for example, the front matter to *Guide Régional Michelin: Normandie* (1933–34).

90. Francon, "Le guide vert," p. 327.

91. Ibid., pp. 76, 311, 313–14.

92. *Guides Thiolier: Bourgogne et Morvan, Fontainebleau et sa forêt* (n.p., 1938), p. 44; "Les guides et les cartes Michelin," *Bulletin Intérieur Michelin,* 20 June 1946, p. 2.

93. The 1926 guide to Brittany had included some 38 itineraries, that of 1928 95, that of 1935 62, and that of 1949 none at all: Francon, "Le guide vert," p. 329.

94. Nordman, "Les Guides Joanne," p. 1036.

95. Dean MacCannell, *The Tourist: A New Theory of the Leisure Class* (New York: Schocken, 1989), pp. 91–107.

96. *Guide Régional: Bretagne* (1926), pp. vi–xxiv.

97. Bertho Lavenir, *La roue et le stylo,* pp. 234–35. I translated *pâté* and *galette* as pie and pancakes because what peasants actually ate may well have more closely resembled traditional shepherd's pie and American-style pancakes than the fine *pâtés* and *galettes* that both the French and foreigners associate with the latter words.

98. Peer, *France on Display.*

99. Curnonsky, *Souvenirs*, pp. 192–98.

Conclusion

1. John F. Sweets, *Choices in Vichy France: The French under Nazi Occupation* (New York: Oxford University Press, 1986), pp. 12, 194–95.

2. Denis Estorgues, "Le syndicalisme chez Michelin de 1930 à 1950" (master's thesis, Université de Clermont-Ferrand II, 1983), pp. 61–66 and repeated in André Gueslin, "Le système social Michelin," in *Michelin, Les hommes du pneu: Les ouvriers Michelin, à Clermont- Ferrand, de 1889 à 1940,* ed. André Gueslin (Paris: Editions de l'Atelier/Editions Ouvrières, 1993), pp. 86–87. On de la Rocque, the Croix de Feu, and the Parti Social Français, see Robert Soucy, "French Fascism and the Croix de Feu: A Dissenting Interpretation," *Journal of Contemporary History* 26 (1991): 159–88; idem, *Fascism in France: The Second Wave, 1933–1939* (New Haven, Conn.: Yale University Press, 1995), pp. 104–203; William D. Irvine, "Fascism in France and the Strange Case of the Croix de Feu," *Journal of Modern History* 63 (1991): 271–95; and Samuel Huston Goodfellow, *Between the Swastika and the Cross of Lorraine: Fascisms in Interwar Alsace* (DeKalb: Northern Illinois University Press, 1999), pp. 137–38.

3. Alain Jemain, *Michelin: Un siècle de secrets* (Paris: Calmann-Lévy, 1982), p. 120.

4. Sweets, *Choices in Vichy France,* p. 85; Gueslin, "Le système social Michelin," p. 87; Jemain, *Un siècle de secrets,* pp. 119–20.

5. Jemain, *Un siècle de secrets,* p. 125; Sweets, *Choices in Vichy France,* p. 194. At the same time, Jean Michelin housed Fernand Boverat of the Alliance Nationale as the latter attempted to ingratiate himself with Vichy authorities, and the Michelin company gave Vichy authority to reprint the prewar pamphlet, "Guide de la jeune mère: Comment alimenter vos bébés" (1941), whose title page read "publié avec l'autorisation spéciale de la Maison Michelin." On Boverat during the war, see Cheryl A. Koos, "Engendering Reaction: The Politics of Pronatalism and the Family in France, 1914–1944" (Ph.D. diss., University of Southern California, 1996), pp. 207–11.

6. On Renault's actions during World War II, see the detailed account in Emmanuel Chadeau, *Louis Renault* (Paris: Plon, 1998).

7. Michael J. French, *The U.S. Tire Industry: A History* (New York: Twayne, 1990), pp. 100–101.

8. Michelin, "100 ans de progrès pour servir," publicity brochure without date or page numbers.

9. Donald N. Sull, Richard S. Tedlow, and Richard S. Rosenbloom, "Managerial

Commitments and Technological Change in the US Tire Industry," *Industrial and Corporate Change* 6, no. 2 (1997): 461–501; Jemain, *Un siècle de secrets*, p. 220; French, *The U.S. Tire Industry*, pp. 100–119.

10. Jemain, *Un siècle de secrets*, pp. 213–14; French, *The U.S. Tire Industry*, pp. 100–119.

11. Sull et al., "Managerial Commitments," 461–501; Jemain, *Un siècle de secrets*, pp. 209–30; French, *The U.S. Tire Industry*, pp. 100–119.

12. Michelin, "Cent ans de progrès."

13. French, *The U.S. Tire Industry*, pp. 131–32; Philip Mattera, *World Class Business: A Guide to the 100 Most Powerful Global Corporations* (New York: Henry Holt, 1992), pp. 330- 37, 458–63; "François Michelin: Le patron ermite," *L'Express*, 16 April 1998, p. 61.

14. The poster, "Michelin XAs nouveau!" appeared in 1965 and has been reprinted in Olivier Darmon, *Le grand siècle de Bibendum* (Paris: Hoëbeke, 1997), p. 106.

15. On the ways that advertisers attempt to reach across class lines, congratulating the rich and promising good fortune to the poor without offending either, see Roland Marchand, *Advertising the American Dream: Making Way for Modernity, 1920–1940* (Berkeley: University of California Press, 1985). On Aunt Jemima, see M. M. Manring, *Slave in a Box: The Strange Career of Aunt Jemima* (Charlottesville: University Press of Virginia, 1998).

16. Jean-François Mesplède, *Trois étoiles au Michelin: Une histoire de la haute gastronomie française* (Paris: Gründ, 1998), pp. 87–90; Herbert Lottman, *Michelin, 100 ans d'aventure* (Paris: Flammarion, 1998), pp. 303–311, 337–45. In the 1960s and 1970s, Michelin also sponsored an array of "beach games" (*jeux de plages*) for a whole generation of French children on vacation.

17. "Connaissez-vous les cartes et guides Michelin?" (1996), a publication of the Michelin Service du Tourisme, no page numbers.

18. Michelin, "Catalogue: Cartes et guides" (Paris: Michelin, 1998), pp. 56–57.

19. Marc Francon, "Le guide vert Michelin: L'invention du tourisme culturel populaire" (Thèse de doctorat du nouveau régime, Université de Paris VII, 1998). For a comparison of the green and blue guides, see also Bernard Lerivray, *Guides bleus, guides verts, et lunettes roses* (Paris: Editions du Cerf, 1975).

20. Michelin, "Catalogue: Cartes et guides," pp. 44–45.

21. Francon, "Le guide vert Michelin."

22. Michelin, "Catalogue: Cartes et guides," pp. 32–33, 48–49; "Connaissez-vous les cartes et guides?"

23. This is the figure frequently cited by Michelin in its publications and in tours of its small museum in Clermont-Ferrand.

24. Michelin, *Bulletin Intérieur Michelin* (March 1945), p. 2; "Cent ans d'industrie: Histoire anecdotique de la Maison Michelin," vol. 2, p. 253, Michelin archives; "La carte Michelin au service des alliés," *Michelin Magazine* (May 1994), n.p.; Michelin, "Une histoire passionnante" (1989), p. 4. Before both the North African and the D-Day campaigns, the British and American armies equipped themselves with Michelin maps, reproduced by their respective war offices: Great Britain, War Office, General Staff, Geographical Section, "Maroc-Algérie-Tunisie" (Washington, D.C., 1942), sheet 151; United States, Army Map Service, *France* (Washington, 1944).

25. Jemain, *Un siècle de secrets,* p. 179.

26. Alain Arnaud, "Michelin: L'expérience d'un éditeur à la conquête du marché touristique nord-américain," *Cahier Espaces* 51 (n.p., n.d.), pp. 70–74.

27. Ibid.

28. Michelin, "Catalogue: Cartes et guides," p. 50.

29. This paragraph has been greatly informed by my conversations with M. Didier Derbal, who oversees the Michelin archives in Clermont-Ferrand.

30. Hervé Le Bras, *Marianne et les lapins: L'obsession démographique* (Paris: Olivier Orban, 1991). On the controversy generated by Le Bras' book, see Andrés Reggiani, "Procreating France: The Politics of Demography, 1919–1945," *French Historical Studies* 19, no. 3 (Spring 1996): 725–54. On the Fondation Alexis Carrel, see also Alain Drouard, *Une inconnue des sciences sociales, la Fondation Alexis Carrel, 1914–1945* (Paris: Institut National d'Etudes Démographiques, 1992).

31. On the Concorde as a French national symbol of modernity, see Rosemary Wakeman, *Modernizing the Provincial City: Toulouse, 1945–1975* (Cambridge, Mass.: Harvard University Press, 1997), pp. 199–205; Kenneth Owen, *Concorde and the Americans: International Politics of the Supersonic Transport* (Washington, D.C.: Smithsonian, 1997). Like aviation, nuclear power came to represent the grandeur or *radiance* of a modern postwar France: Gabrielle Hecht, *The Radiance of France: Nuclear Power and National Identity after World War II* (Cambridge, Mass.: MIT Press, 1998).

32. Richard Kuisel has pointed out the continuity between interwar and postwar notions of Americanization and anti-Americanism: *Seducing the French: The Dilemma of Americanization* (Berkeley: University of California Press, 1993), especially pp. 1–14.

33. France receives, by far, more foreign tourists than any other country. In 1991, the World Tourism Organization reported that 55.7 million foreigners traveled to France; its closest rival, the United States, received only 42.1 million. Donald E. Lundberg and Carolyn B. Lundberg, *International Travel and Tourism,* 2nd ed. (New York: John Wiley, 1993), p. 7.

Note on Sources

Primary Sources

Although many of the primary sources on Michelin are available at the Bibliothèque Nationale de France (BNF), at the Bibliothèque Municipale et Universitaire (BMU) in Clermont-Ferrand, or in assorted libraries in the United States, this book also relied heavily on the holdings of the Michelin company in Clermont-Ferrand, making it possible to get a full overview of various company publications. Michelin has preserved and assembled a large collection of company posters, postcards, and newspaper and magazine advertising. Press clippings from early-twentieth-century newspapers fill twenty-six large volumes with rubrics covering automobiles, aviation, and tourism. Michelin also possesses a comprehensive collection of the company's brochures on aviation, pronatalism, and Taylorism, including the various issues of *Prospérité*. Some of the company's brochures and its newspaper columns and advertisements in *L'Auto, Le Journal, Omnia,* and *L'Illustration* are readily available at the BNF and in some cases through interlibrary loan in the United States. Michelin possesses a small volume of notes by Edouard Michelin, many of which consider Taylorism, as well as internal histories of the firm, most notably those of André Michelin's collaborator, Fernand Gillet. In addition, the company maintains a collection of the red guides, the regional/green guides, maps, and guides to the battlefields, although many of these are also available at the BNF and through interlibrary loan. Michelin's internal publication for employees, the *Bulletin Intérieur Michelin,* renamed *Bib-Revue* in the 1960s and *Michelin Magazine* in the 1980s, can be found in the BMU and the Archives Départementales du Puy-de-Dôme in Clermont-Ferrand.

No historian has yet received unfettered access to internal company memoranda, financial statements, labor records, or the gold mine that the correspondence between Edouard Michelin in Clermont and André Michelin in Paris would most likely contain. Until such evidence is revealed, a definitive traditional comprehensive business history of the firm—however far that kind of work is from the interests of this author—will not be possible. Company custodians of Michelin advertising and touring collections themselves claim no knowledge of any such internal documents. It seems very doubtful that all documents have been destroyed; they are more likely seen by the Michelin family as private information. As Michelin chairman François Michelin approaches retirement, interested historians can only hope that his son and groomed successor, Edouard, believes that the company and French society might profit from greater openness.

The publications of Michelin's fellow travelers in tourism are almost all available at the BNF. Competing guidebooks edited by Karl Baedeker, Adolphe Joanne, Marcel Monmarché (Guides Bleus), and Marcel Thiolier can all be found under the editors' names, as can useful books about early-twentieth-century tourism by Léon Auscher, Louis Baudry de Saunier, Antoine Borrel, Pierre Chabert, and Edmond Chaix. The Bibliothèque Trocadéro, the Parisian municipal library of the sixteenth arrondissement, houses the former library of the now defunct Touring Club de France. It possesses not only the Touring Club's *Revue* and all of the TCF's own guidebooks and publications but also an excellent collection of other groups' and individuals' publications, including a rich array of guidebooks to the battlefields published by a host of organizations. The Archives Nationales has very limited holdings on the TCF, including the bound minutes of meetings, as most of the group's files were lost in the association's dissolution in the 1980s. The BNF and the Bibliothèque Trocadéro maintain extensive collections of publications on gastronomy for tourists, including the books by Curnonsky and Marcel Rouff.

Secondary Sources

Although this book's extreme indebtedness to French sources is obvious in the text and notes, the following sources are disproportionately in the English language for two reasons. First, the contribution of anglophone scholars to French cultural history has been very significant in the past two decades. Second, the following references, organized roughly by chapter, are meant to aid English-speaking students, as well as historians, wishing to delve more deeply into the various topics considered in the book.

CULTURAL HISTORY

The disciplines of anthropology and literary studies strongly influenced the creation of cultural history as currently practiced by a host of historians, particularly those of modern France. Influential works in anthropology include Claude Lévi-Strauss, *The Savage Mind* (London: Weidenfeld and Nicolson, 1966); and Clifford Geertz,

The Interpretation of Cultures: Selected Essays (New York: Basic Books, 1973). Notable literary studies are Roland Barthes, *Mythologies,* trans. Annette Lavers (New York: Hill and Wang, 1972); Jacques Derrida, *Of Grammatology,* trans. Gayatri Chakravorty Spivak (Baltimore: Johns Hopkins University Press, 1976); Michel Foucault, *The Order of Things: An Archaeology of the Human Sciences* (New York: Vintage, 1970); and idem, *The Archaeology of Knowledge and the Discourse on Language,* trans. A. M. Sheridan Smith (New York: Pantheon, 1972). For the articulation of what most historians seem to mean by the term *cultural history,* see the excellent essay by William H. Sewell Jr., "The Concept of Culture," in *Beyond the Cultural Turn: New Directions in the Study of Society and Culture,* ed. Victoria E. Bonnell and Lynn Hunt (Berkeley: University of California Press, 1999), pp. 35–61, as well as the introductions to both that volume and to Lynn Hunt, ed., *The New Cultural History* (Berkeley: University of California Press, 1989).

Many historians have now considered bourgeois culture in the nineteenth century: Whitney Walton, *France at the Crystal Palace: Bourgeois Taste and Artisan Manufacture in the Nineteenth Century* (Berkeley: University of California Press, 1992); Douglas Peter Mackaman, *Leisure Settings: Bourgeois Culture, Medicine, and the Spa in Modern France* (Chicago: University of Chicago Press, 1998); Michael B. Miller, *The Bon Marché: Bourgeois Culture and the Department Store, 1869–1920* (Princeton: Princeton University Press, 1981); and Philippe Perrot, *Les dessus et les dessous de la bourgeoisie: Une histoire du vêtement au dix-neuvième siècle* (Paris: Fayard, 1981). On the city of Paris as a sort of open-air spectacle that mixed as much as it distinguished between social classes, see Vanessa R. Schwartz, *Spectacular Realities: Early Mass Culture in Fin-de-Siècle Paris* (Berkeley: University of California Press, 1998). Leora Auslander traces the evolution of production and consumption through the nineteenth century, organizing her book around the issues of taste, state power, and social hierarchies over the *longue durée* of the seventeenth to the early twentieth centuries (*Taste and Power: Furnishing Modern France* [Berkeley: University of California Press, 1996]).

MICHELIN AND THE AUTOMOBILE INDUSTRY

There has been no cultural history of Michelin (or any other French firm in the twentieth century). Michelin's general business strategies and labor policies have received some attention in André Gueslin, ed., *Michelin, Les hommes du pneu: Les ouvriers Michelin, à Clermont- Ferrand, de 1889 à 1940* (Paris: Editions Ouvrières/ Editions de l'Atelier, 1993); in Lionel Dumond, "L'industrie française du caoutchouc, 1828–1938: Analyse d'un secteur de production" (Thèse de doctorat du nouveau régime, Université de Paris VII, 1996); in Pierre Pascallon, "Michelin et le développement industriel auvergnat," *Revue d'Auvergne* 91, nos. 1–2 (1977), supple-

ment, pp. 3–163; and in the more journalistic works by Herbert Lottman, *Michelin,
100 ans d'aventure* (Paris: Flammarion, 1998) and Alain Jemain, *Michelin: Un siècle
de secrets* (Paris: Calmann-Lévy, 1982). Several coffee-table volumes trumpet the
wonders of Michelin without providing much context: René Bletterie, *Michelin:
Clermont-Ferrand, Capitale du pneu 1900/1920* (Paris: Civry, n.d.); Pierre-Gabriel
Gonzalez, *Bibendum: Publicité et objets Michelin* (Paris: Editions du Collectionneur,
1994); Jean-François Mesplède, *Trois étoiles au Michelin: Une histoire de la haute gastronomie française* (Paris: Gründ, 1998); and Olivier Darmon, *Le grand siècle de Bibendum* (Paris: Hoëbeke, 1997). Despite its smaller size by the 1910s and 1920s, the
competitiveness of Michelin is placed in relief by recent studies of American tire
makers. Michael J. French's *The U.S. Tire Industry: A History* (New York: Twayne,
1990) is a good overview. Mansel G. Blackford and K. Austin Kerr's detailed *BFGoodrich: Tradition and Transformation, 1870–1995* (Columbus: Ohio State University
Press, 1997) portrays both the history of Goodrich and the changing market for
tires. Although a study of labor, Daniel Nelson's book contains much useful information on the American giants Goodyear and Firestone (*American Rubber Workers
and Organized Labor, 1900–1941* [Princeton: Princeton University Press, 1988]).

Traditional business histories of the French automobile industry abound. An
early survey of the industry as a whole is James M. Laux, *In First Gear: The French
Automobile Industry to 1914* (Montreal: McGill-Queen's University Press, 1976). Due
to the early strength of French auto makers, the French context has also received
pride of place in more general histories of twentieth-century automobile industry,
including James M. Laux, *The European Automobile Industry* (New York: Twayne,
1992) and Jean-Pierre Bardou et al., *The Automobile Revolution: The Impact of an
Industry*, trans. James M. Laux (Chapel Hill: University of North Carolina Press,
1982). The early scholarly work of Patrick Fridenson (especially his *Histoire des
usines Renault: Naissance de la grande entreprise, 1898–1939* [Paris: Seuil, 1972]) has
been followed by a host of more popular histories and biographies of the auto makers: Emmanuel Chadeau, *Louis Renault* (Paris: Plon, 1998); Alain Frerejean, *André
Citroën, Louis Renault: Un duel industriel* (Paris: Albin Michel, 1998); Alain Jemain,
Les Peugeots: Vestiges et secrets d'une dynastie (Paris: J. C. Lattès, 1987); Jean-Louis
Loubet, *Citroën, Peugeot, Renault et les autres: Soixante ans de stratégies* (Paris: Le
Monde Editions, 1995); Sylvie Schweitzer, *André Citroën, 1878–1935: Le risque et le
défi* (Paris: Fayard, 1992); and Jacques Wolgensinger, *André Citroen* (Paris: Ar-
thaud, 1996).

ADVERTISING, RACE, AND GENDER IN TWENTIETH-CENTURY FRANCE
Early histories of advertising in France focused on the medium, such as posters, in
an almost celebratory style. Surveys of advertising and posters in France and the

world are not very critical but provide much valuable information (Marc Martin, *Trois siècles de publicité en France* [Paris: Odile Jacob, 1992]; and Alain Weill, *The Poster: A Worldwide Survey and History* [Boston: G. K. Hall, 1985]). Marie-Emmanuel Chessel's *La publicité: Naissance d'une profession, 1900–1940* (Paris: Centre National de Recherche Scientifique, 1998) is an excellent account of the "professionalization" of advertising with an eye for how French admen carefully adapted advertising methods sometimes deemed American, tailoring a message and the media appropriate for interwar France.

On the importance of race and gender in defining social hierarchies in the West as opposed to the East or Africa, Edward Said's *Orientalism* (New York: Pantheon, 1978) and Anne McClintock's *Imperial Leather: Race, Gender, and Sexuality in the Colonial Contest* (New York: Routledge, 1995) are indispensable, and Martin W. Lewis and Kären E. Wigen's work is quite useful (*The Myth of Continents: A Critique of Metageography* [Berkeley: University of California Press, 1997], especially chapters 2 and 3). On French views of Africans in particular, see William B. Cohen, *The French Encounter with Africans: White Response to Blacks, 1530–1880* (Bloomington: Indiana University Press, 1980) and William H. Schneider, *An Empire for the Masses: The Popular French Image of Africa, 1870–1900* (Westport, Conn.: Greenwood, 1982).

Thus far, the sort of sustained analysis of the role of gender, race, and class in advertising that has developed in the historiography on the United States has not been undertaken to the same extent by historians of twentieth-century France. See, for examples, Jackson Lears, *Fables of Abundance: A Cultural History of Advertising in America* (New York: Basic Books, 1994); Roland Marchand, *Advertising the American Dream: Making Way for Modernity, 1920–1940* (Berkeley: University of California Press, 1985); idem, *Creating the Corporate Soul: The Rise of Public Relations and Corporate Imagery in American Big Business* (Berkeley: University of California Press, 1998); Ellen Garvey, *The Adman in the Parlor: Magazines and the Gendering of Consumer Culture, 1880s-1910s* (Cambridge: Cambridge University Press, 1996); and Pamela Walker Laird, *Advertising Progress: American Business and the Rise of Consumer Marketing* (Baltimore: Johns Hopkins University Press, 1998). The case study of Aunt Jemima forms an interesting contrast with Bibendum: M. M. Manring, *Slave in a Box: The Strange Career of Aunt Jemima* (Charlottesville: University Press of Virginia, 1998). On the importance of gender not only in advertising but in the automobile industry and the engineering profession, see Virginia Scharff, *Taking the Wheel: Women and the Coming of the Motor Age* (New York: The Free Press, 1991); Ruth Oldenziel, "Boys and Their Toys: The Fisher Body Craftsman's Guild, 1930–1968 and the Making of a Male Technical Domain," *Technology and Culture* 38 (1997): 60–96; and Ruth Oldenziel, *Making Technology Masculine: Men, Women, and Mod-*

ern Machines in America, 1870–1945 (Amsterdam: Amsterdam University Press, 1999).

TOURISM, REGIONALISM, AND GASTRONOMY IN MODERN FRANCE
Sociologist Dean MacCannell's *The Tourist: A New Theory of the Leisure Class* (New York: Schocken, 1976) has oriented much recent scholarship in the history of tourism. Overviews of tourism in France include Alain Corbin's *L'avènement des loisirs, 1850–1960* (Paris: Aubier, 1995); Marc Boyer, *Le tourisme* (Paris: Seuil, 1982); André Rauch, *Vacances en France de 1830 à nos jours* (Paris: Hachette, 1996); and idem, *Vacances et pratiques corporelles: La naissance des morales du dépaysement* (Paris: Presses Universitaires de France, 1988). More recently, the history of tourism has been successfully mined for much evidence of social hierarchies (that tourism had sometimes been claimed to erase) and as the very embodiment of first bourgeois, then mass, consumer culture: Catherine Bertho Lavenir, *La roue et le stylo: Comment nous sommes devenus touristes* (Paris: Odile Jacob, 1999); Ellen Furlough, "Making Mass Vacations: Tourism and Consumer Culture in France, 1930s to 1970s," *Comparative Studes in Society and History* (1998): 247–86; and idem, "Packaging Pleasures: Club Méditerranée and French Consumer Culture, 1950–1968," *French Historical Studies* 18, no. 1 (Spring 1993): 65–81. Recent international studies of tourism provide necessary perspective for developments in France: Shelley O. Baranowski and Ellen Furlough, eds., *Tourism, Commercial Leisure, and National Identities* (Ann Arbor: University of Michigan Press, 2001); and Orvar Löfgren, *On Holiday: A History of Vacationing* (Berkeley: University of California Press, 1999).

There is comparatively little serious scholarship on the centrality of food and wine to French national identity. Sociologist Stephen Mennell's general survey remains the best overview (*All Manner of Food: Eating and Taste in England and France from the Middle Ages to the Present* [Oxford: Blackwell, 1985]), particularly when complemented by the more general cross-national essays in Jean-Louis Flandrin, Massimo Montanari, and Albert Sonnenfeld, eds., *Food: A Culinary History from Antiquity to the Present* (New York: Columbia University Press, 1999), as well as by the more specific essays on France: Priscilla Parkhurst Ferguson, "A Cultural Field in the Making: Gastronomy in 19th-Century France," *American Journal of Sociology* 104, no. 3 (November 1998): 597–641; Bertram M. Gordon, "The Decline of a Cultural Icon: France in American Perspective," *French Historical Studies* 22, no. 4 (Fall 1999): 625–51; Georges Durand, "La vigne et le vin," in *Les lieux de mémoire*, ed. Pierre Nora, vol. 3 (Paris: Gallimard, 1997), pp. 3711–41; and Pascal Ory, "La Gastronomie," in *Les lieux de mémoire*, ed. Pierre Nora, vol. 3 (Paris: Gallimard, 1997), pp. 3743–69.

Similarly, the importance of regionalism (and not just denial of it) in the construction of French national identity has received relatively little sustained attention, with the notable exceptions of Shanny Peer's *France on Display: Peasants, Provincials, and Folklore in the 1937 Paris World's Fair* (Albany: State University of New York Press, 1998); Anne-Marie Thiesse's *Ecrire la France: Le mouvement littéraire régionaliste de langue française entre la Belle Epoque et la Libération* (Paris: Presses Universitaires de France, 1991); and Laird Boswell's work-in-progress on Alsace.

WORLD WAR I COMMEMORATION AND BATTLEFIELD TOURISM

Daniel J. Sherman's excellent *The Construction of Memory in Interwar France* (Chicago: University of Chicago Press, 1999) considers the commemoration of World War I in French culture, helping to move the discussion about the cultural impact of the war beyond the long-established argument as to whether the war brought "modernity" (as argued by Paul Fussell, *The Great War and Modern Memory* [New York: Oxford University Press, 1975]; Modris Eksteins, *The Rites of Spring: The Great War and the Birth of the Modern Age* [Boston: Houghton Mifflin, 1989]; and Thomas W. Laqueur, "Memory and Naming in the Great War," in *Commemorations: The Politics of National Identity*, ed. John R. Gillis [Princeton: Princeton University Press, 1994]) or consisted of an adaptation of nineteenth-century norms (a view most frequently articulated by Jay M. Winter, particularly in his *Sites of Memory, Sites of Mourning: The Great War in European Cultural History* [Cambridge: Cambridge University Press, 1996]). David Lloyd's recent work on interwar battlefield pilgrimage and the construction of national identity in British dominions offers an essential comparative perspective for understanding the French context (*Battlefield Tourism: Pilgrimage and Commemoration of the Great War in Australia and Canada, 1919–1939* [New York: Berg, 1998]). Although somewhat difficult to find (Michelin has a copy; and the author sent me a copy), Antoine Champeaux's master's thesis contains useful information about the guides: "Les guides illustrés Michelin des champs de bataille, 1914–1918" (master's thesis, Université de Paris IV, 1984).

PRONATALISM

The cultural construction of *dénatalité* has recently become a well-established part of the historiography of modern France, displacing an older literature that accepted contemporaries' alarmist assumptions of "depopulation" at face value. Karen Offen's early "Depopulation, Nationalism, and Feminism in Fin-de-Siècle France" (*American Historical Review* 89 [June 1984]: 648–76) established much of the context before World War I, while Mary Louise Roberts has brilliantly shown the importance of gender, including the central theme of "depopulation" for which women were supposedly to blame, in the cultural definitions of French society in the interwar period (Mary Louise Roberts, *Civilization without Sexes: Reconstructing Gender in Postwar France, 1917–1927* [Chicago: University of Chicago Press, 1994]). For the role of bourgeois men in controlling the reproductive lives of working-class women, see Elinor A. Accampo, Rachel G. Fuchs, and Mary Lynn Stewart, eds., *Gender and the Politics of Social Reform in France, 1870–1914* (Baltimore: Johns Hopkins University Press, 1995). Fuchs also edited a forum on French "depopulation" in *French Historical Studies* 19, no. 3 (Spring 1996): 633–754; contributions of Joshua Cole and Andrés Horacio Reggiani suggested the extent to which the very institutional foundations of French demographic study resulted, and continue to result, from a fear of *dénatalité*. Cheryl Koos further contextualized French pronatalism by showing the very strong links of the movement with the far right in interwar France.

AVIATION

The works of Emmanuel Chadeau have done much to describe both the French aeronautical industry (*De Blériot à Dassault: Histoire de l'industrie aéronautique en France, 1900–1950* [Paris: Fayard, 1987] is the shortened form of his *thèse d'état*) and the importance of flight in the West (*Le rêve et la puissance: L'avion et son siècle* [Paris: Fayard, 1996]). Robert Wohl offers a fascinating account of the centrality of flight in the construction of French (particularly high) culture in the early twentieth century (*A Passion for Wings: Aviation and the Western Imagination, 1908–1918* [New Haven: Yale University Press, 1996]), while Herrick Chapman has used the aviation industry to consider the issues of state capitalism and labor (*State Capitalism and Working-Class Radicalism in the French Aircraft Industry* [Berkeley: University of California Press, 1991]). On Michelin's aviation program in particular, Antoine Champeaux includes useful materials in "Michelin et l'aviation, 1896–1936" (Mémoire de diplôme d'études approfondies, Universités de Montpellier III et Paris I, 1988 [consulted in the Michelin archives]). Peter Fritzsche's study of flight in Germany provides a comparative context that does much to clarify developments in France (*A Nation of Fliers: German Aviation and the Popular Imagination* [Cambridge, Mass.: Harvard University Press, 1992]).

TAYLORISM AND AMERICANIZATION

The definitive study of the adoption of Taylorist rationalization in French industry is Aimée Moutet's *Les logiques de l'entreprise: La rationalisation dans l'industrie française de l'entre-deux-guerres* (Paris: Ecole des Hautes Etudes en Sciences Sociales, 1997), which is an adapted form of her *thèse d'état*. Charles Maier revealed the larger context of rationalization in Europe during the interwar years ("Between Taylorism and Technocracy: European Ideologies and the Vision of Industrial Productivity in the 1920s," *Journal of Contemporary History* 5 [1970]: 27–62). Articles by Moutet and Patrick Fridenson cover the prewar years: Moutet, "Les origines du système Taylor, le point de vue patronale (1907–1914)," *Le Mouvement Social* 93 (1975): 15–49; Fridenson, "Un tournant Taylorien de la société française, 1900–1914," *Annales ESC* 42 (1987): 1031–60. The divergence between French industrialists' definitions of Taylorism and Taylor's actual ideas is put in relief by Daniel Nelson's biography of Taylor (*Frederick W. Taylor and the Rise of Scientific Management* [Madison: University of Wisconsin Press, 1980]).

At least since the First World War, the "Americanization" of France has been an issue barely distinguishable from the "modernization" of France, even among historians of both phenomena. Recently, several authors have considered the "Americanization" of France more critically, with an eye for how images of the United States could be manipulated by different social groups in France for their own ends. For the interwar years, Ellen Furlough's "Selling the American Way in Interwar France," *Journal of Social History* 26 (1993): 491–520 nicely complements Richard F. Kuisel's study of the postwar years (*Seducing the French: The Dilemma of Americanization* [Berkeley: University of California Press, 1993]). Victoria de Grazia has considered the role of American text-oriented advertising in displacing European poster art ("The Arts of Purchase: How American Publicity Subverted the American Poster, 1920–1940," in *Remaking History*, ed. B. Kruger and P. Mariani [Seattle: Bay Press, 1989], pp. 221–57) and the model of American consumption in Europe more generally ("Changing Consumption Regimes in Europe, 1930–1970: Comparative Perspectives on the Distribution Problem," in *Getting and Spending: European and American Consumer Societies in the Twentieth Century*, ed. Susan Strasser, Charles McGovern, and Matthias Judt [Washington, D.C.: German Historical Institute; Cambridge: Cambridge University Press, 1998], pp. 59–83). For the broader international context, see Peter N. Stearns, "Stages of Consumerism: Recent Work on the Issues of Periodization" *Journal of Modern History* 69 (March 1997): 102–17; Victoria de Grazia, ed., with Ellen Furlough, *The Sex of Things: Gender and Consumption in Historical Perspective* (Berkeley: University of California, 1996); and Mary Nolan, *Visions of Modernity: American Business and the Modernization of Germany* (New York: Oxford University Press, 1994).

Prade, Georges, 168
Prix Michelin. *See* prizes
prizes: British Empire Cup, 166; Coupe
    Michelin, 158–64, 179, 186; Grand Prix
    Michelin, 157–67; Prix de l'Aérocible
    Michelin, 167–71; Prix Michelin de la
    Natalité, 139–43; Prix Michelin de la
    Sécurité, 177–79; for wartime aviators,
    91–92
pronatalism, 126–55
Prospérité, 199–223
Prost, Antoine, 113
Puiseux, Robert, 270–71
Puy-de-Dôme, 159, 165, 186, 232–33, 271

Rabelais, 36–37
race, portrayals of, 41–45, 292n 79
racing: as advertising, 17–20; automobiles,
    17–18, 47; bicycles, 17–18
racism, commodity, 41–45
radial tire. *See* tires, radial
Rand, George F., 113
Rand McNally, 278
red guides: diffusion of, 67–68, 246; and
    gastronomy, 245–56; gender and,
    68–69; international editions of, 67–68,
    334n 77; origins of, 56–70; plagiarizing
    of, 70–71; reader participation in, 63;
    size of, 67; stars in, 247–56; after World
    War II, 274, 276, 277–78
regional gastronomy, 235–45, 265–67. *See
    also* gastronomy
regional guides, 256–64, 276–78
Reims, 100–101, 107–10
religion. *See* Bibendum, Commandments
    of; pilgrimage
Remington typewriter, 2
Renault: engines for Breguet planes, 92;
    nationalization of, 271; as purchaser of
    Michelin tires, 192–93; road signs by,
    230–31; and Taylorism, 196
Renault, Louis, 10, 192. *See also* Renault
Renaux, Eugène, 165

La République, 163–64
Rerum Novarum, 144
restaurants. *See* gastronomy
Riom, 232
Risler, Georges, 147
Riviera, 250, 252
road numbering, 83–86
road signs, 79–83, 229–35
Rolls Royce, 231
Romanet, Emile, 144–47
Romans, 106
Ronald McDonald, 3
Rossillon, Marius. *See* O'Galop
Rothschild bank, 180
Rouen, 262–63
Rouff, Marcel, 241–44
Rousseau, Paul, 19
Roussel, Paul, 159–61

sacred way. See *voie sacrée*
Sailland, Maurice-Edmond. *See* Cur-
    nonsky
salaries at Michelin, 201–2
salons: for automobiles, 22; for aviation,
    83, 136, 163–64; for selling tires, 22
Santos-Dumont, Alberto, 158
Saussure, Ferdinand, 12
Savignac, Raymond, 275
scientific management. *See* Taylor, Freder-
    ick Winslow; Taylorism
Sears, 272
Senoucque, Albert, 165
Sewell, William, 11–12
Simca, 218
Social Catholicism, 129, 144
société en commandite par actions, 10
Société Générale, 180
Soissons, 100–101
solidarism, 55–56
Somme, battle of the, 174
Les Sports, 34
St. Barnabas Society, 123
stockistes, 58, 245–46